AF333161

STAR TESTING
ASTRONOMICAL TELESCOPES

SECOND EDITION

A MANUAL FOR OPTICAL EVALUATION AND ADJUSTMENT

Published by

Willmann-Bell, Inc.

P.O. Box 35025 / Richmond, VA 23235 / (804) 320-7016 / FAX (804) 272-5920 / **www.willbell.com**

Published by Willmann-Bell, Inc.
P.O. Box 35025, Richmond, Virginia 23235

Printed in the United States of America

Library of Congress Cataloging-in-Publication Data.
Library of Congress Cataloging-in-Publication Data

Suiter, Harold Richard.
  Star testing astronomical telescopes : a manual for optical evaluation and adjustment / Harold Richard Suiter. -- 2nd ed.
      p. cm.
  Includes bibliographical references and index.
  ISBN 978-0-943396-90-3 (alk. paper)
  1. Telescopes--Testing--Amateurs' manuals. I. Title.
  QB88.S85 2008
  522'.20287--dc22

                                                    2008045792

08 09 10 11 12 13 14 15 16 9 8 7 6 5 4 3 2

Foreword

It is not often that a book opens the eyes of a whole generation of amateur astronomers—but the first edition of Dick Suiter's *Star Testing Astronomical Telescopes* was just such a book. By giving its readers a simple, sensitive, and reliable test for optical performance, he enhanced the observing experience of every amateur astronomer who took its lessons to heart. Whether you were a novice or expert, *Star Testing* was important because it told you how to get the most from your telescope.

What makes the star test so great is that it's both very easy and very sensitive. The star test is so simple that you probably use it already without being aware of it. When focusing an instrument you probably roll the eyepiece back and forth a couple times before settling on the sharpest position. If the telescope is well figured, collimated properly, and the air is steady, best focus lies midway between two identical out-of-focus disks. Without even thinking about it, you just know that appearance means good optics.

To perform the test, you simply observe a star with a moderately high-power eyepiece, giving careful consideration to the image on both sides of focus. The patterns you see in the focused and out-of-focus star images tell you whether your telescope is aligned for maximum performance, whether the atmosphere is steady, when the telescope has cooled, and when it is ready to do its best.

I have known Dick Suiter for about twenty-five years, since I met him at a star party in Ohio and subsequently asked him to write a short article about star testing for *Astronomy* magazine. His ability to meld abstruse theory with practical observing has always impressed me, and now, in this second edition, he has produced once again a spectacular blend of theory and practice.

In the Foreword to the first edition, I told readers how star testing helped me to get top-notch performance from my 20-inch $f/5$ Dobsonian, back in the days when not many people had experience with telescopes of that aperture. Star testing revealed tube currents, so I added a fan; it revealed a bit of spherical aberration, and I had the primary refigured. With the help of star testing, I tuned up my big Dob to give excellent lunar and planetary images. During the 1988 opposition of Mars, for example, not only did I see Deimos and Phobos for the first time, but I also enjoyed the best views I'd ever had of the planet itself.

In the years since this book first appeared, I think that the increasing number of amateurs who routinely star test telescopes has exerted a powerful but subtle influence on the telescope industry. When the first edition of *Star Testing* came out, it was not particularly uncommon, at a star party, to accept a proud owner's invitation to look through a new telescope— only to find yourself thinking, "Oh my gosh, does this person realize the telescope has major problems?" Over time, I am sure that amateurs who routinely star test every instrument they look through have given a few proud telescope owners a few bad nights, and that has resulted in bad days for some dealers and manufacturers. But the net result is that today, the astronomical public expects—and gets—a far better telescope than they did some twenty years ago. All because Dick Suiter taught us what to look for in a star image.

What impresses me most about this second edition is that what was already good has become even better. With great care, Dr. Suiter has improved the visual fidelity of the figures by computing effects for the whole range of wavelengths visible to the eye. He has tested differing computational schemes against one another to verify his results. He has carefully compared the complex aberrations found in real telescopes with the pure aberrations found in theory—and he demonstrates that they match.

Star Testing is a resource that should be on the shelf where you keep your most-often-used astronomy books. Toss it into your observing kit along with eyepieces, Oreo cookies, and packs of instant hot chocolate. Read it; absorb it; read it again. Star testing is an integral part of observing.

You will benefit when you become sensitive to your telescope's optical performance. If you find that it has a few problems—and what telescope does not?—it is best to deal openly with them. Would-be astronomers who refuse to acknowledge such problems tend to stop using their instruments, and eventually their interest in astronomy. Once you recognize that a problem exists, you can possibly use the star test to correct it. Often, it's nothing that careful collimation will not solve. If your telescope has properly adjusted and excellent optics, the star test will confirm that fact—and you are free to turn your attention to observing the splendors of the heavens.

Richard Berry
Lyons, OR.

Table of Contents

Preface

It has been well over a dozen years and five printings since I wrote the first edition of *Star Testing*. (It was published in late 1994, but largely written in 1992.) Perhaps I presume too much, but I believe that star testing has moved to more prominence since the book first appeared.

It is not that star testing didn't exist before. The point-source test was first used to fabricate a telescope around 1720 when John Hadley used it to make the earliest astronomical Newtonian reflector. Hadley employed a variant of the star test, since his source was a nearby pinhole, but all the essential elements of the star test were used. He drew the eyepiece to both sides of the image and judged the behavior of light in focus by what he saw on each side (King 1955).

However, the star test has never enjoyed popularity as a sole method of testing during fabrication. Even after Hadley's example, other tests were preferred. The star test does not yield surface profiles, except in advanced modern forms, and it is not a convenient indoor evaluation technique. That does not mean that its use in telescope making hasn't had a few outspoken adherents, but the star test is most employed as a final evaluation of a completed working telescope.

It was first popularized in a short work by H. Dennis Taylor. In 1891, he published a tiny book with a set of star-test drawings reproduced photographically and pasted in the frontispiece (Taylor 1983).[1] These drawings have been reprinted many times since, and Taylor's descriptions have been summarized in a number of books (e.g., Twyman 1988, Bell 1922, and Ingalls 1976). Optical experts have long employed the star test to qualitatively compare optics in the field (that's what they used at Stellafane all those years). But it had the cachet of privileged knowledge of a few adepts – not for common folks. I wanted to open up the star test to everyone and show people how the secondary mirror changes the pattern, since Taylor's treatise was solely about refractors and practically ignored reflectors.

Since 1995, telescopes that are directed toward amateur astronomers (as opposed to cheap consumer telescopes) have also greatly improved in optical quality. About half of Newtonian mirrors I had inspected up to 1992 had been of literally wretched to bad quality. Since 1995, they have improved steadily until most have at least an acceptable rating and a surprising number are excellent. This change may be caused by a natural mat-

---

[1] More recent editions do not have this plate. They reproduce the photo on a printed page.

uration of the market, but I like to think that the first edition contributed in teaching a method that everyone can use to find superior optics. In the old days, makers of high-quality optics had to sit by and try to compete with makers of cheap optics. Novice telescope buyers really could not tell the difference and nearly always selected the less expensive instrument. No doubt some makers curse my name because they had to deal with obnoxious customers, but they would have had to deal with those people anyway. Had not the star test provided a convenient topic of complaints, difficult customers would have found something else to fume about.

Still, I knew when I wrote the book that I had left some gaps. In particular, few specific designs were calculated. I dealt mostly with generic aberrations by discussing pure Zernike forms. Most of these were unavoidable at the time, caused by the lack of information about the aberration-mix of unique designs and the length of time it took to generate a pattern in those days. Even with speeds of computers today, some of the polychromatic images calculated in the present book took the same 45 minutes required by the monochromatic images in the original book, mainly because they handle 17 weighted colors rather than one.

This book does present some specific designs, basically the most common designs available on the market. This was not done to answer criticisms that I hadn't modeled "real" telescopes, but to show that properly-made real telescopes do indeed follow one or two of the basic Zernike polynomials, and thus resemble the theoretical telescopes modeled in the first edition more closely than is commonly realized.

There is a myth promulgated right now that "complex" telescopes – Maksutovs, Schmidt-Cassegrains, and apochromatic refractors – can break the behaviors demanded by the star test, and still image well. The behaviors targeted by the star test are fundamental to the theory of diffraction and I hope that some of the complex-telescope patterns calculated here will convince you that the star test still works if used properly.

I have designed some refractors and catadioptrics. All commercial refractor and catadioptric designs are proprietary, so they are unavailable to me. Even though I admit that my designs are perhaps not optimized the same way or even as well as commercial objectives, the fact that a sub-optimal design star tests superbly can be used as an existence demonstration for better designs.

Some people may be disappointed that I have not added true color images in this volume. I discussed this years ago when I advised Cor Berrevoets about the program ABERRATOR (an FFT-based freeware star-test simulator). Cor wanted to adapt it to color. I expressed the opinion that this calculation would only be accurate at one range of brightness values because of the color and monochrome sensing systems and logarithmic dependence of the human eye. If accurate in-focus, the image would be inaccurate out-of-focus, and *vice versa*. I suggested filtering the calculations through a CIE weighting. Cor continued with success at computing focused extended images. ABERRATOR's point-source calculations, however, have colors that seem a bit too saturated at significant defocus values and for mirror telescopes. Actual stellar images seem painted on the retina with pastels, if color shows at all.

If I wanted to depict the color dependence of the star test, my task would be compounded by the inevitable image shifts at the book printer. Grayscale is bad enough, color is frightening. I would be judged for the absolute color in the difficult situation of color-on-black printing. Authors learn to tolerate a good deal of color shift in regular astronomical books, but the colors in astrophotos are somewhat fictional anyway. I would have no such wiggle room. Therefore, I determined once again not to render the individual colors in this book, but to calculate them in a composite polychromatic grayscale image[2]. At least the unanticipated printing contrast shifts would carry the whole grayscale and every color with it. This procedure produces less difference with the monochromatic images in the first edition than one would expect, and that difference is confined to only the brightest targets. My only color picture appears on the cover, showing a bad comatic flare calculated for a 17-wavelength distribution. Imagine it with the red and violet ends very much muted; that is how people see images in the real world.

I have always regretted the overemphasis assigned to my method to make a crude estimate of low-order spherical aberration. The correct way of using this estimate is to first ascertain that a fairly pure low-order spherical aberration exists. Then use the ratio of breakouts to determine if the amount is severe enough to demand further study. Finally, the reader should use the defocus numbers in Chapter 5 and the figures of Chapter 10 to narrow down the estimated aberration. In this edition, the breakout test still appears but is less emphasized and the follow-on technique is explicitly described in the same location.

I envision an audience of amateur astronomers with telescopes of small to moderate size (apertures less than 20 inches or so). The book may occasionally be useful to professional astronomers and optical experts, although the former usually have to deal with the imaging peculiarities of huge telescopes and the latter may have already formed strong differing opinions about many of the topics. When faced with the choice of rigor

---

[2] Computed using the commercial optical-design code ZEMAX.

versus clear exposition, I almost always opted for clarity. Thus, the book resembles more a personal viewpoint than an authoritative document.

I was unwilling to completely remove equations from the central discussion, but I try to limit them to elucidating important topics, and I put most of the non-essential math in the Appendices. The reader need not understand the mathematics in the body of the text to use the star test effectively, but fullest comprehension of causes and effects will reward those who have studied these discussions. Appendix B is only for those wishing detailed knowledge of the way these complicated figures have been generated. It is by no means required reading. Appendix D, on the other hand, is helpful to casual readers because it contains the labeling description for the star-test figures. Be sure to consult it when you finally encounter these illustrations. Appendix G is a detailed description of the specific optical designs I made for the book. Some people will be fascinated by Appendix G; most will be bored if they were to read it.

I would like to thank the host of people who have helped me with this book. I can't list all of them here, but I would like to mention the following contributions especially for their help in the first edition: Bob Bunge, Peter Ceravolo, and Bob Gent for kindly supplying me with various references; Ray Lim for locating typographical errors in the difficult mathematical appendices; E. Wolf and R. Kingslake for going out of their way to help me get reprinting permission on one of the figures in Appendix B; skilled mirror-maker Bill Herbert for his Foucault and Ronchi test photographs; and editor Tom Burns for converting my meanderings to reasonably coherent English. Diane Lucas, Richard Berry, Michael Brunn, Richard Buchroeder, Roger Sinnott, and William Zmek aided me by doing detailed technical editing on my manuscript. In the second edition: Gregory Hallock Smith, Bryan Greer, and William Zmek for the tedium of detailed examination of a text that they had read before, and Jack Koester for his non-technical editing. Of course, my wife Anita gets most of the credit for tolerating my second edition rewrite.

Thanks to these people, the book is much clearer. (Of course, any errors that remain are strictly my fault.)

Harold R. Suiter
Panama City, FL
January 2009

Chapter 1
Introduction

1.1 Telescope Evaluation

Astronomical telescopes are almost unique. Many optical instruments exist, from camera lenses to solar collectors to the eyeglasses perched on your nose, but few besides the telescope (and its miniature cousin the microscope) push so vigorously on the fundamental physical limits of what can be seen. As a consequence, the owner of a new telescope or the long-time user of an old telescope wants to ensure that the instrument is working as well as possible. General telescope users tend to be frightened away from performing workshop tests. The elaborate preparation, equipment, and training are formidable. Such methods are not necessary to simply determine that your telescope works at a level reasonably close to the way it was designed.

Telescopes are easy to test. All that is required is a good high-magnification eyepiece, a conveniently placed star or illuminated pinhole, and some experience. In fact, telescopes are so easy to test that I recommend you check the optical train of your instrument *every time you use it*, as a monitor of alignment, proper optical support, and atmospheric conditions.

You will notice that the methods proposed in this book are a great deal different from the workshop methods telescope makers use. Telescope *fabrication*, as contrasted with telescope *evaluation*, requires that corrective measures be indicated. Hence, makers favor methods that lead to profiles of the optical surfaces, or at least the errors in the slope of those surfaces. They can use such information to decide where they must remove slight amounts of glass during the next figuring step.

Such methods demand specialized equipment. More important, they require skills best learned at the elbow of an expert. Optical shop tests tend to require delicate interpretations, and it is helpful if someone physically demonstrates the test to a person unfamiliar with it.

Telescope evaluation, on the other hand, is a test of a finished product. It requires little or no specialized equipment. The evaluator cares little how an optical piece can be improved. The test is purely a yes-no decision: Is the optical train good enough to transmit the image or is it not? The evaluation encompasses the *entire* optical path, even those elements not customarily included in the telescope, from the top of the atmosphere to the eye of the observer.

The star test evaluates telescopes in their final configuration, doing precisely what they were intended to do. The star test is not easily reduced to numbers (though it can be by careful professionals, see Fienup 1993 or Roddier 1991), but it is very sensitive. A telescope that "passes" the star test need not be evaluated with a bench test. It has already met the most stringent criterion necessary to deliver beautiful images.

The star test is an examination of the image of a point source, most commonly a star, both in focus and on both sides of focus. The power of the star test is contained in this simple motion of the eyepiece to examine the expanded diffraction disk on both sides of focus. Not only is the expanded spot bigger and hence easier to see (the focused diffraction disks of most astronomical telescopes are tiny, even at high power), it "unfolds" into a unique representation of the aberrations that caused it. In particular, out-of-focus circular images appear similar at equal small distances inside and outside the focal point *but only if the optics are excellent.*[1] The expanded disks will appear different if any aberration is degrading the system. The sensitivity of this test is phenomenal. My first mirror was a 200 mm *f*/6 Newtonian reflector that regularly shows excellent images of the planets, yet reveals a slight inclination towards overcorrection, roughness, and a turned edge in the star test.

A telescope is automatically prepared to do the star test and the test is conducted with all optical elements in place. Many of the optical defects discussed in this book have nothing to do with errors on the glass, and would not even have been detected in shop tests during fabrication. For example, the 200-mm reflector mentioned above is carefully held in a 9-point mirror support, yet still shows some slight evidence that the mirror sags in its cell.

As you continue to use the star test, you will use it not only as a test of the instrument itself, but as a check on alignment and seeing conditions. A quick turn on the focuser is all that's needed to verify that misalignment is not disturbing your images. You'll come to depend on the appearance of outside-focus images to see what is happening in the upper atmosphere. When good seeing comes and goes, you can shift that evening's observing plans to take advantage of superior tranquility.

---

[1] At this point, terminology should be defined to avoid subsequent confusion. In all further discussion, an image is said to be "out of focus" if it is defocused in *either* direction. "Inside focus" means that the eyepiece's focal plane is placed between the main optical element and focus; "outside focus" indicates that the eyepiece is withdrawn beyond focus. "In focus" and "focused" are used as synonyms, meaning that a pointlike object is focused to its minimum size.

1.2 Testing the Surfaces

You may rightly suspect that telescopes are not easy to make. The objective, or main optical element, of an astronomical telescope has tolerances a thousand times smaller than the usual accuracies of a metal-cutting lathe.

As a means of comparison, let us imagine the surface of a common 200-mm (8-inch) telescope mirror expanded to 1 mile across (1.6 km). If this mirror had the usual thickness ratio of small mirrors, it would be 880 feet thick (268 m). In common metal-shop practice, it is normal to machine such an 8-inch surface to a thousandth of an inch, or to a scaled accuracy of 8 inches. A wavelength of light expands from 0.000022 inches to 0.17 inches (4.4 mm) at this scale. The maximum optical error tolerable on such a surface would be only 0.55 mm, or about 0.02 inches. Premium optics would be made to an accuracy of less than 0.01 inches (0.25 mm)—a playing-card thickness error on a disk a mile across and 300 yards high.

Clearly, testing such surface accuracy is not trivial. Common calipers, a measuring device of machinists, have a maximum accuracy of only about 10 micrometers (μm), or about 20 wavelengths of green light. Even if sufficiently accurate calipers were possible, one would have the additional problem of repeatedly placing them on the curved surface. The spread of measurements would greatly exceed the inherent accuracy of the measuring tool.

Obviously, some device or technique which can sense micro-deformations on the surface is needed. Light itself is the most appropriate tool. Less certain is the precise manner in which to use light to bring out these defects without demanding the same stringent measurements as gauging the surface profile by physical contact.

For example, it is not difficult to come up with completely useless ways to measure surfaces (and I have seen some demonstrated at telescope-making conventions). One could place a point source of light (say, a pinhole) to one side of a mirror (see Figure 1.1) and reflect it to a spot on a screen. By moving the mask around over the objective, we could see how the spot was forced to shift. In principle, this method would contain all the error information, but it is not the easiest way of proceeding. The spot is fuzzy, the system is difficult to align and focus, and the measurements are difficult to reduce because they are taken far from the optical axis. The position of the spot is dominated by the overall curvature of the surfaces and the motions of the mask, rather than the interesting deviations from the curvature. The way to perform an accurate and easy-to-interpret test eliminates first-order difficulties such as the innate curvature of the surface. It is best done along the optical axis either near the focus or the center of cur-

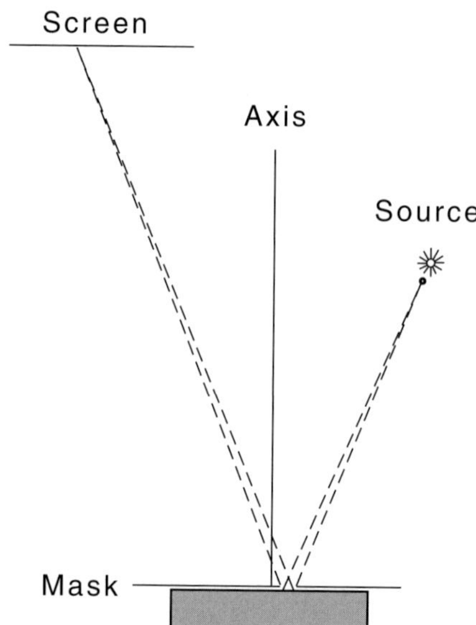

Fig. 1.1 A poorly-designed testing arrangement. Alignment, focusing, and data reduction would be a nightmare even though the test contains all the information required to infer the surface shape of the mirror.

vature of the lens or mirror.

Accurate tests are possible when the strengths of the testing geometry are exploited. The useless test in Figure 1.1 becomes the very accurate caustic test when the image spot is moved near the center of curvature and the sensor is fitted with a knife edge or wire. A variation in which many openings are used leads to the Hartmann test (see Appendix A for more detailed descriptions).

1.2.1 Sources of Errors

Accuracy of surface shape is a function of the stiffness of the material and the definition of just what "surface accuracy" means. For example, most machinists' gauge blocks and flat surface plates, the absolute standards by which they check their measuring tools, are rectangular chunks of steel. Steel is a fairly inflexible material, and machinists think of steel as capable of holding its shape. These blocks are more than accurate enough for the kind of precision demanded in machining as long as the temperature of the room does not vary too much. For optical use, however, the scale of precision is much smaller.

Microscopically, a block of steel expands under heating. Its linear dimensions depend on the ambient temperature. This change wouldn't ordinarily affect optical performance—after all, a sphere is still a sphere even if it has a slightly different radius. But as temperature varies, portions of the metal change faster and internal stresses build up. The surface slightly buckles as the temperature of the piece non-uniformly follows rapidly changing outside temperature.

Likewise, glass deforms under heating. Although it swells less than metal, the thermal conductivity of glass is lower and it has a problem getting rid of interior heat. Mirrors are typically coated on one side with metal, which complicates the way they radiate energy. The temperature of glass is a complex function of thermal radiation, convective air cooling, and conduction through the relatively small area at which the optical disk is supported.

Glass not only changes shape under heating, it deforms under pressure. A helpful way of viewing glass disks at the wavelength scale is to think of them as sheets of rubber. If you push on the top, the surface goes down. If you incorrectly support the bottom, the whole disk sags and the upper surface will deform. Thin pieces are harder to hold flat than thick pieces, large pieces more difficult than small ones. The cells that hold optics must not pinch or warp them.

No process is as stressful to optics as fabrication. Simple grinding even has a few pitfalls. Imagine that a mirror disk has a slight amount of cylindrical curvature on its rear side—it rocks on the backing instead of sitting flat. Under the pressure of grinding on too hard a surface, it will deform away from its "backbone" and suffer astigmatic curvature when the pressure is relieved.

The fabrication stage at which most errors originate, however, is polishing. It takes place on polishing pitch, a material that no one fully understands. Some experienced opticians have learned the limits and general behavior of pitch, but even after years of experience, they are often surprised by its unstable nature.

Pitch is a highly viscous fluid used in a thin layer (4 to 8 mm) covering a disk used as a tool. This lapping surface (or "lap") is usually cross-hatched with grooves that allow the pitch to spread and conform more readily to the polished surface. Pitch will behave more-or-less as a solid at fast speeds and as a liquid at slow speeds. For example, if you hit it with a hammer, it shatters. Lay a stick of pure pitch across the lip of a bowl, however, and eventually you'll find it has run to the bottom.

During polishing, powdered abrasive grains sink into the pitch surface where they are held as microscopic scrapers. Certain polishing agents are more effective than others, and cause more heat to be generated in the lap. The resistance to deformation in pitch varies strongly with temperature. Its characteristics on the outer portion of the disk will vary markedly from those of the inside because wet pitch on the periphery is exposed to the air and cools more rapidly by evaporation. If polishing is stretched too long, the pitch tool dries out, becomes overheated, and loses its shape. Bad polishing habits will result in excessive wear at the edge of the optical disk, giving it a "run-down heel" appearance under sensitive testing. Not varying the stroke when using a machine may cut shallow circular channels in the optics. In the case of fast aspherical optics, more polishing must be applied on the center of the disk. Clearly, many opportunities exist for errors to find their way to the optical surface.

1.2.2 Measures of Optical Quality

One popular way of gauging optical quality is to measure the peak-to-valley wavefront error at the pupil of the objective lens or mirror. A wavefront might be thought of as a line traced along the crest or trough of a wave. In regions far from focus, a wavefront is perpendicular to the direction of wave motion. Using a convenient example, a wavefront is the crest of a surfer's wave, parallel to the beach. The wave is moving toward the beach, at right angles to the wavefront crest.

A perfect converging wavefront is part of a sphere with its center at the focus. The light from a point source converges to the minimum spot size, called the "diffraction disk." After passing through an optical system with errors, a wavefront departs from the spherical shape, and the image spot will be larger and less intense.

Imagine two spheres with a common center at focus, somewhat like the layers of an onion. The outer layer touches the point that lags furthest behind the real wavefront and the inner one touches the point closest to the sphere's center. The different radii of these spheres define the total wavefront error. (See Figure 1.2.)

J.W. Strutt (Lord Rayleigh) formulated an often-quoted rule: If the total wavefront error—peak-to-valley—exceeds ¼ wavelength of the center of the bandpass for human vision (yellow-green or 550 nm), then visual optics begin to noticeably degrade. The reason that the image begins to fall apart is simple—a significant portion of the converging wavefront now has a phase mildly "disagreeing" with the majority. Rayleigh's rule is not a hard limit. For the most common smooth error, low-order spherical aberration,[2] some people don't easily perceive diminished quality until total wavefront error exceeds ⅓ wavelength. Others are more discriminating,

---

[2] Explained in Chapter 10.

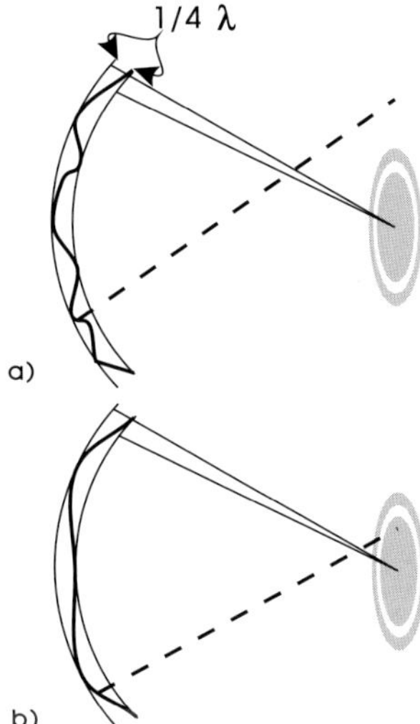

Fig. 1.2 The difference between adequate and very good optics. Both of these wavefronts are contained within two shells ¼ wavelength apart, but the rougher one in a) scatters light farther from the radius of the minimum-size diffraction spot. (Not to scale.)

detecting degradation at ⅛ wavelength and below. Much of the sensitivity to optical faults depends on the type of observing, the type of error, and the sophistication of the observer (Ceravolo *et al.* 1992; Texereau 1984).

Nearly every telescope user has a hazy memory of reading about the ¼-wavelength Rayleigh tolerance. At one time, there was an "exaggeration race" on, where different manufacturers found tricky ways to restate the quarter-wavelength tolerance in various forms, disguising the fact that their specifications were basically the same as their honest competitor's ¼-wavelength mirrors.

These days, the accuracies of telescope optics are less often specified in artificially exaggerated peak-to-valley numbers. A measure of the reduction of the diffraction disk intensity is preferred now, with a central intensity reduced to 80 percent that of perfect optics being the commonly-used cutoff.[3] This ratio of central irradiances is called the *Strehl definition*, or more often the *Strehl ratio* (Born and Wolf 1980). The Strehl ratio has the feature of being proportional to the square of the surface deformation,

---

[3] Some in the adaptive-optics community define a *system Strehl ratio* that includes transmission effects, because they are worried about having enough photons to make their corrective optics work. In this book, energy is normalized in the image, and that central intensity is defined as a purely *diffractive* Strehl ratio. This definition, I believe, is closer to that of the original intent and fits in well with other aspects of modern optics. There will be more on this topic in later chapters.

and so it is scaled to follow effect on the images more precisely. In other words, a ¼-wavelength deformation affects the image four times more than a ⅛-wavelength deformation, not twice as much, as it would seem.

Wavelength claims of either type, however, are often made from a measurement along only a diameter instead of the whole surface, which unravels some of the superiority of the Strehl ratio. Strehl ratios of this type are weighted by the radius, and so are often strongly dependent on the poorly-sampled outer zones of the mirror.

These differing claims of surface accuracy are not inherently dishonest, as long as they are given in sufficient detail that one can pick apart their meanings. The Strehl-ratio tolerance is superior to the Rayleigh limit because it quantifies the fraction of the wavefront that is bent away from a perfect sphere, and it is the most useful one-number criterion if the measurement has sampled the whole wavefront rather than a simple diameter. However, the Rayleigh tolerance is far from useless, particularly because both are typically measured by amateurs using broad zones of the mirror surface. For smooth, cylindrically-symmetric errors, i.e., defocusing, low-order spherical aberration, and refocused higher-order spherical aberration, the ¼-wavelength peak-to-valley and 80-percent Strehl tolerances are functionally and quantitatively identical.

Accuracy specifications in various forms are seldom written with decoding instructions, and consumers are left to wonder what the claims mean, even if they know the distinctions. In recent years, a certain fraction of quality telescope makers have avoided the whole question of assigning numbers to their optics. They merely state that their optics are diffraction-limited and let it go at that. They let their reputations carry the rest of the load. Such a designation is better than artificially inflated numerical claims. "Diffraction-limited" conventionally means that a Strehl ratio greater than 0.8 has been met (Schroeder 1987), but it may also mean the quarter-wavelength Rayleigh tolerance is satisfied, depending on the habit and training of the person stating the number. High-quality makers generally have unstated tolerances even higher than these minimum limits, though they will not tell the public what they are.

Another factor often neglected in statements of optical quality is the slope of the error. If sharp channels, turned edges, or roughness appear on the optics, the overall wavefront error can often be contained within the Rayleigh or Strehl tolerances. The anomalous slope does not persist over the whole aperture, but the sharply-sloped fault still diverts some light out of the central spot to contaminate the quality of the image. According to either the Strehl ratio or the peak-to-valley Rayleigh tolerance, the optics are still "officially" good, yet a noticeable error exists in situations requiring detection of a dim target next to a bright source. This failure of the specification numbers results from lack of sensitivity to the size scale of the error in any of these single-number wavefront criteria. *Simply put, any one number cannot contain sufficient information to completely describe an optical surface. It must neglect certain aspects of quality in favor of others.*

A. Danjon and A. Couder addressed this topic in their book *Lunettes et Télescopes* (1935, pp. 518–522). They noticed that some instruments could slip by Rayleigh's limit yet possess enough surface roughness to scatter a hazy glow through lunar and planetary images. They stated that optics could not be judged as "good" until two conditions were simultaneously met:

1. Over the greatest part of the aperture, the wavefront has a mild slope and does not divert light rays outside the diffraction disk.

2. The Rayleigh ¼-wavelength tolerance is everywhere obeyed, and over most of the aperture, deviations should be appreciably less.

Condition #2 is just the Rayleigh limit, with a verbal warning having the same goal as limiting the root-mean-square (RMS) deviation used in most Strehl-ratio calculations. After stating these two conditions, Danjon and Couder point out that condition #1 on the slope of the mirror is typically more difficult to meet than the ¼-wavelength condition.

Incidentally, for aberrations that smoothly change over the whole aperture (such as the error sketched in Figure 1.2b), the maximum wavefront departure that leads to condition #1 is closer to $\frac{1}{6.5}$ wavelength peak-to-valley. Thus, optics that truly satisfy both Danjon and Couder conditions are not only good, but excellent.

It has become fashionable recently to disparage the Danjon and Couder criteria, with those doing the criticism claiming that condition #1 is unnecessary, and that all analysis techniques based on it are invalid (such as the method described in Millies-Lacroix 1976[4]). They seem to express a distaste for any method based on transverse aberration (or how far rays are diverted to the side in the image). However, condition #1 contains a perspective on optical quality that is inherent in neither the Rayleigh peak-to-valley specification nor the Strehl ratio. True, such ray-trace criteria don't describe the shape of the diffraction image, but neither do wavefronts inferred at the pupil. The diffraction behavior may be indirectly

---

[4] The first appearance of the "tornado plot" was in a publication by Yoder *et al.* (1953), who there referred to the effect of 3rd-order spherical aberration on the pupil. It is identical to the Millies-Lacroix plot except that the tolerances are somewhat looser. The fact that these authors are talking about the wavefront, and not limiting the discussion to transverse aberration, emphasizes the way that wavefront and transverse aberration are closely linked.

calculated from either. Additionally, the transverse aberration is a convenience to customers and fabricators. Keeping rays at the pupil directed at less than the diffraction spot radius is a workable way of specifying optics. Whether the D&C conditions are still the best way of stating the specification is arguable, but that disagreement does not reduce the quality of the many fine mirrors made through the years using criteria based on transverse aberration.

Another widely-used criterion is to specify the encircled energy ratio at some radius (or more than one radius) from the center. If you were contracting for a 1-meter telescope, you might be interested in keeping a certain fraction of the light within a very good ⅓ arcsecond as determined by the *seeing* at your site, than in reaching the seldom attainable diffraction resolution of $\lambda/D = 0.11$ arcsecond. You might demand 90% of the available light within 0.33 arcsecond, and 98% within 2 arcseconds, for example. Those two specifications would require not only control of the overall surface figure, but additional control of the smoothness.

Alternatively, some professional observatories will specify a global Strehl or encircled-energy ratio criterion to govern the overall resolution and figure of the mirror. Further they specify a tolerance value of what is called the *bi-directional reflectance distribution function* (or BRDF) at a couple of field radii that govern the scattering from middle and fine-scale roughness. Such detailed guidelines, however, are far beyond most amateur optics, amateur measurement techniques, and especially amateur budgets.

1.2.3 The Modulation Transfer Function

The most complete way of indicating optical quality is to present the detailed *modulation transfer function* (or MTF), which is the ability of an optical system to preserve the contrast of oscillating bar patterns of various spacings. It will be the method preferred in this book, because it can be calculated for any imperfect system. No optical difficulty can escape the MTF. Dusty optics, pits on the optical surface, spider diffraction, telescope vibration, microripple, aberrations, and obstructions of any sort reveal themselves in a lowered transfer function. MTF charts have the advantage of giving the spacing of detail that the optical problem attacks. The meaning of the MTF will be discussed in Chapter 3.

1.3 The Star Test—A Brief Overview

Observers rightly regard an out-of-focus instrument as nothing more than a problem to be cured. A telescope is either in focus or is almost useless—

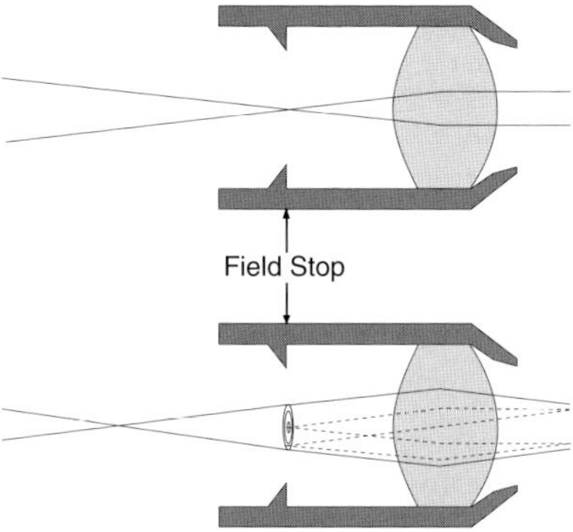

Fig. 1.3 The use of the eyepiece for the star test. The top arrangement shows the usual proper setting of an eyepiece. The eyepiece on the bottom shows the star test's inspection of the defocused disk as an equally valid setting.

at least for the job it was meant to perform. When used properly, a telescope must be focused as accurately as possible.

Implicit in the customary use of a telescope is the *fixed objective* assumption, which regards the image produced by the objective as the whole purpose of the telescope. The eyepiece is relegated to the secondary status of a mere accessory, a magnifier. It slides along the optical axis and has only one correct setting.

The star test uses the telescope in a new way. We must assume the eyepiece has a fixed position. From this viewpoint the eyepiece's field plane is seen as the whole purpose of the exercise. The eyepiece is regarded as *always* being in correct focus and examines whatever occupies its field plane. The field plane is usually constricted by a sharp-edged mask called a *field stop*. If you invert an eyepiece and look in the bottom, this stop is usually visible as a ring inside the base. The field stop is the crisp edge you see in an eyepiece. This edge has nothing to do—as it first seems—with the boundary of the objective.

Figure 1.3 shows an idealized eyepiece. We regard the eyepiece as fixed and the objective as mobile.

If, as in the top of Figure 1.3, the objective is located at precisely the right distance to put an image of a star on the eyepiece's field plane, the instrument is said to be *in focus*. The light coming from a point source leaves the rear of the eyepiece in a parallel bundle. Figure 1.3, bottom, shows an *out-of-focus* instrument. Here the path of the rays depicted by solid lines exits the tube in a converging bundle that is not focused properly on the retina unless the eye's powers of accommodation are very large.

The "looking path" of two points on the out-of-focus disk is denoted by dotted lines. One imagines that the bundle of light is sliced neatly at the focal plane of the eyepiece and that the eyepiece images this slice perfectly. If the eyepiece is moved to and fro across the position of focus, each slice can be examined in turn, and the memory of all such slices forms a collective record of the behavior of light near the focus. We are allowed to use this viewpoint because the in-focus location of the eyepiece is no more special than an out-of-focus location.

1.3.1 Diffraction Rings

Almost everywhere, the complicated situation of a converging wavefront can be approximated by replacing the wavefront with little "arrows" moving perpendicular to it. These arrows are called *light rays* and the intensity of such a beam can be calculated as the cross-sectional area of the ray bundle. However, elementary geometry used on a converging light cone leads to an important breakdown in the ray approximation. If a certain amount of power is incident on an area of aperture, the intensity can be calculated as this power divided by the area.[5] Halfway to focus, the area of the cone has shrunk to ¼ of its value right against the aperture, but it still contains the same power—so the intensity has increased 4 times. Move halfway again, and intensity further increases 4 times to a factor of 16 greater than it was at the aperture. You can double it again and again until you get to focus. What happens there?

The ray description has the area of the cone go to zero as the light approaches focus. This area must be multiplied by the intensity to make the power the same as it was all along the path. Because the speed of light through air is uniform, the energy content of the beam neither increases nor decreases. The ray approximation says that if the optics are perfect, the intensity of an image point is infinite. Needless to say, infinite intensity is impossible.

During the two hundred years between the invention of the telescope and the final acceptance of the wave theory of light, people actually believed there was no limit on optical quality. If optics were made of exquisite quality, the central spot would shrink in size—or so opticians thought. They must have agonized when their optical masterpieces, on which they had worked so diligently, still showed that disk surrounded by a system of rings.

We now know that there is a fundamental limit to imaging. Diffraction softens the image in the region of focus. For a given telescope focal length, the central spot (called the *Airy disk*) decreases linearly in diameter for larger apertures. The formula for the visual radius of the Airy disk with

---

[5] Radiometrically, this quantity is not the "intensity" at all, but should be called the "radiant flux density." However, the term "intensity" is used informally among physicists.

aperture diameter D and focal length f is

$$r_{\text{Airy}} = \frac{1.22\lambda f}{D}. \qquad\qquad \textbf{1.1}$$

(Here λ is the wavelength.) Thus, the diameter of the diffraction disk is 11 μm (0.0004 inches) for a 150-mm (6-inch) $f/8$ and 5.5 μm for a 300-mm (12-inch) $f/4$. Since 4 times the light was intercepted by the larger telescope and it was squeezed inside ¼ the area, the larger telescope has a central image intensity 16 times as bright. A focused diffraction image appears on the left side of Figure 1.4—the whole square is $(20\lambda f)/D$ across.

To those accustomed to purely symbolic equations, the factor 1.22 in the expression for the Airy disk seems messy and inexact, but it is unavoidable. Its origin is the circularity of the aperture, and the introduction of Bessel functions. If we were to make a square objective of distance x on a side, the brightest portion of the diffraction spot would be a little square with closed-form side dimension $(2\lambda f)/x$. Similarly, apertures with aberrations and obstructions have their own unique diffraction spot sizes. The Airy disk happens to be the perfect diffraction disk of the circular window that so many real optical devices resemble.

Diffraction, at any one wavelength, is an angular effect. The angular blurring of the image is not lessened by increasing the focal length of the telescope, or equivalently, the magnification. If one doubles the focal length, the Airy disk also doubles in size. Until this independence of fundamental image blurring on focal length was appreciated, telescopes were commonly specified by their focal length instead of their apertures. Nowadays, such terminology appears quaint.

The central spot is not the whole story. Thin, ghostly rings encircle the bright spot. With large instruments under ideal conditions or bright sources, they are observable out to 3 or 4 rings, but stars in small telescopes might show only one ring readily.

The out-of-focus pattern of Figure 1.4 is channeled by some circular furrows. These are diffraction rings, too, although their location and magnitude are not a simple matter to calculate. The first conjecture would be that the dark lines were the previously hidden structure of the diffraction image that is now exposed because the expanding disk filled it with light. However, this reasonable guess is wrong. The images don't behave that way at all.

As the focuser is racked in or out, the grooves continue to appear at the center and move outward, like ripples spreading from a pebble dropped in a pond. That central point successively dims to blackness and then lightens to a new ring. The one seemingly unchanging feature of the expanding

focused 　　　　　　　　　　　**out-of-focus**

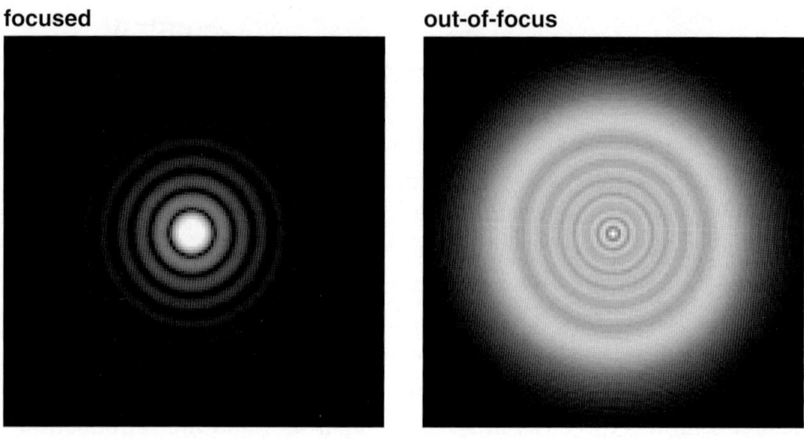

Fig. 1.4 The theoretical diffraction image of a perfect telescope, both for the in-focus and one defocused case. The direction of eyepiece travel does not matter; the out-of-focus pattern looks the same on either side. The frame at the left is magnified 4 times that of the right.

spot is the broad outer ring. It seems narrower as the focuser is turned, but it never goes away.

Notice another feature. The expanded disk is channeled, but its average brightness is more-or-less constant. The outer ring is somewhat brighter to average out that dark ring just inside of it. Still, except for the slightly brighter outer ring, the disk is remarkably flat. This principle is even more true for white-light diffraction images. Each contributing color exhibits a different number of rings in its expanded disk. The minima of one color sit on top of maxima of another color, and the net effect is largely to wash out any variation in the interior of the disk. Far out of focus, unobstructed telescopes display a flat disk with a clearly defined outer ring separated from the interior by a dark ring. Only a hint of grooved structuring exists inside, with maybe a central bright spot. One may think of this behavior as the transition from diffraction performance to the simple light-ray description of the defocused spot as a featureless disk.

Finally, a last characteristic is fundamental to the star test. The inside-focus disk is very nearly identical to the outside-focus disk. No circularly symmetric aberration can appear the same on both sides of focus. If the patterns are closely similar, and if they are circular, the optics are almost perfect.

Figure 1.5 shows an actual photograph of the defocused situation calculated in Figure 1.4. The contrast of the pattern has been increased by using one very pure red light from a helium-neon laser reflected in a small reflective sphere. A hole punched in metal and placed over a small refrac-

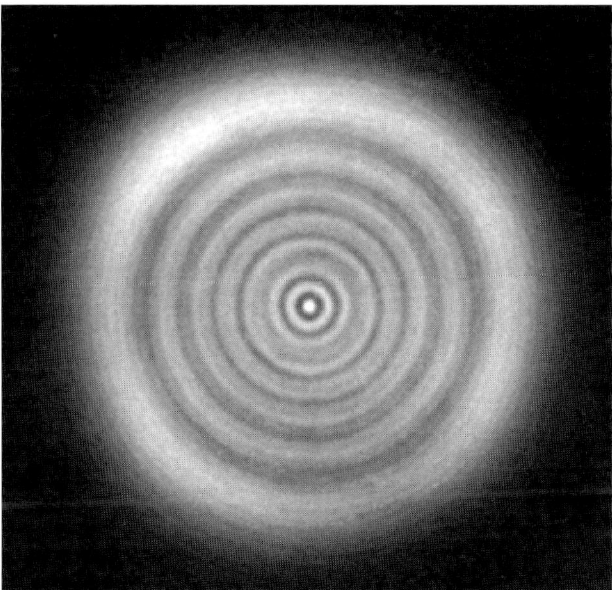

Fig. 1.5 Photo of actual defocused situation calculated in Figure 1.4. The image was taken with a strongly stopped-down refractor viewing a laser reflected in a tiny sphere.

tor created the aperture. The theoretical pattern reproduces the real behavior, even showing the terracing of the outer parts of the disk.

1.4 The Reason for Star Testing

The appearance of a defocused image as seen in Figure 1.5 is rare. Aberrations or other optical difficulties conspire to destroy this perfection. By moving the eyepiece inside and outside focus, you can detect and possibly identify the problems that affect your telescope. Comparing images seen at the same distances inside and outside focus is particularly powerful. The differences between these patterns will betray one of the most common optical errors, spherical aberration.

You will seldom see any one aberration unadorned by a mixture of other optical effects. The diffraction pattern is hard to diagnose using the star test without external information. Diagnosis is not the point, however. You may decide after inspection that one of the difficulties described here is dominating your system. If the errors are correctable or external to the telescope, you should try for identification of the problem as a guide for telescope or site modification, but don't be surprised if you cannot pin the error on a single cause.

As you read this book, you will learn the appearance of the best image possible, both in and out of focus. That appearance does not vary. It may be modified by obstructions, but the effects of a secondary are predictable.

A good stellar image has a short list of identifying characteristics:

1. The in-focus stellar image is circularly symmetric; it has a dim ring hugging the outer perimeter of the diffraction disk, and rings beyond that are vanishingly dark. (Bad optics have rings too, but they are bright and too many of them can be counted. They are often asymmetric.)

2. The out-of-focus image is circularly symmetric; it is almost precisely identical for equal distances on either side of focus. (The images of aberrated apertures can be one or the other, but they are not both identical and round.)

3. The out-of-focus image has a fairly flat distribution of intensity along the radial direction, except for a slightly brighter outer ring. It is divided by diffraction grooves, but they are of very low contrast and are sometimes suppressed by the mixture of colors.

4. If a central obstruction is used, its shadow reappears during defocus at equal distances on either side of focus. In the case of perfect optics, it appears to be the same fraction of the total diameter at equal distances on both side of focus.

You will also learn a systematic way of identifying optical errors and see image intensity patterns calculated for known amounts of aberration. Using this information, you may be able to estimate the size of your own telescope's aberrations and act on them if they are severe.

It is also useful to change viewpoints. Many useful concepts and procedures in modern wave-based physical optics (as opposed to old-fashioned ray optics) can be used by observers to more fully understand their instruments. The first and greatest of these concepts is to view the telescope as a generalized filter. We can then use ideas developed for the electronics industry, especially those electronics having to do with audio, with a few modifications changing the terminology to optics. The second concept is the idea that light is a wave. Diffraction imposes fundamental limits on image quality. Using the viewpoints of wave optics and filtering theory,[6] we are forced to rely less on the questionable store of folk wisdom, mythology, and belief that has accumulated around telescope use.

---

[6] This way of looking at the process of imaging is called Fourier optics.

Chapter 2
An Abbreviated Star-Test Manual

Below, the reader will be introduced to the star test. Many people have misinterpreted this tutorial as encapsulating all they have to know about the star test, with the unfortunate consequence of unfairly praising or disparaging their instruments. There is nothing in this chapter that allows an observer to assess the amplitude of the errors. For that information, the chapters discussing specific optical problems in turn have been written, and Chapter 5 aims at enabling people to attempt quantitative estimation. Those who read this chapter (and nothing else) know just enough to be dangerous. The same is true of various magazine articles that have appeared on star testing, including my own.

Nevertheless, this tutorial is a valuable map of the landscape before we get down to the somewhat myopic viewpoint of surveying each optical error in great detail.

In this chapter, the reader will learn a number of things:

1. The focal-length eyepiece range to use with the star test.
2. The source magnitude of stars to view while star testing.
3. A classification of optical errors into those characteristic of the design, errors that may be repaired, and errors that are mistakes in fabrication.
4. The appearances of the pure-form optical errors, as well as a brief introduction as to what causes them.

2.1 Some Necessary Preliminaries

You will need a high magnification eyepiece to conduct the star test successfully. The ideal power to use varies from telescope to telescope and depends on what you are trying to see, but it is around 1x per millimeter of aperture. For a commercially-available 200-mm Schmidt-Cassegrain telescope, magnification should equal 200. Since the focal length of such an instrument is 2000 mm, the eyepiece focal length should be around 10 mm. In other words, the eyepiece focal length in millimeters should really be lower than the telescope's f-number. A telescope with an aperture ratio of $f/6$ demands a 6-mm eyepiece; an $f/15$ refractor needs an eyepiece focal length less than 15 mm, and so forth. This rule is not hard and fast. If you have a little longer focal length eyepiece in millimeters than the aperture ratio, but can still see details in the expanded disk, please go ahead and use it. If you want to examine small details of the image when defocusing only a small amount (especially useful in detecting astigmatism), then you will need somewhat higher magnification than the 1 power/mm recommended here. If you are testing richest-field telescopes or finders, which are to be used at low power exclusively, you needn't use such a high magnification eyepiece.

The appropriate eyepiece is one with a minimum of optical surfaces. Try to obtain a good-quality Plössl or orthoscopic ocular and place the star precisely at the center of the field of view. More elaborate, multi-element eyepieces are fine for actual observing, but for assessing the primary optics these wide-field designs may be too sophisticated. I prefer that you put as little glass as possible between you and the image because eyepieces are themselves subject to compromises and faults of fabrication. If there is any question about the optical quality of the telescope, please try another eyepiece before downrating the instrument. If you are using a coma corrector or star diagonal, please remove it. You might want to include a Barlow lens, coma corrector, or super-wide angle eyepiece after establishing the baseline quality of the telescope to see the effect of the additional optical elements, but don't use them to judge the objective.

A major consideration in the selection of power in conducting the star test is reduction of the influence of the eye's own aberrations. Using the same focal length eyepiece as the telescope's aperture ratio results in the ocular yielding a 1-mm exit pupil, which serves as an aperture stop on the eye regardless of the latter's actual iris opening. Using only the center millimeter or less of most eyes allows them to work much better. At such a magnification over a target as large as a defocused star disk, the telescope's aberrations dominate the unremovable aberrations of human vision. On the other hand, if you use too high a magnification, you may get interference from floaters or other flaws in the eye. It is usually pointless to conduct a star test at a magnification much past 2 per millimeter of aperture, or 50 per inch of telescope diameter, and it may even be detrimental.

Set up your telescope so that it doesn't overlook nearby roadways or house roofs, because those structures hold the day's heat and slowly release it at night. Parts of the star test can be done peering through a small amount of turbulence, but it is best to avoid additional sources of aberration when possible. Also allow your telescope enough time between setup and star test so that it can approach ambient temperature (at least an hour with a small instrument, longer with larger ones). The instrument need not be transported to a dark-sky site since the test target is a fairly bright star.

The test is immeasurably aided by a well-aligned instrument. If possible, collimate your telescope before testing it. One of the sections below teaches you how to recognize severe misalignment, but further testing will temporarily be halted if the telescope is not at least geometrically aligned (see Chapter 6).

It is unnecessary to wait for full darkness, but you must be able to find a star. The best star for 6- to 8-inch telescopes is first or second magnitude (Polaris is good for higher latitudes if you don't have a tracking mount), but larger instruments may require correspondingly dimmer stars. Small refractors may demand a first- or zero-magnitude star. It all depends on what you're looking for and how much you defocus. If all you're wanting to do is defocus a trifle (for example, to detect a small amount of astigmatism) you may need a low brightness star. Daytime or evening tests can also be performed using the glitter of the Sun reflected in a smoothly spherical Christmas-tree ornament. Place it 60–300 meters away[1] with the intervening path over grass. Try to hang the bulb so that it is viewed against a smooth background.

If you wear eyeglasses, leave them on. This warning should only apply to those suffering from astigmatism, but refocusing a strong correction normally handled by glasses may force the whole optical system to add anomalous aberration. It's easy enough to prevent this extra source of error. Besides, you only need to inspect the center of the field of view and can withdraw your head enough to slip the eyeglasses between your eye and the eyepiece.

Finally, before you begin to evaluate your instrument, decide what you realistically expect it to do. Does it have a low aperture ratio for that type of instrument? A Newtonian telescope of $f/4.5$ or below is making a compromise between corrected field-of-view and physical size. It should not be judged to the same standards as a Newtonian mirror with an $f/8$ aperture ratio. Similarly, a richest-field achromatic refractor with focal ratio of $f/5$ should not be held to the rigorous tolerances of a long lunar-planetary refractor.

2.2 Optical Problems in Turn

The star test can be used with any telescope if you keep in mind only one of the principles discussed in the last chapter: *If the optics are perfect, the*

---

[1] See **Section 5.2** for tables of source diameter and placement.

*defocused images are fairly uniformly-illuminated disks. They appear the
same at similar distances inside and outside of focus.*

The first procedure to try is called the "snap" test. Sadly, this crude
evaluation is often sufficient to judge a telescope as bad. Using high magnification, wiggle the focuser back and forth through best focus. Good telescopes will snap into crisp focus quickly. Bad ones will display a range of
equally acceptable focus positions, none very good. You will be uncertain
about precisely where to stop your hand (Suiter 1990).

Don't condemn the instrument too quickly on the basis of poor performance on the snap test, however. This quick method has several practical difficulties. Often, the instrument is not rigidly mounted, and it is
difficult to tell the condition of the image while your hand is disturbing the
telescope. In addition, some observers have eyes that are capable of large
and fast focus accommodations. Eye focusing is not always under con-
scious control. As the focus is approached, the eye locks in and the human
focusing mechanism subverts the test.[2] Finally, the speed of focus depends
on the type of focuser and the aperture ratio of the instrument. A perfect
$f/15$ instrument has 14 times the depth of focus as an $f/4$.

The perfect circular aperture has only one appearance for the diffraction disk at each value of defocus. All modifications to this pattern represent various levels of optical problems. Such difficulties exist in three
general categories:

1. Baseline characteristics of the optical system, such as
    a. spider vanes in front of the mirror or mirror-clip blockages,
    b. secondary mirror or diagonal obstruction of the optical path,
    c. transmission variations (vignetting and imperfect coatings),
    d. residual color variation in the refraction of the lenses,
    e. irreducible and insignificant residual geometric aberrations of the
       design (such as higher-order spherical aberration in unretouched
       Maksutovs).

2. Transient or repairable problems, such as
    a. misalignment or abnormal tilts of the optical elements,
    b. atmospheric turbulence effects,
    c. artifacts caused by temperature differences near or within the tele-
       scope,
    d. unusual strains in the optical elements (such as pinching or sagging
       of thin mirrors),
    e. dust or dirt on the mirror or lenses.

3. Errors on the glass, such as

---

[2] A way to counteract this automatic focusing will be discussed in Section 5.3.

    a. low-order spherical aberration (a failure of the converging wavefront to conform to a sphere),

    b. rough optical surfaces (causing light to be scattered from the central diffraction spot),

    c. zones (light circular hills or trenches in the mirror, including turned-down edges of the glass disks and dimples or mounds at the center),

    d. chromatic aberration (uncorrected color errors),

    e. astigmatism (non-uniform stretching of the image along an axis).

Some of these errors are discussed briefly below, and some are saved for the detailed discussions later. What you must keep in mind is that there is a proper ordering for consideration of optical errors. You have to learn to recognize the characteristic out-of-focus star patterns that are a consequence of your choice of optical system (category "1" above). For example, nothing will get rid of the obstruction in non-tilted reflectors, so you must expect its shadow in the defocused star image. Also, the slight irreducible zonal spherical aberration remaining in many Maksutovs may be visible, so you must be prepared to see it.

Next, you must learn strategies that lead to minimum interference from the temporary or curable difficulties of category "2." You cannot make a valid check for slight astigmatism if your instrument is hopelessly out of alignment. If you take a warm telescope from the house to the cold outdoors, the images will dance and swirl. Your tube ducts the heat off the mirror, causing the stretching characteristic of tube currents. Often, optics are cruelly bound into their cells, and the warping caused by such stresses masks true optical errors ground into the glass. Finally, you must determine if the atmospheric seeing is good enough to do a valid test. While the star test is less sensitive to turbulence than the double-star resolution evaluation, it is best done under quiet skies.

Once you have determined that category "2" errors are sufficiently small, you can go on to look for errors in the shape of the glass. These errors are the most debilitating because nothing can be done about them, save refiguring the optics. Even here, the order is important. You should look for simple spherical aberration first because it is the most common and likely error.

Finally, you should not defocus the image either too far or too little. If you do, you will either find no errors at all or unfairly exaggerate insignificant errors. If you are testing a fast f/4 Newtonian, you will probably see most of these errors best with eyepiece motion less than $\frac{1}{20}$ inch (1.25 mm). If you are testing a slow f/10 Schmidt-Cassegrain, the same sequence of images is visible with eyepiece motions of $\frac{1}{3}$ inch (about 8 mm). These values are taken from the behavior listed in Table 5.1. Simi-

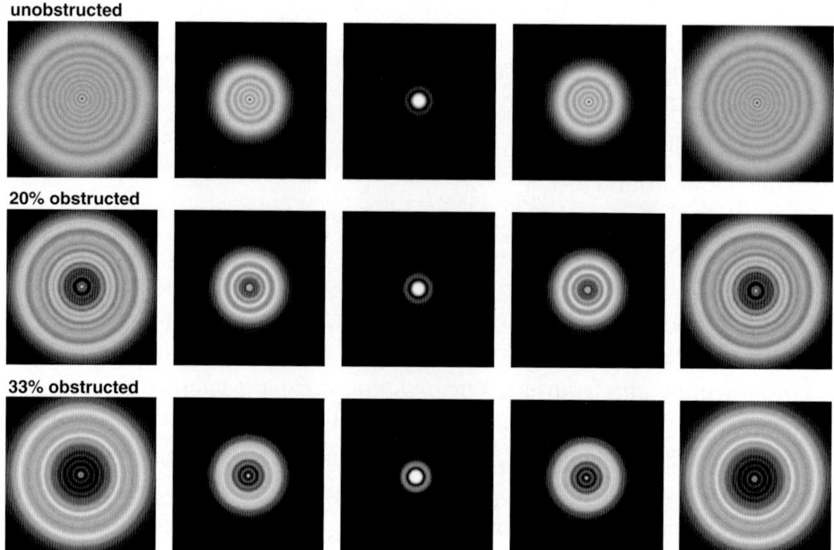

Fig. 2.1 Central Obstruction. Monochromatic images from three perfect telescopes are shown from inside focus (left) to outside focus (right). Typical behaviors of (top) refractors, (middle) long-focus Newtonians, (bottom) commercial Schmidt-Cassegrains.

larly, you can use very small defocus motions to identify the offending aberration, but be sure to consult the information in Chapter 5 lest you detect a tenth-wavelength of built-in residual aberration and unfairly accuse a maker of delivering an unsound instrument. Some catadioptric telescopes are designed with an inconsequential amount of higher-order aberration as a trade-off for other desirable properties. It is not the maker's fault that the star test is so sensitive that you can detect residual aberrations of this size, nor the star test's fault for detecting them. It is your responsibility to accurately assess the magnitude of the errors.

2.2.1 Secondary Mirror Obstruction

The inside-focus pattern is the same as the outside-focus pattern even for obstructed instruments. Though the presence of a diagonal or secondary mirror degrades the optical quality a little, the appearance of the aberration-free image on either side of focus is the same for good optics. As an example, an otherwise-perfect aperture is shown in Figure 2.1 as the obstruction is increased to 20% and 33%.

The most remarkable change from the unobstructed behavior is the coarseness of the pattern. We have traded the fine tracery of the unobstructed pattern for the unsubtle variations caused by the secondary. Only two wide bright rings are shown in the lit region outside the secondary's shadow in the 33% pattern. Fewer dips in intensity are seen across the disk. This is especially true of a white-light image, where the middle portions of the unobstructed pattern cease to have much detail at all.

The sole difference between these three rows is the size of the secondary. Why should a tiny obstruction—at most only $\frac{1}{9}$ of the surface area—have such a profound effect?

First of all, it doesn't really cause that big a difference. Those dips are extremely delicate and they are altered by nearly anything (as in mixing together colors in white light). Except for the minor brightenings and the shadow at the center, the disk is still fairly uniform, and it is still identical on opposite sides of focus.

For the purpose of diffraction calculations, we are allowed to pretend that the shadow is a miniature aperture, only out of phase with the larger opening (called Babinet's principle—Hecht 1987, p. 458). This tiny blockage has a much larger diffraction pattern, just as if the telescope were stopped down to the secondary's size. But remember, it is imbedded as a tiny correction to the larger pattern surrounding it. Thus, the small shadow's diffraction pattern is not visible as a distinct entity, but it causes enough coarse changes to destroy the low-contrast variations seen in the out-of-focus disk of the unobstructed aperture (see Figure 1.4 or 1.5).

The other feature of note in Figure 2.1 is the little bright spot at the center of the out-of-focus shadows. Interestingly enough, the successful observation of such spots is part of the experimental confirmation of the wave theory of light. When Fresnel first presented his paper on the wave theory to the French Academy in 1818, one of the listeners was S.D. Poisson, a vehement opponent of the light-as-waves description (he was a proponent of the light-as-particles, or corpuscular theory). He used Fresnel's new theory of diffraction to show that a bright spot should appear at some locations in the shadow of a circular obstacle. Poisson thought that such a ridiculous conclusion would settle the nonsense about light waves once and for all. Imagine his chagrin when this spot was found soon after. In fact, it had been reported long before, and had escaped notice by optical theorists. This little bright patch is still called *Poisson's spot*. When turbulence is influencing the image, you may not see Poisson's spot (it is very small), so its presence or absence does not indicate anything, but look for it all the same and you may occasionally detect it.

The spot appearing here is a composite of the central spot expected from the full aperture and the inner negative aperture. However, a Poisson spot is observable without even using an outer aperture just by carefully arranging a circular obstruction in a beam of light diverging from a pinhole.

focused OB=30%   **out-of-focus OB=30%**

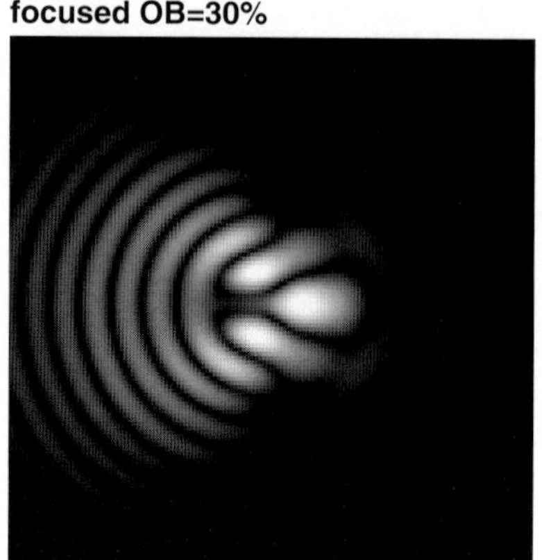 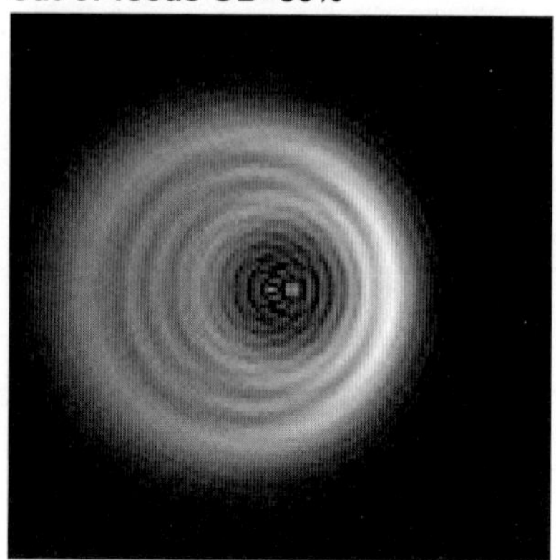

Fig. 2.2 Misalignment. A severely misaligned reflector with a secondary obstruction of 30%. On the left is a focused image and on the right is the same image defocused. The magnification of the focused image is 6 times that of the defocused one.

2.2.2 Misalignment

Misalignment can be easily exposed by star testing. In fact, before testing can be continued much further, the optics must be properly collimated. This procedure is described in more detail later, but Figure 2.2 depicts a typical misaligned Newtonian telescope. The region of good imaging may be outside a high-magnification eyepiece's field stop in a misaligned telescope. Misalignment aberrations can be straightforwardly corrected, so you should not greet their appearance with a feeling of dread. Anything that can be done in the alignment procedure may be undone. Assuming nothing is wrong with the glass, such aberrations will vanish in the center of the field with proper alignment.

The focused image of the Newtonian has a small amount of astigmatism but mostly shows the effects of coma. The word *coma* comes from Latin, meaning "hair" and is also the root word of the familiar astronomical term *comet*. Coma, when severe, stretches the image out into two wing structures.

2.2.3 Atmospheric Motion and Turbulence

Another troubling image aberration is caused by wavefront passage through the many atmospheric turbulence cells occupying the long cylinder of air in front of the instrument. This tube of air is always part of an earthbound optical system. Refraction by air is very small, but it does exist. It changes with temperature and pressure.

Cold air is denser than hot air, and the ability of air to refract light

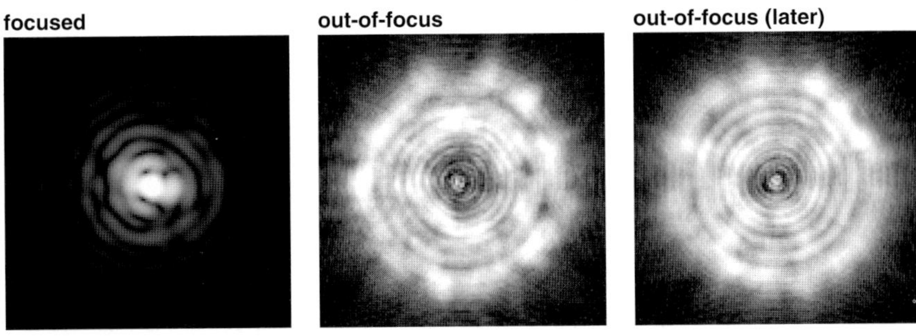

Fig. 2.3 Air Turbulence. A frozen moment of a turbulent image is shown in a focused pattern (left) and a defocused pattern (center). A delay of only a moment can show the example pattern at right. The focused image is magnified 5 times that of the defocused images.

increases with density. Density differences also result in a fundamental instability of the atmosphere. Sunlight chiefly heats the ground, leaving little energy in any particular volume of air in its passage through the atmosphere. The hot ground warms the air above it, and this warmer air becomes less dense. The cool air above it falls, and the warm air rises to occupy the former position of the cool air, making a temporary vortex. At some point, the decrease in pressure cools the rising warm air, and the falling cool air warms for the opposite reason. The two volumes of air have just changed places. The ground is now slightly cooler, and both volumes of air are slightly warmer. Perhaps the upper volume of air will become unstable with respect to a still higher volume of air.

Inefficiencies of the process make the difference in temperatures of the two layers less profound, and without further heating of the ground, the process will wind down like an old clock. Such motion is an invisible cyclical heat engine, transporting energy from the hot ground to the cool upper atmosphere. The energy of sunlight percolates upward by convection through a cauldron of air.

During the daytime, this heating continues. Convection cells persist and sometimes grow. Leavened by the efficient heat transport properties of water and taking advantage of the shade provided by clouds, convection structures can expand to colossal dimensions—beyond little wisps and dust devils, beyond even thermal currents exploited by soaring birds. They can grow to summer afternoon thunderstorms. At the margins of the major convective structures are fine-scale cells, and at the edges of those little cells, even finer scale cells, until the swirls disappear into seething turbulence at microscopic scales.

It is no surprise that solar observation is usually best conducted in the morning, before the ground has heated much. A daytime test using the image of the Sun in a reflective sphere is best done in the very early morning. Also, turbulence is strong anywhere near clouds, since clouds are a flag that indicates one of these gigantic heat engines is operating.

It is nighttime that mostly concerns us, however. Without sunlight to drive the process, convection must rely on residual ground warmth or currents in the atmosphere caused by masses of air at different temperatures. At some time during a clear night, the ground cools by heat radiation to have a lower temperature than the air above it. Because cool air is denser than warm air, this result is more stable. The convective cells on clear nights are less spectacular. Through near-pinholes, such as the human eye, this effect is seen as a slight spreading or bunching of light—stellar "twinkling." Through larger apertures, such as a telescope, the effect seldom appears as changes in brightness. In small telescopes, the image jumps around; in larger instruments the image is fixed but blurred.

In the outside-focus star test image, such cells look like nothing so much as the dappling of sunlight on the bottom of a swimming pool. Driven by high altitude winds, they wash across the aperture.

Figure 2.3 models a snapshot calculation of such a roughened diffraction disk. Look particularly at the focused image. The diffraction pattern varies so rapidly that such appearances aren't often directly observable. At least you can follow the changes in the outside-focus pattern as the cells sweep rapidly across the front of the telescope. The most visible change in the focused pattern is the angle at which splinters of light appear.

2.2.4 Tube Currents

Yet another atmospheric effect concerns the telescope user. Small telescopes are often carried from a warm house to the cold outdoors. These days, they are frequently transported to dark-sky sites in a warm automobile for some distance before reassembly. Even permanently mounted observatory instruments are seldom maintained at precisely the outside temperature at which they will be used. As a consequence, portions of the mounting—and most notably the mirror itself—must cool.

Glass has a high heat capacity and the accompanying low heat conductivity. In other words, glass optics have both a lot of heat to dump and the inclination to hold on to it a long time. Large, full-thickness mirrors take hours to cool.

Air convection is responsible for a large fraction of the energy transport. In the presence of gravity, the air warmed by higher temperature surfaces in the tube is displaced by falling cool air. In an open-tubed reflector

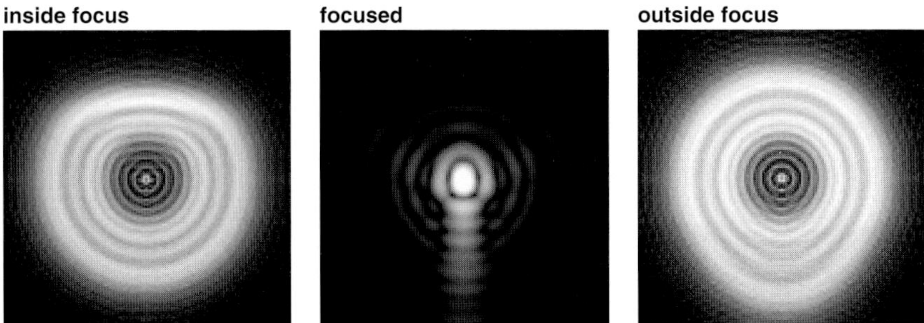

Fig. 2.4 Tube Current. A common tube current appearance is the squeezed or herniated lobe on one side of the disk, and a flattened look on the other. Magnification of the focused image has been increased 6 times.

that is tilted at an angle, the result is a tube current, as seen in Figure 2.4. Other cooling effects are visible in windowed reflectors and even refractors, but they may not take precisely this form.

The heated air exits the telescope as it would a chimney. It generally hugs the upper side of the tube as it is ducted out. As light passes more quickly through the less-dense warm air, the wavefront there curls up like a page being turned. The light passing through the curled wavefront has already reached a sort of quasi-focus and started to diverge again, appearing below the unaberrated in-focus image like the moon setting over a quiet sea. After the telescope cools, these effects go away.

2.2.5 Pinched or Deformed Optics

Particularly common with overly-tight reflector cells or thin mirrors bending under their own weight is the aberration that has the appearance of Figure 2.5. Details of the deformation will change the precise out-of-focus

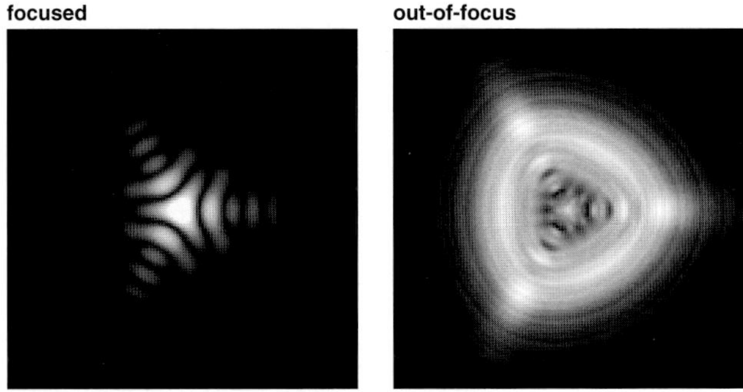

Fig. 2.5 Deformed Optics. The three-lobed pattern that results from too-tight mirror clips or a thin mirror that is inadequately supported. Left: focused pattern. Right: one appearance of the slightly defocused disk. Focused pattern is expanded 2.5 times.

pattern. This deformation depends on which clip is tight, how many support points hold up the mirror, and whether the optics are supported on the edge or on the bottom. It will change with different mounting designs and telescope elevations. In focus, the particular deformation modeled here results in a three-sided spike pattern.

2.2.6 Spherical Aberration

If abrasive is placed between two disks of glass and they are rubbed together with the orientation and stroke length completely random, what can be expected? At the end of thousands of such rubbing motions, part of two spheres must result—one convex and the other concave.

Thus, spheres (or more accurately, tiny bowl-shaped portions of spheres) are easy to make, at least compared with other three-dimensional surfaces. Unfortunately, spheres do not image properly. If a hemispherical reflector or a single spherical lens surface is imagined, it is not difficult to see why.

Let's say light rays are incident on such surfaces as in Figure 2.6. As the impact point deviates from a center-on-center direct hit, the focus wanders from a single point. Clearly, this aberration is an ever-present danger.

The mechanisms that evolved to correct this problem are fascinating because they point out the essential differences between refractive and reflective astronomical optics. They are almost two separate lines of development.

In the case of refractors, the spherical lens was retained and refined. In fact, most of the effect of spherical aberration can be corrected by the correct choice of curvature of the lens front surface relative to its back. In

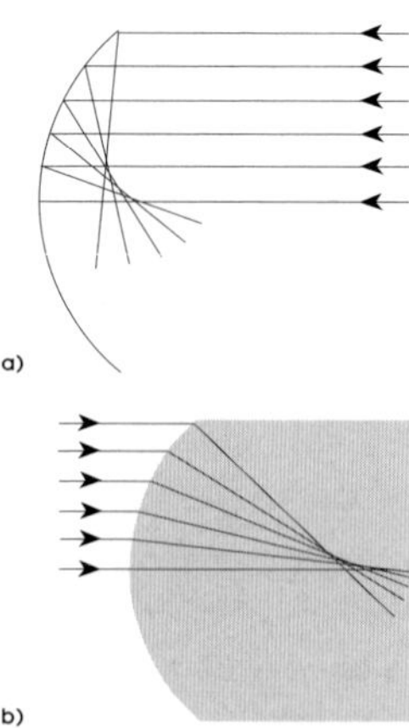

Fig. 2.6 Spherical aberration in reflectors (a) and refractors (b).

such systems, optical surfaces can be placed close together without getting in each other's way. More elaborate optical systems could be designed by invoking trade-offs between the curvature and separation of closely-spaced lens surfaces. The first optical designer to attack aberrations systematically was Joseph Fraunhofer. His masterpiece instrument, the great Dorpat achromatic refractor, was corrected not only for spherical and chromatic aberration, but off-axis coma besides. This instrument became the pattern for telescopes of the 1800s in much the same manner as John Hadley's reflector telescope had been the fundamental design of the 1700s. Fraunhofer accomplished this task with spherical surfaces (with a small amount of retouching) and clever design. Variations of Fraunhofer's air-spaced 2-element doublet refractors are sold today and still give outstanding images.

Reflecting telescopes took another path. If you place a lens very near a mirror, light traverses it twice, which may or may not be useful. Rear-surface mirror telescopes have been designed (indeed, one was suggested by Newton himself in *Opticks*, p. 105), but they are usually specialty instruments that are difficult to construct. All-spherical reflectors that correct low-order spherical aberration are not impossible, but they had to wait for sophisticated mixed lens/mirror systems (such as the Maksutov telescope). Makers of reflectors turned to aspherical optics quite early.

The aspherical surface for a one-mirror astronomical telescope is a 2-dimensional parabola that is spun on its axis like a top. This three-dimensional surface is called a "paraboloid." Unfortunately, the paraboloidal surface does not happen randomly. The maker must take the more-or-less statistical process that forms a sphere and control it (with pressure, special strokes, or small tools) to wear the surface to a paraboloid. Two-mirror and catadioptric systems (mostly Cassegrainians) have been devised that demand not only paraboloids, but every sort of conic section from hyperboloids to oblate spheroids.

Aspherizing requires good bench testing and understanding of materials and methods. As might be suspected, some telescope makers are more conscientious with such operations than others. Pitch, which is used for polishing and shaping the surface, is one of the most cantankerous materials used in any process. Fabrication can go badly when pitch is used without respect.

In my experience (up to 1992) of testing approximately 100 nominal paraboloids, about half of the commercial mirrors had been marginally undercorrected. They had a surface somewhere between a sphere and paraboloid, barely within tolerance or slightly outside of it. A quarter of the

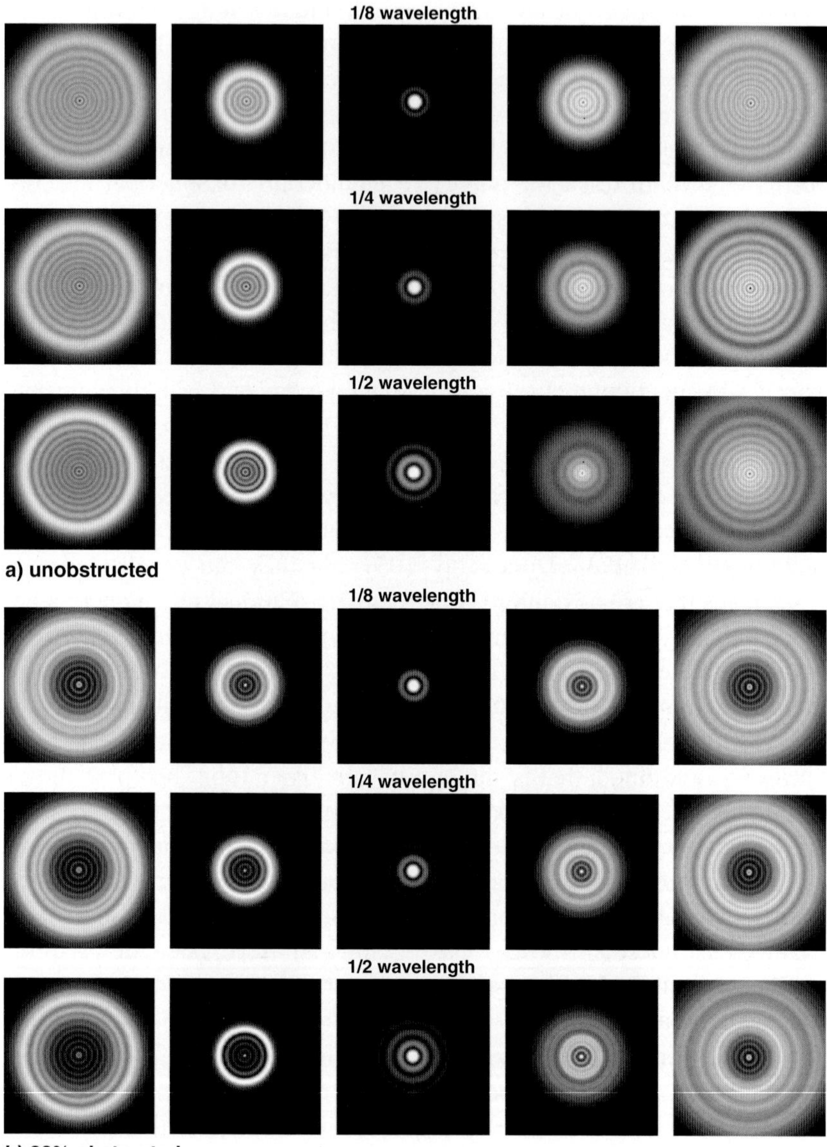

Fig. 2.7 Low-order Spherical Aberration. Undercorrection appearing in a) unobstructed apertures, and b) 33% obstructed apertures. Inside focus (toward left): much of the light is collected in a strong outer ring, leaving a dim center. Middle: The focused image steadily worsens as aberration increases. Outside focus (toward right): the outer side of the image fades away and the missing light is found near the center. (Focused-image magnification is 4 times that of the defocused frames.)

mirrors had been severely undercorrected, and about a quarter were figured within acceptable limits. Most of the undercorrected mirrors were of the short-focus variety, between *f/4* and *f/6*. Nearly all of the *f/6* to *f/8* paraboloids were adequately figured. Since 1992, my experience with Newtonians has improved considerably. Three quarters are figured acceptably and a third of those are excellent. Whether this represents changes in the market or just my testing fewer bad mirrors, I do not know.

I have seen few commercial Newtonian mirrors that were overcorrected—although many amateur-made mirrors seem to be. For experienced opticians, paraboloidal mirrors are not all that difficult to make. Because Newtonian telescope mirrors tend to be undercorrected, the makers must be spending as little time as possible on each one. This practice can perhaps be explained (if not justified) by the continued low prices of consumer mirrors. The makers are attempting to figure mirrors to the closest edge of the tolerance, minimizing time and costs. The inevitable statistical spread means that there will be the possibility that some fraction of such mirrors will be unacceptable.

Undercorrection is shown in Figure 2.7. Inside focus, much of the light is bunched up into the outside ring. Beyond focus, it has been pushed to a fuzzy patch in the center or to the outside of the secondary shadow. This is the signature of undercorrection. Memorize these patterns. If you have an opportunity to test Newtonians, you will see undercorrection repeatedly, and it is common enough in other types of telescopes. Lower-order spherical aberration is perhaps the only unadulterated glass error that you will ever see. The rest are usually mixed together.

Some people may misread the overall tone of these comments as a condemnation of Newtonian reflectors. Nothing of the sort is intended. The two best telescopes I have ever seen—and that includes refractors— have been exquisite Newtonians. The crisp star images of a well-made and well-aligned paraboloid are a beautiful sight. It is a shame that more of them do not perform as well as they can.

2.2.7 Rough Surfaces

Another common aberration afflicting telescopes is surface roughness. It results from using rapid polishing materials and maintaining insufficient contact between the pitch polishing lap and the optical surface being worked. The polishing operation is a strange and delicate process. Transient conditions on the surface of the lap can modify the contact between the tool and the optics. If the surface is polished by hand, one can feel the tool grab and kick almost like a living thing. Manual workers know immediately that something is wrong, and they can rewet the lap with polishing

focused   **out-of-focus**   **smooth**

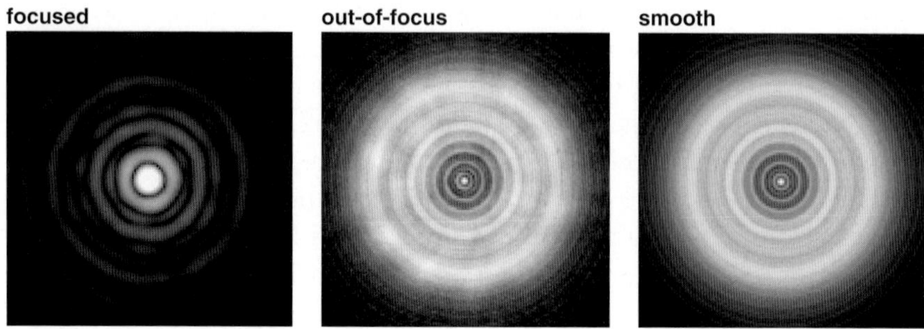

Fig. 2.8 Roughness. A small amount of surface roughness. Left: the focused image. Middle: defocused rough surface image. Right: a smooth defocused image for comparison.

compound and press until contact is once again established.

As a simple matter of economics, optics aren't worked much by hand. An 8-inch telescope mirror probably requires between 2 and 4 hours of the optician's time. Such work must be prosecuted rapidly with machines or the cost of telescope mirrors would quickly escalate out of the reach of consumers. When a lap begins to seize, the machine neither notices nor cares. It has power enough to roll over the lap's squeaking complaints. If the optician has not worked out a standing procedure to avoid such difficulties, one possible consequence could be rough surfaces.

Figure 2.8 is an example of the behavior of light in the diffraction pattern from an excessively rough surface. Of course, only one such appearance is illustrated. The pattern depends on such details as the scale of the roughness as well as its graininess and periodicity. These patterns, if the page is placed at a sufficient distance, exhibit a weak spiky effect. These spikes can be distinguished from the roughness induced in the atmosphere by their motionless aspect. Atmospheric turbulence shifts and changes. As a result, spikes shoot out from one side of the image and then another. Rough surfaces, however, are coldly rigid. The sky isn't usually steady enough to test for this problem on an actual star; most often this is an error best judged with an earthbound artificial source.

2.2.8 Zonal Aberrations

Part of what makes a successful optical surface is the effect of statistical averaging. One of the most paradoxical features of optical work is that the best surfaces are the result of superficially sloppy practice. Behind the variations, however, are carefully delineated boundaries.

Machines are less random. The operator must make adjustments to add a pseudo-random component to the stroke. If insufficient artificial

Fig. 2.9 Zonal Defect. Zonal aberration on an unobstructed aperture caused by a trench at 60% of the radius of the disk. Far inside of focus, the zone appears as a bright ring on the uniform disk, and outside it becomes a dark ring. On the right is an unaberrated pattern.

variations are imposed on the machine, it will tend to dig circular furrows or wavy deformations, called zonal defects, in the optical surface. Zones can also be the result of employing small polishers on a larger optical surface. Use of small polishers without sufficient blending can cause zones in both handmade and machine-fabricated optics. See Figure 2.9.

Zonal defects are common in small optics at the center. They appear as an indentation or a bump (the photograph in Figure A.3 of Appendix A is an example). Zones at the center are less harmful than zones appearing at other radii. In obstructed reflectors they may be confined largely to the shadow of the secondary mirror and are thus rendered harmless. Even if they show, defects at the center occupy only a small part of the surface area.

2.2.9 Turned Edges

Turned-down edge is a defect where the outer portions of the optical surface curl over gradually. This special case of zonal aberration results in a surprising amount of damage to the image because the edge of the aperture has a large fraction of the total surface area. It deflects more light.

Turned edge comes either from polishing pitch that is too soft or from applying incorrect pressure when the optical surface is extended over the edge of the polishing tool (Texereau 1984). Once generated, it requires a lengthy process to remove, and fear of it causes some telescope makers to overcompensate with an extremely hard grade of pitch. Others add adulterants that change the pitch's readiness to flow. Stiff or waxy pitch often worsens the problems with rough surfaces, though, so the net effect is a trade between two noxious aberrations.

In an otherwise perfect reflector, turned-down edge appears as a softening of the ring structure inside focus and a corresponding hardening of

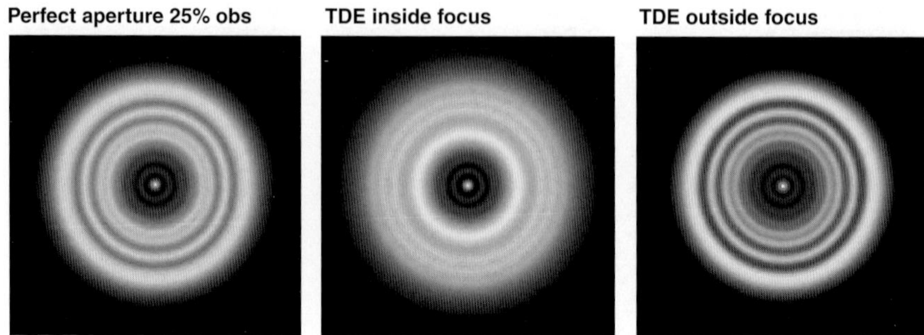

Fig. 2.10 Turned-Down Edge. Turned edge in a 25% obstructed aperture. A normal 25% aperture appears at the left. The inside-focus disk shows loss of contrast and a diffuse glow surrounding it. The outside-focus disk seems less affected in terms of light distribution, but the contrast is increased in the rings. Polychromatic image.

the ring structure outside of focus. To avoid confusion, look for this effect through a strongly colored filter. Figure 2.10 shows two 25%-obstructed apertures, one normal and the other with a turned edge. The light misdirected from a turned-down edge appears as a hazy glow in the vicinity of the image on the inside of focus (remove the filter when looking for this halo). It often is difficult to distinguish from overcorrection, particularly for broadly rolled edges.

On both sides of focus, a narrow turned-down edge displays a fairly flat distribution of light in the disk.

2.2.10 Astigmatism

Pure astigmatism can occur even in compound telescopes designed to eliminate coma (particularly refractors) if the system is not aligned properly. The cure is simple, and the aberration disappears quickly upon the telescope's collimation. Astigmatism unmixed with coma also appears in Newtonian telescopes that have curved secondary mirrors. Because of the 45° tilt of the supposedly flat mirror, a bulging or concave diagonal will be exhibited as astigmatism in the image.

Astigmatism in the objective itself is caused by three fabrication errors. It can result from pressing the disk against a cylindrical surface—for example, the rear surface of the disk may not be flat. Another cause is too rapid cooling of the glass disk when it was poured, freezing unrelieved stresses into the disk. This error is scarcely ever seen in deliberately made optical disks, but is common enough in portholes or other undocumented glass. And last, failing to rotate the optical disk with respect to the tool forms a cylinder into the disk directly.

In all cases, astigmatic images will show two oval patterns at 90° to

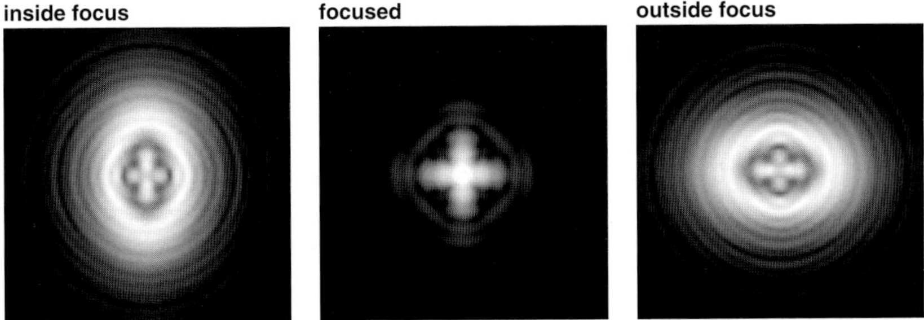

Fig. 2.11 Astigmatism. Appearance of astigmatism immediately on either side of focus. Middle: at best focus, the pattern is a cross. Out of focus, the profile is stretched into an oval, with the direction of stretch changing a quarter turn on opposite sides of focus.

one another on either side of focus (see Figure 2.11). If astigmatism is severe, certain focus positions will give stellar images that look like straight lines.

At best focus, astigmatism will show the indicated crossed pattern. Small amounts display a thickening of the first diffraction ring along the cross directions.

2.3 Concluding Remarks

This chapter is meant only as a brief atlas of the star test. Do not confuse this nodding acquaintance with real expertise, any more than reading a terrestrial atlas is a substitute for travel. You will continue learning new features of the star test years into the future, and the best trainer is experience in evaluating many telescopes.

The size of the preceding sections is one measure of their relative importance given as optical problems. If one topic can be recommended for study, it is low-order spherical aberration—i.e., simple correction error. Become an expert in its detection. You will see this error again and again.

One other thing must be emphasized. You will see marking numbers on some of the image squares in the chapters that follow. At the center bottom is the value of defocus in wavelengths. The units of this defocus can be decoded to actual eyepiece movement by using Table 5.1. At the upper right corner is a number related to the magnification used to view the pattern; in these units the tightest diffraction spot is always the same size. Please see Appendix D for a translation of what these numbers mean when you encounter the star test figures that use them.

Chapter 3
Telescopes Are Filters

The telescope is a device that reproduces an image of reality. It does not write on paper like a copy machine, but its images are just as unreal. It creates its reproduction on a tiny image plane only a few centimeters from the tip of the observer's nose. The scrutiny is done using a powerful magnifying lens called an eyepiece. Only through the multiplicity of imaged values (location, color, brightness) and the selective visual processing power of the human brain is this reproduction interpreted as reality.

Similarly, the visual image projected onto the retina is only representative of reality. Those with keen vision are able to derive more information about the external world than people with weak vision, but each individual tends to give equal value to that perception, regardless of its absolute quality. We tend to ignore errors.

Before going farther, look at Figure 3.1. Think of a filter as a process that degrades information contained in an image or signal. Filters are not only objects such as the colored disks of glass you might attach to the eyepiece. They are anything which removes information from the image, even relatively subtle factors such as the limitation on aperture and the wavelength of light viewed. This concept, which I call the "wobbly stack," represents a partial list of the filters between the observer of an image and reality. Some of the filters depicted are not independent. Eyepiece aberrations may, by good fortune or design, partially cancel the main instrument's aberrations. For example, wedge dispersion in the atmosphere was often compensated by slightly offsetting fields-of-view of the simple eyepieces common in the old days.

The effects of each of these filters can be lessened, but not all may be removed. For example, one can avoid atmospheric turbulence by going into space or using adaptive optics. One can lessen aberration errors by building near-perfect optics. One can even avoid the mushy errors of the eye-brain system by using the more predictable filtration of photography. So why wouldn't it be possible to put an absolutely perfect image onto a sheet of photographic film?

The most important form of filtration cannot be removed or diminished—a filtering caused by the *aperture*, or the finite extent of the window through which the telescope looks. Let's imagine that we have some sort of super-film that records everything that is projected on it. Further imagine that the instrument has absolutely perfect optics. We might naively suppose that if a photograph of Jupiter taken with such a system is inspected with a microscope, we would have a perfect image of everything at the distance of Jupiter—volcanoes on Io, tiny ice crystals in the atmosphere, the smallest cloud-belt whorls, etc. If we can do it for a full-sized telescope, we can do it for a little one. It might even be easier to make perfect small optics than large ones. If the optics are tiny, what then?

We have reached a fundamental limit. Small lenses don't image as well as large ones. The sharpness of the image of every optical system is limited by the presence of the aperture. If we rip a representation of the universe through a tiny hole, we should expect this image to be a bit scuffed by the passage. Passage of light through larger apertures results in less damage.

3.1 Perceptions of Reality

Figure 3.1 shows many limitations that we can do nothing about. Although other receptors are often used, most people interpret eyeball vision as the best representation of reality. Thus, we all send the signal through an unavoidable filtration system that happens to be attached to our heads. Some might argue that this filtering is always present and can be viewed as a sort of baseline. True, we are probably accustomed to our own vision, but we are most familiar with its performance under bright lights, using two eyes. In a telescope, with low illumination and monocular vision, many of the errors that would otherwise be incidental are worsened until they cause significant and unexpected information loss.

The only things that can be done about such unavoidable errors is to be aware of them and use strategies to lessen their effects. For example, the most commonly used corrective measure for the awkward distribution of low-light sensors on the retina is not to look directly at dim objects, i.e., to use averted vision. The unbalance caused by monocular viewing can be eliminated with binoculars.

The filters covered in this book concentrate on the center of the stack, from the atmosphere down to the image inspected by the eyepiece. This is not to say that these are the worst sources of error, but they are filters most associated with the telescope, its environment, and its use. They are the forms of filtering that we are most able to affect by corrective actions.

Eyepieces are neglected here. Although changing the eyepiece changes the optical system, it is not usually the worst source of optical dif-

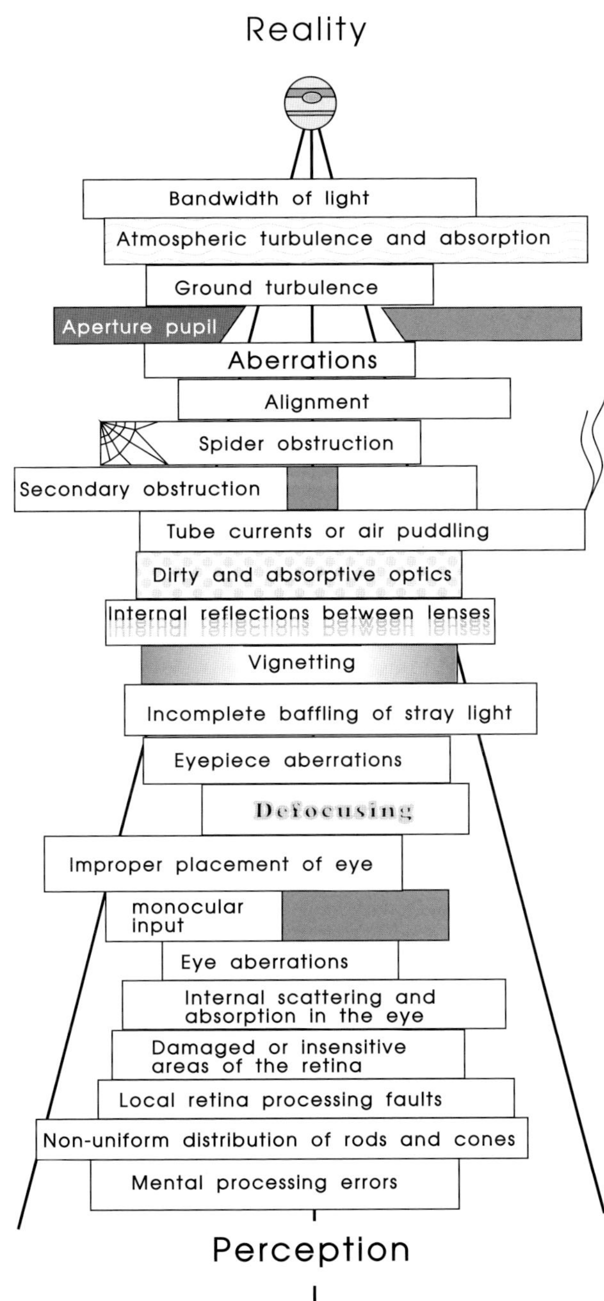

Fig. 3.1 A diagram depicting some of the aberrations, obstructions, misalignments and processing errors that can degrade an image: the "wobbly stack."

Table 3.1

Generalized telescope and associated equipment filtrations that behave
in an analogous manner to another reproduction device, the high-fidelity
sound system.

FILTRATION (telescope/eye/camera)	FILTRATION (sound system)
aperture diameter	size of speakers
colored filters	equalizer filters
image processing	signal processing
scattered light	audio hiss or other noise
spatial response	frequency response

ficulty. If a high magnification is used, the errors of the primary optics will dominate unless the eyepiece is entirely defective or it is used for the unusually steep light cone of a low focal ratio telescope. Also, a poor eyepiece can be identified by trying it in several instruments.

Aberrations that may seem less important are covered because they are generally caused by construction details peculiar to the particular instrument or repeated poor habits of use. For example, if you store your telescope in a hot shed and only use it in the early evening, you will be plagued by tube currents and local seeing effects. Telescopes that are misaligned seldom become better aligned the next time they are used. Optics don't tend to heal by themselves.

Clearly, the most desirable image would be a one-to-one mapping of points on the real object to points on the image. Telescopes already violate this principle by compressing much of the three-dimensional universe to a two-dimensional plane or slightly curved surface. Only for nearby objects are the images stretched out into a three-dimensional image space. Thus, a gas cloud 500 light-years away is imaged on top of a giant star visible 200 light-years beyond it. People are accustomed to this effect and tend to ignore it. Only where we have independent knowledge of the three-dimensional placement of the objects does this compression becomes objectionable. Many people are familiar with a similar perspective compression effect when they use binoculars at a sports event. If they are far enough from the action, the field seems flattened to a playing area a few steps deep.

For most purposes, the filtration caused by squeezing the image onto a sheet is not harmful. In fact, this and other forms of filtering can be quite useful. Astronomical telescopes are not troubled with the vanishingly small depth of field so bothersome in microscopy, where only part of the object is clear at any one focus.

Another positive use of filters is the way emission nebulae pop out from skyglow using narrow-band nebular filters. Although such filters diminish the interesting signal slightly, they profoundly reduce the superimposed skyglow. The observer is happy to take the somewhat weaker true image as a trade for eliminating artificial light.

3.2 A Comparison to Audio

Because generalized filtering is a difficult concept at first, let's use an example where the filtration ideas appear in our common vocabulary—the commercial sound system (see comparisons in Table 3.1). Because electronics are more recent than telescopes, and the terminology for audio was invented by engineers brought up on signal-processing mathematics, many modern words used for sound systems tend to have filtering concepts built right in. Many people are familiar with audio hardware or have a superficial understanding of the words. Comparisons, or at least some strained analogies, can be made to similar patterns in telescope filtration. In the audio descriptions, a frequency of 20,000 cycles/second or 20,000 Hz equals 20 kilohertz, abbreviated as "kHz."

3.2.1 Aperture Diameter/Size of Speakers

A typical stereo system has a cascaded set of sound reproduction devices called speakers, more technically named *transducers*. They convert electrical energy to the air compressional waves of sound. A typical stereo has subdivided speakers into at least two ranges, "woofers" and "tweeters." Low-frequency woofers are quite large, while high-frequency tweeters tend to be very tiny.

Low-frequency speakers, having to cycle a large quantity of air, occasionally move large distances and have enormous diameters. High-frequency speakers cannot be visually perceived to move at all and seem to work well in small sizes. Most people figure the reason the frequency range is broken up into multiple transducers is because no single speaker can cover the range, which is mostly true. However, we can easily imagine, if not build, a single speaker that would reproduce the entire frequency range with equal facility. Why would we still expect speakers to be divided into multiple units?

Flat speakers (and optics) transfer energy in an angular range according to how many wavelengths fit across the aperture. The speed of sound is about 330 meters/second. A sound at 880 Hz, or 880 cycles/second, would have a wavelength of $330/880$ or $3/8$ meter. If we had a speaker of this size, about 15 inches, it would put out most of the 880 Hz energy in a 120° cone. In other words, twice the 60° angle of an equilateral triangle 1 wavelength across the speaker at the base, 1 wavelength on each side. This size is effective for stereo imaging, because sound can spread widely into a room. So we're done designing a speaker, right?

Consider our imaginary, one-size-fits-all single speaker for a moment. It works fine at 880 Hz, but we can hear to frequencies about 20 times higher, to around 17,600 Hz. Young people might hear higher pitched sounds, older ones lower. Now the wavelength at 17.6 kHz is 18.9 mm (about ¾ inch). The emergent cone of energy from this same speaker is twice the narrow angle in a long, skinny triangle 20 wavelengths at the base and 1 wavelength high—less than 6°. At high frequencies, the angular spread of our single speaker is so narrow that we need to carefully aim such a source to hear it directly. Even worse, the mix of frequencies depends on whether we are directly in front of the speaker or sitting to the side, because each frequency has its own cone. The "sweet spot" of best stereo effect would be hard to find. High-pitched music from such a speaker may only be detectable in one ear at a time and may vary with motion of the head. Sound engineers still design multi-speaker systems of different sizes because no single speaker would emit sound into the optimum angle at all frequencies. They also design speakers in other shapes for this reason. Flat is sometimes not the best form for a speaker.

What is annoying in a sound system is desirable in a telescope, however. A typical telescopic aperture is 200 mm, or 360,000 wavelengths across. We still use the crude estimate of the angle of the energy cone similar to the speaker above—twice the narrow angle of a skinny triangle one wavelength high and 3.6×10^5 wavelengths across the base. This cone is less than 1.2 arcseconds across. Thus, most of the energy of a star detected by a 200-mm visual telescope can be found in an angle confined to less than 1.2 arcseconds.

One can also see from these resolution arguments why the location of a subwoofer (a very low frequency speaker) makes no difference. At a typical frequency of a subwoofer—say, 33 Hz—the wavelength is 10 meters. Any subwoofer smaller than a railroad car cannot even pretend to be able to direct the sound in any particular direction. Subwoofer location matters little because it radiates sound in all directions. The listener also isn't capable of hearing it in a unique location; the sound has such low frequency that normal two-eared perception is subverted by acoustic transmission directly through the head.

Similarly, radio telescopes have to be huge to offer any resolution. A 20-cm (8-inch) telescope has an aperture of only about one wavelength of the 21-cm line often observed by radio astronomers. If one were foolish enough to build a single-element, 8-inch radio telescope, a point image of 21-cm radiation would occupy a broad, fuzzy, 120° angle.

3.2.2 Colored Filters/Equalizer Filters

Equalizer filters are often added to sound systems to compensate for the room damping or reverberation. The rough equivalent in a telescope is to add color filters to the optical system. The color filters perform a similar shift or emphasis in the frequency spectrum.

It is no whim or accident that the sonic signature of an individual room is called the room's *coloration*. Sound spaces with unusually high amounts of upholstery or curtain absorption are called "warm," meaning that they absorb high frequencies much more strongly than low ones. Red and orange colors are also called "warm." The cure for a warm sonic space is to boost high frequencies. Similarly, when we use a blue filter on Mars, we can see high dust clouds more easily because they reflect more of the white sunlight than the ruddy Martian surface.

The brain itself acts as a compensating filter for the external world. Since the time of Edison, listeners have repeatedly declared their contemporary audio technology to be perfect, even when it was scratchy and indistinct. Our light perceptions are equally questionable. The brain will automatically adjust nearly any color balance it sees to the color balance it experiences beneath the Sun.

A good example is illumination under artificial lighting. Look into a fluorescent-lit room from the outdoors. Your color balance is held fixed by the brighter natural sunlight, so the room light you see inside is greenish. Mercury vapor lighting is composed (mostly) of two pure colors, a green line and a blue-violet one. Even under these extraordinary conditions, the eye manages to fool us most of the time. Only when the eye is confronted with just one color does it give up. Even then, I suspect if you gave the eye-brain system enough time, it would eventually adjust to perceive any single color as a kind of washed-out gray.

3.2.3 Image Processing/Signal Processing

So many techniques to process audio signals and images exist that they all cannot be included here. Let's look at an example—oversampling.

Sixteen-bit numbers are imprinted on a Compact Disc (CD) at the rate of 44.1 kHz, and they are read off the medium at the same rate. The values are then delivered to a digital-to-analog converter (DAC), which takes the numbers and "undoes" them back into a voltage that one hopes is a reasonable simulation of the original recorded signal.

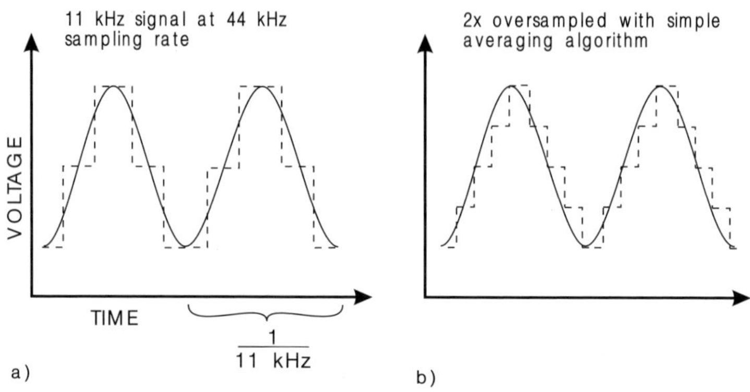

Fig. 3.2 Sampling of digital audio signals.

One would think that 44.1 kHz would be more than fast enough to reproduce sound of less than half that frequency, but it turns out that this is only barely sufficient. If the original signal is 11 kHz, a sampling rate of 44 kHz yields only four values per wave (see Figure 3.2a). The DAC is attempting to produce the dashed stair-step pattern, which does not simulate the original sinusoidal tone very well. The stair-step is rich in higher frequencies, called harmonics, of 22 kHz and higher. More exotic combinations of waveforms could actually create artifacts that would spill into audible frequencies.

One could use electronic analog filters to reject DAC frequencies 20 kHz and higher, but it takes extremely sophisticated (i.e., expensive) electronics to produce a precipitous cutoff beyond a certain frequency. Inexpensive analog lowpass filters attenuate sound at a rate of, for example, 12 dB/octave, or a factor of 16 with each doubling of frequency. In other words, a perfect amplifier feeding one of these lowpass filters with an intensity of 1 at 11 kHz has an intensity of $\frac{1}{16}$ at 22 kHz and $\frac{1}{256}$ at 44 kHz.

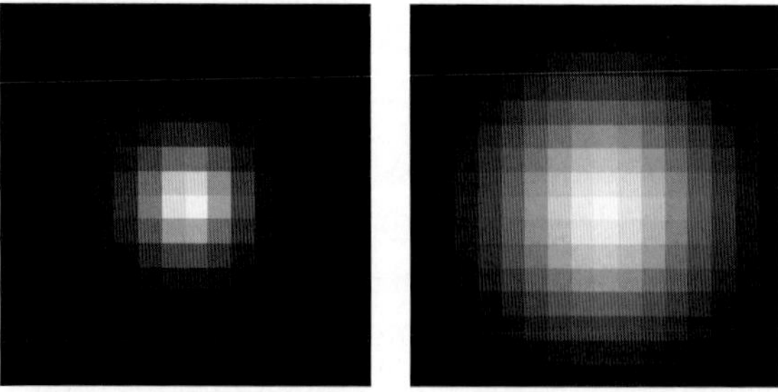

Fig. 3.3 Using higher magnification is a form of oversampling. Here, the object seen at higher power is spread over four times the area of the retina, allowing more receptors to participate in averaging the image.

The high frequencies are folded back into the low ones in a phenomenon called "aliasing." Also, some of the attenuation also leaks into audible frequencies. The response curve is unacceptable for high-fidelity sound reproduction.

The solution often used is to digitally "oversample" the signal. A DAC that operates at 88.2 kHz is used, and the signal is sampled at twice the usual rate. Using a digital filter with an interpolation algorithm (algorithm is a fancy word for *recipe*), one can achieve something like the dashed line of Figure 3.2b. The analog lowpass smoothing can then be readily applied with a cutoff frequency of 44 kHz instead of the more objectionable 22 kHz. Of course, actual CD players probably use much more sophisticated filters and algorithms than this one. This scheme has been presented only to give you a feel for the processing (Strong and Plitnick 1992, p. 440).

The simple act of choosing a higher-powered eyepiece than necessary to barely resolve the object is itself a form of oversampling (see Figure 3.3). Here, you are giving up the outer portions of the image area in favor of expanding the signal over more retinal receptors. Areas of the image can be identified that have a fairly constant signal strength and coloration, but if they are smaller than a retinal receptor, that intensity will be mixed together with a nearby area.

The retina, like any other light sensor array, is non-uniform. One mechanism the visual system uses to suppress this non-uniformity is to consider receptors in batches. A detail that shows on only one receptor may be interpreted as noise and ignored unless it is very bright or contrasty. Going to higher magnification allows you to expand small areas until they cover more individual receptors. The average over many detectors triggers this batch detection, so lower contrast details can be seen. This procedure works on low-contrast details even when the magnification expands the blurring of the telescope beyond the best resolution of the eye, which makes it a true form of oversampling. The eye has a maximum resolution of about 1 arcminute, but our perception continues to improve until magnification drives the radius of the Airy disk beyond 4 to 8 arcminutes.

3.2.4 Scattered Light/Audio Noise

Every sound system is troubled by noise. In some cases, this noise can be objectionable (as in scratchy old phonograph disks) and in other cases the noise can be imperceptible (as in modern digital systems). The absolute level of the noise is less important than its *signal-to-noise ratio*, or SNR. When the music is soft, a constant low level of noise can become unacceptable because it now sounds relatively strong compared with the interesting signal. However, the same level of noise goes undetected when the music is loud.

The signal-to-noise ratio is often compressed logarithmically to the *decibel* scale,

$$\text{SNR} = 10 \log_{10}(I_s / I_n), \qquad\qquad \textbf{3.1}$$

where I_s is the power of the signal and I_n is the power of the noise. For the strongest signals in a typical CD, the digital SNR is somewhere above 96 dB. That translates to a digital noise power of only 1 part in 3.98 billion (of course, there can be other sources of noise). For a high quality analog cassette tape, a similar number might be around 55 dB or an SNR power of 316,000 (Strong and Plitnick 1992, p. 441). Of course, the degradation of quality caused by passage through the rest of the sound reproduction system lowers these SNRs a good deal. Noise begins to become objectionable when it is 20 dB (1/100th the power) below the interesting signal and offensive when it is 10 dB, or 1/10 the signal power. When SNR reaches 0 dB, or conditions where noise and signal have equal intensities, people have a difficult time recognizing spoken single words, only catching about 70% of them (Kinsler *et al.* 1982, p. 284).

A rough analogy to noise in optical systems is scattered light.[1] Of course, scattered light is not really noise in the sense that it doesn't change with time, but it originates from such a multiplicity of sources that no one is going to trace them back to their causes. It is expedient to treat scattered light as a random process. Assume that the surfaces of your optics are dirty, or they are rough on the wavelength scale. Some light diffracts from the tiny irregularities and is scattered beyond the image of the object. If you are trying to observe a very dim target right next to a bright object, the smearing of the light, even though it is a very tiny fraction of the bright object's light, can be strong enough to render the dim object unobservable. The "noise" floor, or background, has risen enough to be objectionable. Contrast in the vicinity of the target is appreciably reduced.

We can calculate the scattering from a single round piece of dust, 1/1,000 of the diameter of the aperture across (for a 200-mm telescope, the speck of dust would be a 0.2 mm disk). One millionth of the energy incident on the aperture would hit the rear side of the speck and be absorbed or reflected. However, we would see a fairly normal Airy disk that has *two-millionths* of the energy missing (van de Hulst 1981). The other millionth part of the energy has been scattered throughout the field of view. If we

---

[1] Since the first edition, more people have become conversant with *photon noise*, a Poisson-distributed noise derived from individual photoelectrons in CCDs. Photon noise is no analogy, but true noise.

assume that we are looking at a large extended object and that none of the scattered light has been lost outside the region of interest, then the signal-to-noise ratio, or more accurately, the signal-to-background ratio is as high as 1,000,000 or $10 \log_{10}(1,000,000) = 60 \text{ dB}$, still far below the noise level of good magnetic tape. At 1,000 specks, we would find that our background could be well on the way to becoming noticeable at −30 dB or $\frac{1}{1000}$ the signal, which is approximately 7.5 magnitudes. Thus, if we were looking at a first magnitude star, we would see the scattered light in a fuzzy glow with a total brightness of 8.5 magnitude, not very bright at all.

Scattered light only damages the image in specialized observing situations. A background that is 24 dB down is only about as bright as the second ring of a perfect diffraction pattern. Such light would only be troubling if it covered something dim, a situation that is seldom the case in dark-field observation. Scattering is a worse problem in solar and lunar observation, or in the perception of very low contrast detail on planets.

Not mentioned before is the fact that the scattered light from 1000 dust motes is distributed into a huge fuzzy glow with an average width 1000 times the Airy disk. So even though the scattered light from 1000 dust particles has the integrated brightness of $\frac{1}{1000}$ the energy passing through the aperture from the primary source, its brightness at any one location the size of the Airy disk is still only a millionth the brightness of the star.

3.2.5 Spatial Frequency/Audio Frequency Responses

A good sound system is supposed to perform sound reproduction between the frequencies of 20 and 20,000 Hz (20 kHz). A frequency of 20 Hz is a rumble you almost feel instead of hear; 20,000 Hz is a fingernails-on-blackboard squeak. Outside these limits, it was once thought, humans do not hear well enough to make reproducing tones in recorded music worthwhile. More recently, audiologists have discovered that the situation is somewhat more complicated. Nevertheless, high fidelity sound reproduction is conventionally contained between 20 Hz and 20 kHz. Tones lower in frequency than 20 Hz are called *infrasound,* and tones higher in frequency than 20 kHz are called *ultrasound.* The analogies to "infrared" and "ultraviolet" are obvious.

Think about what the 20–20,000 Hz limits mean. As long as a tone has a frequency between these limits, the audio electronics will reproduce it without too much loss or unusual gain. Once frequencies go into the infrasound or ultrasound range, the electronics are allowed to fail badly and the sound is degraded or vanishes. In other words, even if a higher frequency tone were put in the front end of the audio electronics, little or nothing is reproduced at the speaker end.

A similar effect can be seen in telescopes, but in optical systems the most interesting effect does not take place in the domain of optical frequencies, or "colors." It occurs for angle.

When a telescope is pointed at a white picket fence not too far distant, the gaps can be viewed crisply against a darker background. If, instead, you tape a small piece of paper with alternating white and dark bars on the same fence and look at it through the telescope, you are less likely to see the bars reproduced well. At some fine scale determined by moving the telescope closer or farther away, you will see the bars printed on the paper dissolve into a gray blur. The instrument is no longer reproducing the reality that may be verified by moving closer to the object. You know that the telescope is looking at bars, but it is no longer transmitting them as distinct stripes. Here, the effect is very close to the inability of audio electronics to reproduce too high a frequency, only in this case a portion of the object cycled between dark and light and back to dark again as different angles were considered.

What is the analogy to frequency here? In the case of an audio tone,

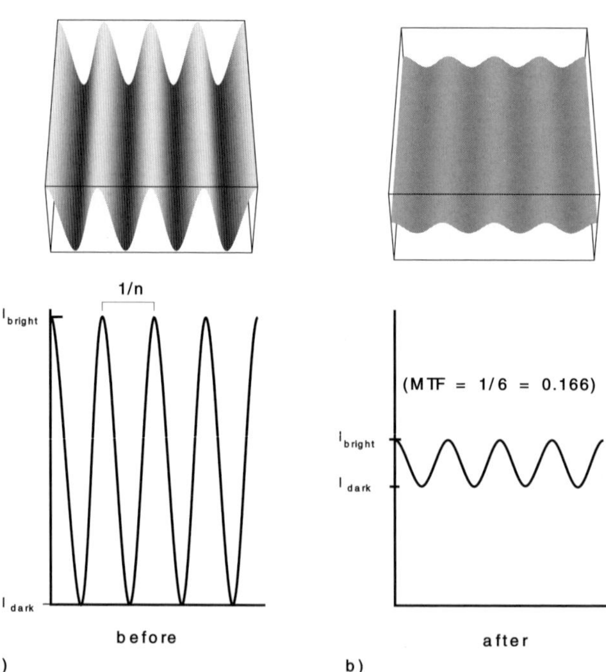

Fig. 3.4 Target intensity of modulated brightness: a) before filtering, b) after filtering. Note: the energy of the image is still the same, but in b) it is less modulated. The target is not actually corrugated—this is just a way of indicating surface brightness.

the standard units of frequency are cycles/second or Hz. In the case of astronomical images, the units are cycles/arcsecond. The resemblance to frequency is so unmistakable that the quantity used to describe the transmission of detail in an image is called *spatial frequency*. Spatial frequency has units of cycles/angle, but it is sometimes stated in cycles/distance in the focal plane—something like "200 line-pairs/mm" or just "lines/mm." It usually appears without the more precise "in the focal plane" trailer.

One might wonder if a low frequency cutoff exists in optical systems analogous to the 20 Hz cutoff of audio systems. You might think that none exists because dark and white bars are easier to see at low frequency. An effective cutoff is given by the limitations of the field, however. Once fewer than one line-pair shows in the lowest magnification eyepiece, the spatial frequency cannot really be said to be transmitted. A visual telescope working at the Schmidt-Cassegrain's typical aperture ratio of 10 has a maximum spatial frequency of 182 line pairs/mm (550 nm) at the focal plane. If that focal plane has a width of 10 mm inside a large eyepiece, the lowest spatial frequency transmitted is greater than one line-pair per 10 mm, or 0.1 line pairs/mm. The ratio of these two numbers is about 1820. Thus, the telescope has about the same 3 orders-of-magnitude bandwidth of the typical 20–20,000 Hz audio system.

3.3 The Modulation Transfer Function (MTF)

Many of the audio-to-visual comparisons above are interesting but have no practical application. Spatial frequency response of optics is a core issue, however, and there is a very important reason it is used. This response is the most objective measure of the quality of an optical system.

Spatial frequency response is generally written using the filtering concept of a transfer function. If the target displays a sinusoidal modulation—changing gradually from bright to dark—and we view the optical system as a black box filter, then it exits with a lesser variation. Perhaps the exit signal varies only from light gray to dark gray.

The spatial frequency target is not a light and dark bar pattern as in the picket fence example above but a smoothly varying pattern such as that shown in Figure 3.4a. Strictly, such a pattern should extend infinitely to either side, but practical use limits it to a few bars. The common 3- or 4-bar resolution chart is not a valid target field in the sense of either being infinite or sinusoidal, but it is such an easy target to use that it commonly serves to estimate optical quality anyway. One other requirement is that the illumination of the bar pattern is completely incoherent, usually a very easy requirement to fill.

If C stands for the modulation contrast and v is the spatial frequency, the way we will define the modulation transfer function (MTF) is

$$C_{\text{after}}(v) \; = \; \text{MTF}(v)C_{\text{before}}(v), \qquad\qquad \textbf{3.2}$$

where $\text{MTF}(v)$ is always less than 1.

Let's look at this simple equation and see what it means. If one has a target pattern with a certain value of modulation contrast, the transfer function of the optical system always acts on that contrast to lessen it. Readers familiar with filtering concepts know that, in general, the transfer function can change the phase of the signal (suggesting that the most general transfer function is complex). Here, we will be concerned only with its amplitude, the modulation transfer function. The modulation contrast (both before and after) is measured from the intensity levels at the darkest place on a dark bar and the brightest location on the light bar.

$$C(v) \; = \; \frac{I_{\text{bright}} - I_{\text{dark}}}{I_{\text{bright}} + I_{\text{dark}}}, \qquad\qquad \textbf{3.3}$$

where the intensities are measured as in Figure 3.4. Note that if the "before" pattern has a dark intensity of zero, the modulation transfer function becomes the modulation contrast itself (Hecht 1987, p. 507).

The MTF is all-encompassing and powerful. Even optical difficulties not originating in wavefront errors but in obstruction and non-uniform transmission find a way of being expressed in the modulation transfer function.

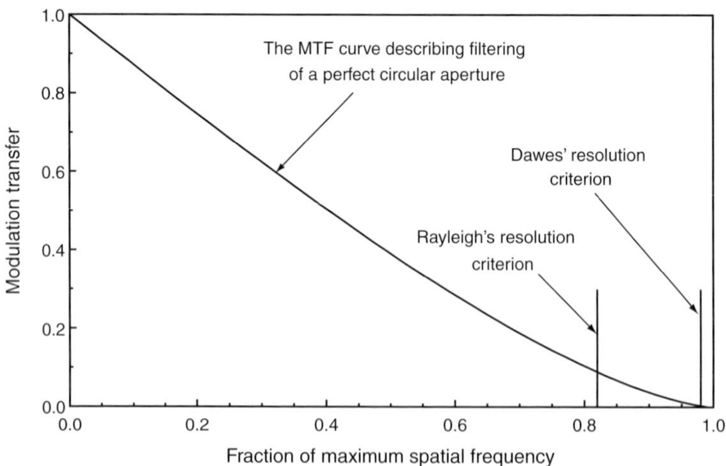

Fig. 3.5 The best-case unobstructed circular MTF. Incoherent lighting of the target is assumed.

What form does it take? One would think that in perfect optical systems, the value of the MTF would be 1 for all spatial frequencies. No such optical system exists although a large planar mirror comes very close. Consulting Figure 3.1 depicting the "wobbly stack" of filters, we can ask which filters are active even under ideal conditions. Clearly the atmosphere can be neglected—assume that the telescope is under perfect skies, or in space. The eye and its processing errors also will be ignored. Assume alignment and cleanliness are perfect and that vignetting can be ignored. This process can be continued until the stack is made as short as possible.

What is left? Remember, even a perfect optical system has these two filters: 1) the wave nature of the light used to make the image, and 2) the limited aperture. This irreducible minimum is enough to force the MTF to a value less than unity and determine where it goes to zero. The perfect circular aperture's MTF is depicted in Figure 3.5. It is a more-or-less linear fall to the maximum spatial frequency. This frequency is identical to the inverse of the maximum linear resolution of a sinusoidal pattern. The circular aperture causes a little curl at the highest spatial frequencies, but mostly the ideal MTF is a uniform slope. Figure 3.5 also shows the location of the inverses of the Rayleigh and Dawes criteria of resolution. The extremely tight Sparrow criterion is somewhat off the chart. The reason that these various criteria are different from the MTF resolution is because they refer to various types of point resolutions rather than sinusoidal bars.

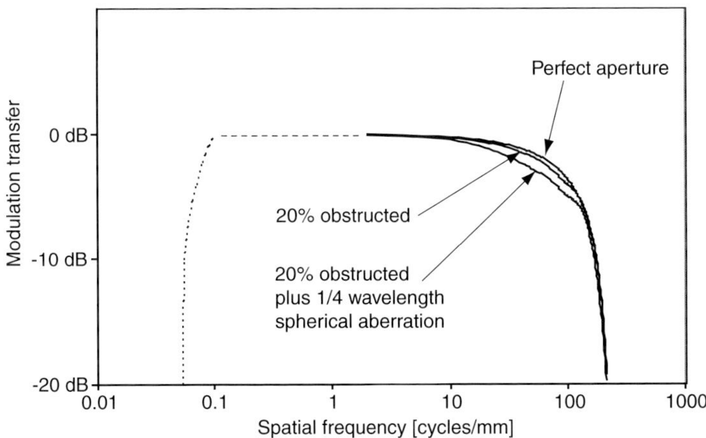

Fig. 3.6 A perfect optical MTF is shown with a 20% obstructed pattern and spherical aberration added. The dashed line is a sketch of the anticipated low frequency cutoff. Because we are most interested in the upper end of the chart, we do not plot optical frequency distributions this way, but rather as it is in Figure 3.5. Audio tones of a few hundred hertz are more important than the shape of the audio response at high frequency, so the logarithm emphasizes them.

Admittedly, this frequency-response curve doesn't much resemble the flat curve of the typical audio system, because it is plotted on a linear scale and extends somewhat beyond the spatial frequency bandwidth. (See Figure 3.6 for a sketch of an optical MTF plotted the way it would appear in audio specifications.) The logarithmic decibel scale used for music tends to compress the spectrum to a wide, flat-topped appearance ($0.5 = -3$ dB, $0.25 = -6$ dB, etc.). If this plot were of an audio system, the bottom axis would go from 10 to well beyond 10,000 Hz. In fact, most audible sounds occupy the lowest $\frac{1}{10}$ of the frequency spectrum. Much interesting detail in sky objects is also at low spatial frequencies where the transfer function is high, but the response curve of the high frequency end is where the resolution resides. Even so, this plot shows how similar the two systems are. The reason the audio and optical transfer functions appear so different is that the optical transfer-function is plotted to emphasize the highest frequencies where the transfer is likeliest to be eroded by aberrations.

The maximum spatial frequency (1.0 in Figures 3.5 and 3.7) is

$$S_{max} = \frac{1}{\theta_{min}} = \frac{D}{\lambda}[cycles/angle] \qquad\qquad \textbf{3.4}$$

with θ_{min} representing a separation angle slightly narrower than the $(1.22\lambda)/D$ angle associated with the radius of the diffraction disk, also known as the Rayleigh criterion. Here, a bar separation equal to the radius of the diffraction disk would occur at $\frac{1}{1.22} = 0.82$ of the maximum spatial frequency (marked with bar in Figure 3.5).

If we wish to use the notation that is more associated with cameras, the maximum spatial frequency at the focal plane (where F is the focal ratio) is

$$S'_{max} = \frac{1}{F\lambda}[cycles/length] . \qquad\qquad \textbf{3.5}$$

For example, if the aperture has a diameter of 152.4 mm, the aperture ratio is $f/8.12$, and the wavelength considered is 560×10^{-6} mm, then S'_{max} is 220 cycles/mm or lines/mm. S_{max} in angle notation for a 152.4 mm aperture would be about 272,000 cycles/radian or 1.32 cycles/arcsecond.

3.4 The MTF in Use

Astronomy authors claim that an aperture obstructed by about 30% is about as bad as an unobstructed aperture with $\frac{1}{4}$ wavelength of spherical aberration. How do they know this? Seemingly, they are comparing two widely different phenomena. Are the images equally degraded?

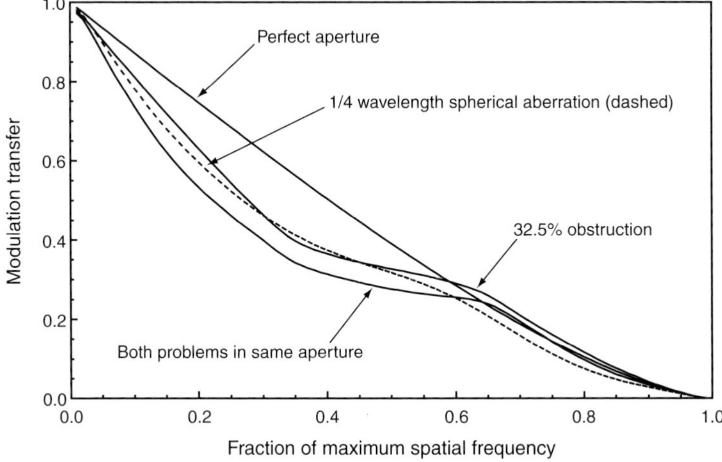

Fig. 3.7 Comparing the filtration of simple spherical aberration and central obstruction.

Everyone who has experience with evaluating images knows that these writers couldn't have derived that 30% number from experience. Differences between optical images are subtle, and this number is too precise. There are many ways to compute this quantity, such as specifying the amount of energy in the central disk or diminution of central intensity. For our purpose here, we calculate the energies contained inside the diffraction spot of an unobstructed aperture with ¼ wavelength of spherical aberration and several obstructed apertures that are otherwise perfect. We choose the obstruction that most closely matches the energy loss of spherical aberration. The answer is an aperture about 32% obstructed.[2]

However, we must examine the frequency responses of these two situations to see where the blanket statement of equivalence breaks down. The corresponding MTFs are plotted in Figure 3.7. We see that they are roughly the same for spatial frequency of about 0.3. For 200-mm telescopes, these two optical situations show periodic planetary detail having spacings of roughly 1 to 2 arcseconds equally well.

The responses at higher and lower spatial frequencies are different, however. At the high end, the obstructed aperture delivers better contrast than ¼ wavelength of spherical aberration. Surprisingly, it even performs better than a clear aperture. (See Section A.3 for more about the improvement of double-star resolution with obstruction.)

At the lower end, the obstructed aperture has a higher MTF for spatial frequencies until bar separations are more than eight times the Rayleigh

---

[2] See Table 10.1.

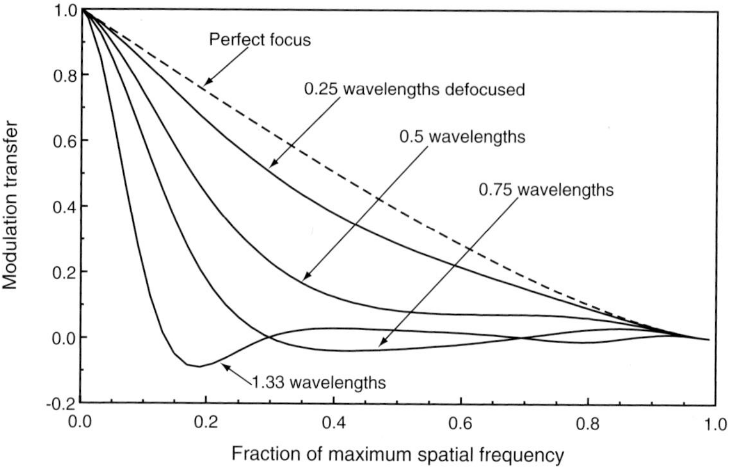

Fig. 3.8 MTFs associated with various amounts of defocusing.

criterion. For a 200-mm aperture, details separated between 2 and 5 arc-seconds are shown better by the obstructed aperture. On coarse details separated more than 5 arcseconds the two contrasts are pretty much the same, but still slightly less than the perfect aperture.

Thus, for low spatial frequencies, ¼ wavelength of spherical aberration is slightly worse than the obstruction. For double-star separation at or slightly beyond the Rayleigh criterion, the obstruction wins again. Only for intermediate spatial frequencies is the frequency response about the same or inferior in the obstructed instrument. Typical extended objects are composed of all spatial frequencies, however, and the net effect would be to reduce the quality of the image (Russell 1995).

The obstruction may be adjusted until the transfer functions are a little closer (the shadow would be nearly 35% of the full aperture), but all-inclusive comparisons should not be made without caution.

3.4.1 MTF Associated with Defocusing

The simplest way to degrade the image is by defocusing it. Figure 3.8 shows the way MTF droops with increased defocus. At ¼ wavelength defocus, performance begins to suffer noticeably; by ¾ wavelength it is unacceptable.

These transfer functions have an extremely interesting property. At higher values of defocusing, the curve goes negative for some spatial frequencies. If this behavior is traced back through the contrast equations above, it is seen to result in a switch of bright and dark image intensities. The sinusoidal bar targets look like negatives of themselves.[3]

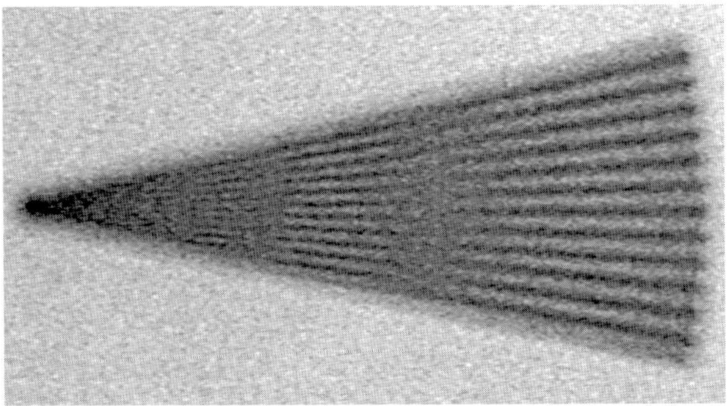

Fig. 3.9 Defocused radial bar target. Look from the narrow end at glancing incidence to the paper. Some of the bars reverse dark-to-light. Photographic-film grain detail causes the noise.

Does this mean that at certain focus positions the star will look dark and the night sky light? Of course not. The image of a truly point-like star is composed of all spatial frequencies. The contrast is transferred with the correct sign at plenty of other spatial frequencies. This bizarre effect does not affect normal images in this manner—it usually just contributes to the overall appearance of blur. However, if two or three cloud bands on Jupiter have the right separation, a slightly defocused telescope can result in a spurious band appearing between them.

Just because it is rare does not mean that an arrangement cannot be made to allow observation of this effect. In Figure 3.9 contrast reversal is readily seen. This picture was taken with a severely defocused camera lens pointed at a radial bar target. These bars aren't sinusoidal but they are periodic. As the pattern passes through a completely unresolved gray area, it emerges on the other side with the opposite intensity. Only near the edge does it fail to act this way (Goodman 1968, p. 126).

3.4.2 Stacking of MTFs

As long as the optical difficulties are independent of one another (that is, changing one does not affect another), the total MTF of the system is just the individual MTFs multiplied together according to the following recipe:

$$\text{MTF}_{\text{total}} = \text{MTF}_{\text{perfect}}\left(\frac{\text{MTF}_1}{\text{MTF}_{\text{perfect}}}\right)\left(\frac{\text{MTF}_2}{\text{MTF}_{\text{perfect}}}\right)\cdots\left(\frac{\text{MTF}_n}{\text{MTF}_{\text{perfect}}}\right). \qquad \textbf{3.6}$$

Each quantity in brackets is known as the *degradation factor* of the

---

[3] The MTF can't be negative under its strictest definition, but here it is allowed to carry the sign of the full optical transfer function when the phase is 0 or π. See Appendix B for more details.

optical error denoted by 1, 2, etc. The numerator of each degradation factor is the MTF as it would have been measured if the aperture suffered from that optical problem alone.

Of course, no two optical difficulties are truly independent, even if they seem to be unrelated at first glance. From Figure 3.7 we may reason that spherical aberration on the surface of a mirror has nothing to do with a secondary suspended above it, but the secondary casts a shadow on the mirror and thus changes the contribution of the aberration to the wavefront. The middle spatial frequencies of the "Both" curve of Figure 3.7 are not as bad as expected from this degradation equation.

A particularly fortunate result of this lack of independence is the effect on a zonal defect at the center of a mirror. If the diagonal is sufficiently large, it will completely shade the error and you will not be able to detect the aberration.

Some optical devices, particularly those without interchangeable eyepieces, have optics designed as a unit. Aberrations inherent in the objective are canceled by oppositely directed aberrations in the eyepiece or other tailpiece optics. When separated, they perform poorly or not at all (common in military surplus equipment). When trying to use the degradation equation for such optics, the evaluator must be careful to circumscribe the whole system—not just part of it.

For most astronomical telescopes, the unknown nature of what will be

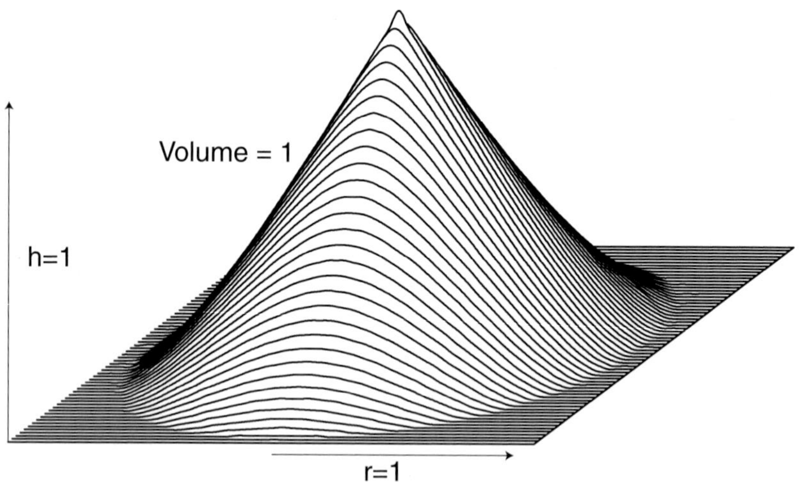

Fig. 3.10 The MTF is a solid with independent transfer functions stretching in every direction. For perfect, symmetric apertures, it takes on a cone shape. For aberrated and unobstructed apertures for which the maximum spatial frequency is set equal to one, the volume under the cone is equal to the Strehl ratio.

attached at the back (whether it is a camera or an eyepiece) demands that the objective alone do good imaging. An unwritten protocol in the design of astronomical telescopes demands that the designer encapsulate the maximum number of desirable features at the focus of both the eyepiece and the objective.

3.4.3 MTF In Every Direction

The bars in Figure 3.4 only point in one direction. The effect of pointing them in another direction, especially in the presence of asymmetric aberrations like astigmatism and coma, can change the value of the contrast transfer, depending on the angle at which they are perceived. A bar pattern only measures the MTF at right angles to it. If you change the direction of the bars, the MTF in that direction could also change. As the bar pattern turns a full circle the MTF maps out a solid exemplified by the perfect MTF of Figure 3.10.

The ideal form of the MTF has a distorted cone shape. In the case of symmetric apertures, the MTF patterns are the same whether they refer to spreading in the horizontal or in the vertical. See the dunce cap shape in Figure 3.10. It turns out that if the maximum spatial frequency is normalized to unity, as in Figures 3.5 and 3.7, the volume under the cone is equal to the Strehl ratio for unobstructed apertures. The diffraction Strehl ratio (normalized for transmission) is this volume even for obstructed apertures.

3.4.4 The Black Box

Lots of amateurs tend to view optical problems as either hopelessly bad or admirably perfect. For them there is nothing in between. What I'm going to try to induce is a shift in thinking from difficult and expensive rigid demands, to the point of view of redesigning the instrument using all inputs, including cost. The common way that amateurs select or specify instruments is to demand the best behavior for the *aperture*. Professional engineers, on the other hand, tend to the viewpoint that the end result is the only criterion, and you design backwards from there. They view the engineered system as a "black box." They construct a set of technical goals, and decide how to weight them, and furthermore construct a set of ergonomic and cost goals and decide how to weight them too. They input all of this into a design model and look for the telescope (or whatever) that will give the highest benefit with the least cost or other factor. Some of this is subjective, of course, but it gives you a feel for how professional workers who are spending a great deal of precious money make their decisions.

Nearly always, for example, observatory designers choose the folded Cassegrain-style reflector. It gives the most light and resolution for the

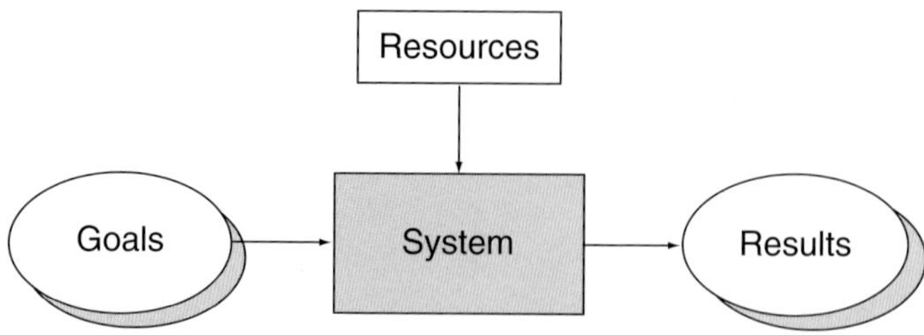

Fig. 3.11 The degradation function may be represented as a "black box" system. The least costly (or some other criterion) aperture/filter is selected to provide the MTF that meets the goal.

least money, and they don't have to suspend heavy sensors at the top of the tube. This is not because there is anything optically superior or inferior about the Cassegrain-style reflector, but this choice gives them the most aperture and the most mounting convenience for the resources expended. That is why you will never hear of a professional observatory specifying a super-large apochromatic refractor.[4] Exotic telescopes are not cost-effective solutions.

All this talk about black boxes and the justifications for observatory reflectors is good, but there is one thing I want you to take from this chapter. That is to view your telescope as an optical *system* producing an image. What goes on inside the black box does not make any difference as long as it delivers an acceptable final product.

MTF is depicted throughout this book as a fraction of the maximum spatial frequency for the aperture. This obscures one fact. The limiting spatial frequency goes up with telescope aperture. Equation 3.4 says a 12-inch telescope has a maximum spatial frequency of about 2.6 cycles/arcseconds at 560 nm. A 20-inch telescope has a limiting spatial frequency of about 4.4 cycles/arcsec. This maximum is no surprise, really. Neglecting atmospheric errors, a half meter aperture allows finer resolution than a 300-mm one.

Now consider what the MTF in Figure 3.12 is showing us. If the only goal is a resolution of slightly better than 0.4 arcsecond, we could have specified an ultra-precise 12-inch design, either for bragging rights or transportability. Reputable firms could no doubt produce such a product at an extraordinary price. If the goals are a combination of 0.4-arcsecond resolution and light-gathering power, but resources are limited, then we can perhaps find a better fit to our systems model with a ¼-wavelength 20-inch. The presence of slight obstructions or mild aberrations need not be viewed as fatal errors but as inputs to a system-design model.

---

[4] Unless, of course, there is an eccentric rich person who would want to endow such a thing.

3.4.5 Shape of the MTF Curve Related to Form of the Image

Most of the aberrated or obstructed apertures in this book show a characteristic spatial frequency at which the MTF turns the corner and starts recovering. Typically, the aberrated or obstructed MTF hangs below the perfect case in an elbowlike droop. How can we judge the effect on the image by seeing just the MTF?

First of all, we have seen how the MTF has differing maximum spatial frequencies as the aperture increases. Imagine, if you will, an average of the perfect MTFs for both the 12-inch and 20-inch apertures of Figure 3.12. Follow the curve halfway between these two perfect apertures and the result is very similar to one of these elbow shapes. The effect of aberration covering a great deal of the aperture behaves similar to this composite MTF concept. It is as if the aberrated aperture is two separate surfaces, each contributing to the image, and the MTF is a mixture of each.

The MTF may be inferred a number of ways. It can not only be computed as the measurable contrast of a sinusoidal grid as it was earlier in the discussion, but also from the aberrated aperture or from the image directly by using different mathematical procedures (for more detail, see Appendix B); Thus, if we construct an artificially diminished image with the excess power appearing as a halo around it, we can calculate the MTF and see how it changes.

Power must be conserved, so any intensity diverted into the halo must

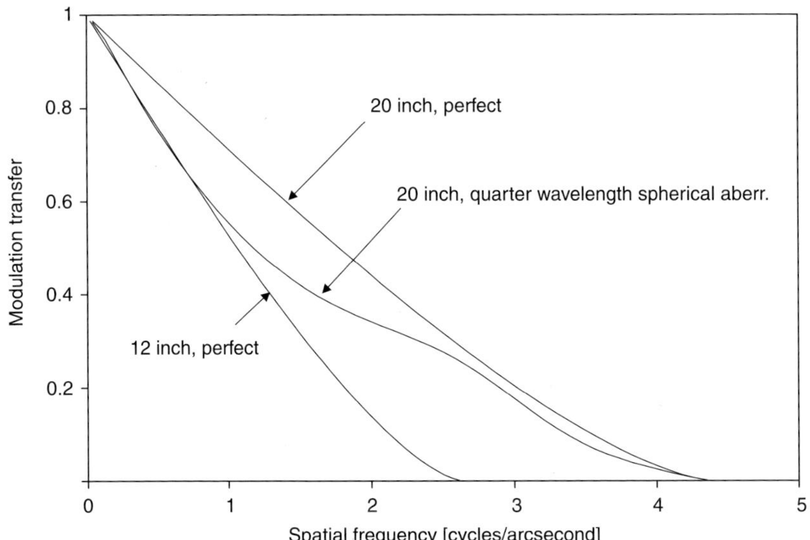

Fig. 3.12 Three MTF curves in actual units showing a perfect 12-inch, perfect 20-inch, and a 20-inch with about ¼ wave of spherical aberration (obstruction not considered).

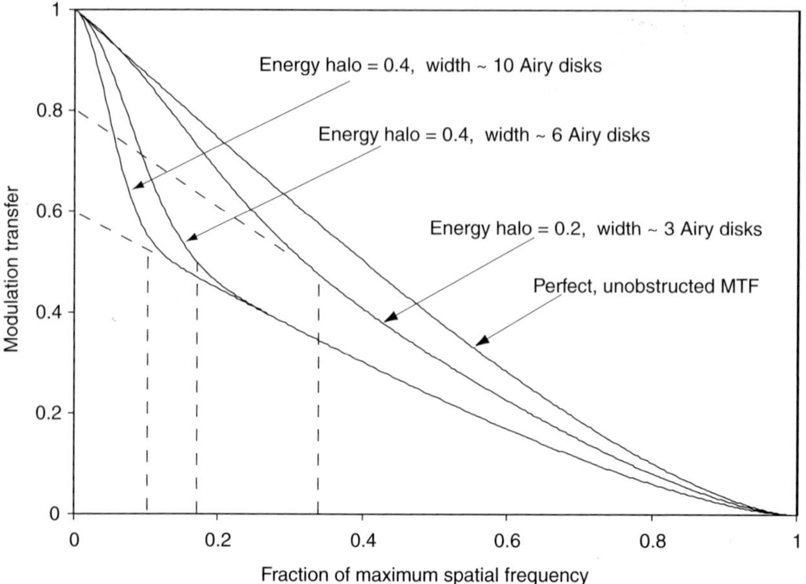

Fig. 3.13 Power or energy considerations in the shape of the MTF. A diminished conventional image pattern is superimposed on a Gaussian (or bell-shaped) halo of the missing energy.

be accounted for as diminishment in the main image, and this MTF calculation will bring out the intimate connection between the MTF and power accounting in the image. It is not always obvious in the MTFs appearing elsewhere in this work, because there is some amount of wave optics interference in the calculation, but every MTF has this power summation as a function of image radius as the basic cause of its shape.

In Figure 3.13 is shown the result of MTF calculations derived from various image contours. There are two parameters of importance in calculations of the halo. One is the width of the halo, and the other is the fraction of the total power passing through the aperture that reappears in the halo. In the figure, the highest MTF is the normal perfect pattern with no halo around it. The next lowest (and one most similar to the MTFs appearing elsewhere in this work) is for a halo about three times as wide as the Airy disk and containing 20 percent of the total power. From later curves appearing in this book, we see that mild amounts of whole-disk aberrations like correction error, coma, astigmatism, and even defocusing divert the energy within 2 to 4 Airy radii. The next lowest is from a halo 6 times as wide as the Airy disk and the final 10 times as wide. Both these curves take 40 percent of the power from the primary diffraction pattern.

Notice how the curves to the right of the elbows, when continued back to zero spatial frequency by the dashed lines, intercept at the power value remaining in the primary image. It is as if the lower parts of the curves (to the left of the elbow) are sitting on top of normal patterns, each diminished by the amount reappearing in the halo. Notice also that the approximate position of the respective elbows is the inverse of the width of the Gaussian halos. The elbow of the halo 10 Airy disks wide is about at $\frac{1}{10}$ the maximum spatial frequency, the halo 6 Airy disks wide is about $\frac{1}{6}$ the maximum spatial frequency, and so forth.

From the behavior of these curves, it is apparent that the low-resolution features of the image are transferred by the low-frequency part of the MTF curve and the high-resolution details of the image are transferred by the upper part. In other words, the smallest diffraction patterns are represented by the highest spatial frequencies in the MTF, and the largest diffraction patterns are represented by the lowest spatial frequencies.

Suppose we have dust on a mirror that removes, say, $\frac{1}{100}$ of the energy and diverts it into a halo 500 times as wide as the Airy disk. We might then expect the curve to drop to a value of about 0.99 at $\frac{1}{500}$ the maximum spatial frequency, and then continue on normally.

3.4.6 Measuring the MTF

Methods exist to experimentally measure the MTF curve for a set of optics. However, later chapters will not emphasize them since they involve a great deal of extra equipment (Baker 1992, Murata 1965). Particularly good is the reference Berkovitz 1968, which shows how MTFs are measured in camera systems by using the spreading of an edge. Telescope MTFs gathered this way would require additional magnification on the tiny image to expand at least four pixels over the blur. The associated aberrations of the magnification device are difficult to separate from the aberrations of the telescope, though not impossible. Most measured MTFs in diffraction-limited systems are indirect results of whole-disk wavefront aberration measurements such are produced by interferometers. The way these MTFs are calculated by an autocorrelation of the aberration is mentioned in Appendix B.

This book concentrates on the theoretical description of ideal cases. Pay close attention to the MTF curves in the chapters that follow. Even though these curves may appear uninteresting and similar to each other, they encompass all optical problems. If you learn to interpret them, you will better understand the way your optics are degraded.

Chapter 4
Diffraction

Literally, diffraction means a "breaking up" of a wavefront, which is disturbed by passage of the wave near an obscuring body. Diffraction is much more than the process that generates rings around a stellar image. While it is true that diffraction makes these rings, it is possible to shade the edges of the aperture to make such ringing arbitrarily small. Diffraction, though, still blurs the central spot. Images without rings still display diffraction.

For people learning diffraction the first time, the various viewpoints presented in this chapter may be overwhelming and the main conclusions can be lost in the detail. Several elements are crucial for this discussion, however, so let's look at them at the beginning:

1. Diffraction is caused by the wave nature of light. Because light cannot be localized, images are necessarily fuzzy. The quantum mechanical concept of the photon as a particle does not help fix its position.

2. Diffraction is a consequence of a limited aperture. It is an *angular* blurring that is independent of the focusing power of the instrument.

3. Diffraction is not just a phenomenon observable in the rings. The spread of the central spot is caused by diffraction.

4. Quasi-static light and dark regions exist because diffraction can be modeled as a standing wave, with the aperture providing the boundaries.

5. A limit exists to optical quality that is difficult to exceed.

Once readers have a working knowledge of diffraction, the star test is more easily understood. Diffraction itself is not often explained well, and many people believe a kind of folklore about what it is and what causes it. Authors either sidestep diffraction entirely, or they take the too-easy (for them) refuge of mathematics. The scarcity of introductory instructions is unfortunate, because the concepts underlying diffraction are not all that difficult to understand. More important, some of these ideas are the most beautiful and fundamental in physics. Diffraction touches on notions as varied as the limits of what is knowable and concept of the particle.

Say that we were in a universe where light consisted of infinitely tiny particles that could be traced from a star, through the telescope, and finally to the eye. Diffraction would not exist in such a world. Either the particles continue undisturbed beyond the obscuring body, or they hit and are stopped. Close approach would make no difference. The region outside the illuminated cone near focus, where you wouldn't normally expect light, is called the *geometric shadow*.

When optical designers ray trace optics, they are using the "light-as-particles" assumption that light will go only where geometry allows. The concept of light moving in particles is called "ray optics." In ray tracing the system, they are assuming that light will go through the optics as if it were a tiny elastic sphere, smacking into optical surfaces, being deflected according to empirical rules of refraction and reflection. Surprisingly, this ballistic model will predict the course of light fairly well. To first order, it does a fine job.

Unfortunately, the behavior of light has other characteristics not explicable by modeling it as particle motion. Light seems to be able to turn corners. By itself, this feature is not too surprising. Any enterprising 17th century physicist could have postulated, for example, that a particle of light has some finite size. As it brushes near a surface, it could be pushed back into the lit region. But experiments done near straight edges showed that light was deflected into the shadows of the obscuration. Instead of banging its shoulders into the obscuring body and being thrown into the lit region, the light ray seemed to be grabbing the edge of the aperture and swinging around. Strangely, some of the shadow also seemed to bleed into the lit region.

Some angles of deflection are preferred. Moving away from the position of best focus, the illumination dims as expected. But here's the strange behavior—it then starts to brighten again. With higher and higher angles, the illumination goes through many such oscillations, which cause the familiar diffraction rings around stars. The mechanical hypothesis of light-as-particles has no way of understanding such behavior, except by supposing a defect in the optics. At focus, one doesn't expect light to be anywhere except at a dazzling point in the center.

The eventual explanation came by saying that light did not resemble particles so much as waves. Many of these effects can be observed in a puddle—or other shallow pool of water—by watching the behavior of the water waves in the surface. Lay a piece of wood across part of the puddle so that it interferes with the steady course of waves, and start a wave by dropping a pebble on one side of the pool. Notice that as the wave passes the end of the wood, it curls around into the protected region. This is diffraction.

Water waves are very interesting, but they don't immediately seem to relate to optical quality in a telescope. However, we will see how the behavior of a wave near an obstacle or edge results in degradation of an image. Diffraction is the mechanism that produces unavoidable imperfection in telescopes.

In the old light-as-particles model, the image could be perfect as long as the surfaces were perfect. Unfortunately, the real world of optics doesn't permit infinite precision. Waves are not localized. It's much like finger-painting while wearing mittens—drawing fine lines or dots is impossible with such pudgy fingers.

People have often thought that diffraction could be suppressed by obscuring or diffusing the edge. I can scarcely count how often I've seen the incorrect assertion that feathering the edge of the diagonal would eliminate its diffractive effects. The aperture is always in the optical system, so modifying its edge affects the course of how waves are broken up, not the fact that they will be disturbed. Tapering the diameter of the secondary merely decreases the size of the aperture and hence increases the amount of diffraction. We can shove diffraction effects around like a child rearranging vegetables on a dinner plate by changing the transmission as a function of radius, but we are unable to make them go away.

4.1 The Coordinates of Light

Where is a light wave? In Figure 4.1, try to point out the location of the light wave. One might conjecture that the light wave is one of the long crests that is depicted in the figure, but this conjecture mistakes the graphical portrayal for truth. In depicting the progress of light, it is convenient to trace out the crests or troughs, but those locations have no special implication. The points halfway up the wave are no less part of it. In pointing out a wave, we must be careful not to allow human interpretation to color the answer.

How could we choose one crest over another? We might mark a wave crest corresponding to a feature of telescopic interest, such as the one just passing through the aperture. Such an assignment has no meaning, however. Light doesn't care what is planned for it; it has no memory or motivations.

The answer to the original question is easy but unsatisfying. The wave isn't at any specific location. In fact, a better way of describing a wave is to recognize it as a dynamic process.

The important thing to remember is that a wave has sources (lamps, lasers, stars, etc.) modified by boundaries. The sources are the points that

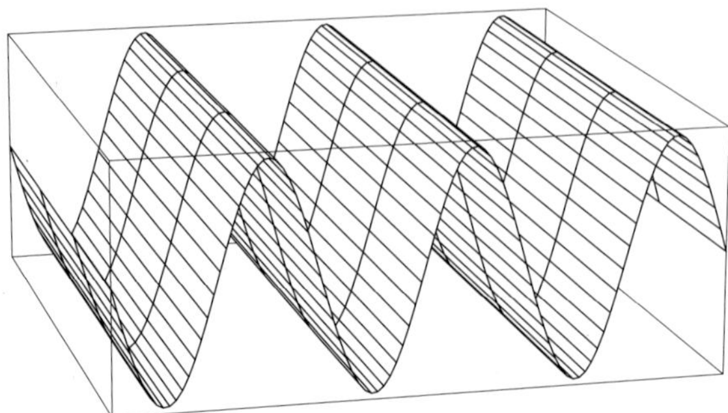

Fig. 4.1 Part of a wave.

can be enclosed in volumes observed to be emitting more energy than they are receiving. The boundaries are comprised of the diffracting obscurations, the edges of the aperture, the reflecting surfaces, and the transition surfaces between refracting media. The behavior of the wave at those surfaces (whether it is absorbed, reflected, or transmitted at a slower speed) must be specified. To describe diffraction properly, we must solve what is called a *boundary-value problem* for the wave equation in the region of interest (Baker and Copson 1950).

Where a wave is coming from, how strong it is, and where it is going are more important than where it is. Our wave is like a choreography script, with each tiny location in space having its own dancing instructions. The wave we perceive is the totality of all of these dancers moving in concert. The very fact that we can't point at a particle of light and say "it is there" puts some uncertainty on light's ability to be directed. This softness in light's location induces the blurring of the final image.

The quantum revolution revealed that light is packaged into discrete particles called photons. Since the quantum theory is a refinement of earlier methods, doesn't that mean that light is composed of little rays after all?

The quantum theory is strange and non-intuitive. To understand it, we must be willing to abandon our experience of the macroscopic world. We are just too big to use what we know about the world in an effort to visualize the quantum universe.

In quantum mechanics, a way one can describe the position of a photon before it has been detected is to give its probability density; i.e., the likelihood of encountering the photon at a particular location. A particle's probability density is related to its wave function. Where the wave function is strong, the probability density is high. But the density is only measurable by running many photons through the optics.

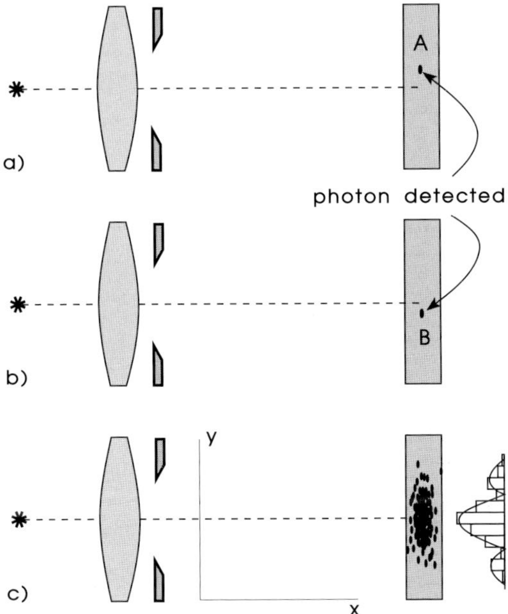

Fig. 4.2 a) One hit at A; b) another hit at B; c) many hits with histogram.

Perhaps the best way of explaining this principle is to imagine the following experiment. Say we had the aperture of Figure 4.2a, and a single photon was coming from the left. A photsensitive array is at the focus. The photon strikes the array at location A, triggering a response in a single pixel of the array. If we were to stop here we would already have contradictory evidence. On one hand, the lit pixel isn't located precisely at the position of geometric focus, which is where the dotted line intersects the film. But the array was only struck at that one point, leaving us with the impression that a particle was detected. "Maybe the alignment is off," we might say, and carefully run the experiment again. Now it hits at B.

For the next trial, we expose the array long enough to gather many photon hits, and plot them to the right as in Figure 4.2c. Now we see a trend—the diffraction pattern is beginning to show. But we already know that such behavior is explained by a wave.

So, we return to the original question, is light best described by particles or waves? On one hand, it is clearly observed that something impacts the array at individual locations. You can dim the intensity of the light until the reception rate into individual pixels is arbitrarily slow. That seems to indicate tiny particles. On the other hand, lots of such individual particles are statistically distributed in the pattern expected for a wave.

We took it as an article of faith that light has no intelligence or motives. Here we seem to find that photons apparently can remember where other photons have already been detected. Record-keeping and communication are two more powers that are difficult to ascribe to light particles.

What we have encountered here is one of the seeming paradoxes of quantum physics. The whole idea of particles (whether they be photons, electrons, quarks, or whatever) whizzing around is an example of one of those concrete real-world ideas that doesn't apply at the quantum level. We cannot impose our kind of macroscopic reality on the microscopic situation. We would like particles to be neatly spherical, to be colored with different hues, and all to have a bright, rectangular highlight where a window reflects on the upper left side. But the quantum world doesn't present us with such comforting pictures.

The first physicist to put this in mathematical form was Werner Heisenberg, who expressed what is known as the uncertainty—or better, "indeterminacy"—principle. Far from being a discouraging limit to what we can know, this principle tells us new things about the dynamics of particle motion. One can even use the indeterminacy principle to derive an approximate expression for the radius of the diffraction disk.

What is unknown about this situation is where the photon (that is, the "ray") entered the aperture. We know where it ended up and where it started, but we are uncertain about the intervening path. The Heisenberg indeterminacy principle is stated roughly as follows: The uncertainty in the momentum of a particle in a certain direction times the uncertainty of its position is approximately a very small constant, or

$$\Delta p_y \Delta y \approx h \qquad \textbf{4.1}$$

where Δp_y is the fuzziness in momentum in the y direction, Δy is the fuzziness in position y, and h is that small number, called Planck's constant. We can rewrite this approximation to find the spread of the detection angle after many counts:

$$\frac{\Delta p_y}{p_y} \approx \frac{h}{p_y \Delta y} \approx \theta_{\text{diffraction}}. \qquad \textbf{4.2}$$

The expression for the momentum of a photon is the energy divided by the speed of light, or E/c. The photon's energy is $E = h\nu$. The constant c is the speed of light, ν is the frequency of the light, and λ is the wavelength; $c = \nu\lambda$ describes the way these quantities interrelate. The uncertainty in position Δy is just the diameter of the aperture D. Thus, the uncertainty in angle becomes

$$\theta_{\text{diffraction}} \approx \frac{hc}{\nu hD} = \frac{\lambda}{D}.$$

4.3

This result is close to the $1.22\lambda/D$ expression for the angular radius of the Airy disk.[1]

4.2 The Consequence of Filtering

We have already seen in Chapter 3 how the telescope has a limited spatial frequency response analogous to the 20–20,000 Hz frequency response of audio equipment. What is the implication of that filtering? Can knowledge about the limitation on spatial frequencies be used to derive interesting facts about diffraction?

The two-dimensional circular surface of an aperture injects some non-instructive mathematical complications into the problem, so it will be easier to turn once again to the simpler situation of audio electronics. In this case, we are interested only in variations in time or frequency.

The audio equivalent of a star or a point optical source is an electronic pop. It could be caused by a scratch or fault on an old vinyl LP disk. If you've ever tried to listen to a distant radio station through an intervening storm, you are probably familiar with the staccato crackle as a lightning strike takes place. One thing that you probably did not notice, however, is that every one of those pops sounds identical, varying only in amplitude.

This fact seems innocent enough, but it gives us an insight into sound reproduction. Every little crackle sounds precisely the same through a given audio system. If you listened to another audio system having a greatly different configuration—say a cheap, hand-held transistor radio compared with your stereo—it may have a different tonality. Yet each pop produced by that other radio is the same as well.

The correct interpretation of this uniform behavior is that you aren't listening to details of the electrical activity in the storm or the source. You're hearing some sort of signature of the reproduction device itself. The crackling noise is merely its best attempt at reproducing an abrupt jump.

At the beginning of the 19th century, physicist Joseph Fourier discovered that complicated, even discontinuous, functions could be simulated to arbitrarily high precision by the sum of a series of sine and cosine functions of higher and higher frequency. For example a "pop" could be described by this infinite series:

---

[1] Equation 4.3 is simplified. Those who are interested in this topic can find more in any elementary quantum mechanics text, e.g. Park 1974.

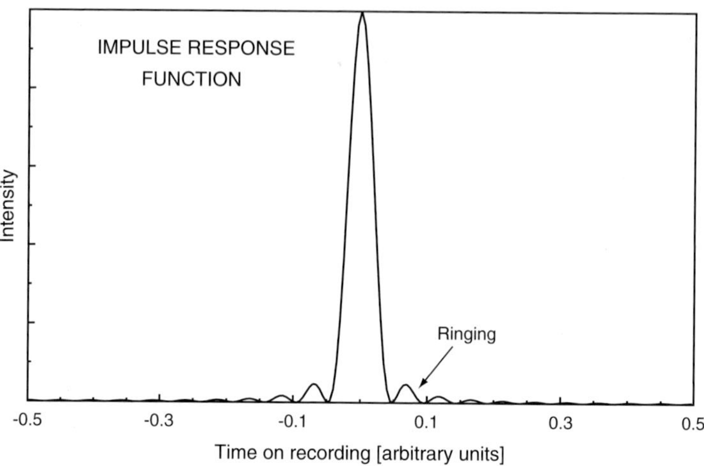

Fig. 4.3 The intensity of an audio crackle is simulated as a truncation of the series in Equation 4.4.

$$a(t) = 1 + 2(\cos(t) + \cos(2t) + \cos(3t) + \ldots), \qquad \textbf{4.4}$$

where $a(t)$ is the amplitude as a function of time. The sharp crack we actually hear is the intensity $I = a^2(t)$.

Being a filter, the audio system doesn't transfer frequencies to an arbitrarily high value; it starts failing at about 20,000 Hz. Thus, we're going to have to truncate this series somewhere. Figure 4.3 shows the effect on the intensity if we lop off the series at 10 cosine terms. The sharp spike of the crackle has been reduced to quite a different function.

This response function resembles a diffraction pattern, complete with rings. In fact, it is identical to the diffraction pattern of a slit or rectangular aperture. Diffraction is a result of the optical system's failure to transmit arbitrarily high spatial frequencies, just as the ringing of Figure 4.3 is an effect of the limited audio bandwidth. A point-like star acts like a sharp audio impulse. A star's diffraction pattern is similar to an impulse response function. It is called a *point-spread function* because most stars are far away and occupy insensibly small angles. Not only do rings appear in Figure 4.3, but also the infinitely sharp spike has broadened into a narrow peak of measurable width.

Astute readers probably realize that the symmetry of the pattern means the response curve oscillates long before the pop on the record actually happens, but that such an oscillation isn't possible. The name of this limitation is *causality*, or the requirement that cause precede effect. Causality is a central issue of special relativity, but it even has implications here. The symmetry of Figure 4.3 is an artifact of the simple-minded way this process was modeled. The phase shifts of the individual cosine terms in the equation above were ignored. We implicitly assumed that all signals go through the system at the same speed with the same phase. An actual system delays every frequency a little and adjusts the phases in such a way that no signal could be detected before it happened. The pattern would become asymmetrical with most of the ringing coming after the sharpest spike. Each audio system has its own characteristic slurring of this pattern.

Fortunately, optical diffraction patterns aren't calculated from time-based frequencies and don't have to obey causality. They can be nicely symmetrical. You are free to look at either side of an optical pattern, but you can't go back in time.

4.3 The Huygens Model

The next and most important way we will look at diffraction is a very old theory originally proposed by Christian Huygens (1629–1695) to explain why light is observed in the shadow region.

The term *wave propagation* is used without much attention to its origins. Propagation is a word originally associated with botany, where it meant to make new plants from cuttings. Its meanings include the multiplication, reproduction, or dissemination of plants and animals. It has come to mean the spread of ideas or even misinformation (hence, "propaganda"). But the concept of reproduction is inherent to propagation and gives a clue as to what Huygens suggested.

Huygens could explain light appearing in shadows by assuming that every point on a wavefront was emitting its own spherically diverging wavelet. The next instant of the wave would be given by the sum of all of these tiny wavelets. Only in the direction the wavefront was traveling would the wavelets not cancel each other. In other directions, a sideways-moving wavelet would cancel another point's wavelet that was out of phase with it. Similarly, if a hedge tries to grow sideways, it finds that a leaf on another branch has already taken the sunlight. The only direction in which growth is free is upward and outward.

An instant later, each new point of the wavefront would emit yet another spherically diverging wavelet. Similarly, in the hedge, inner leaves wither and are replaced by outer leaves. Thus, we could describe wave progression as a rebirth at each new wavefront, a "propagation" in the old botanical sense.

The situation of Figure 4.4 does not require a consideration of propagation, but in Figure 4.5 we see a reason for it. When the wave propagating

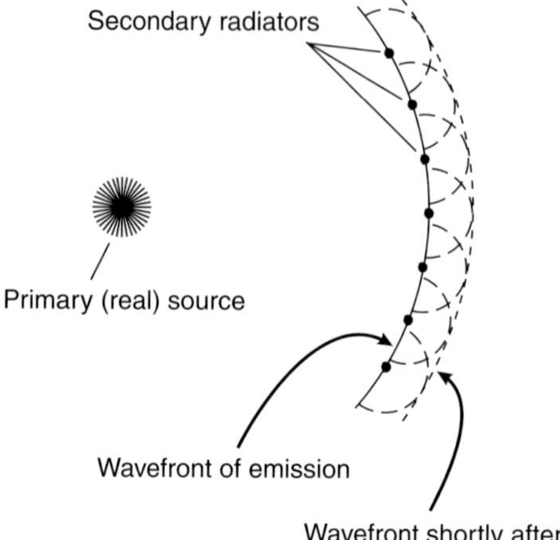

Fig. 4.4 Free-field divergence of a spherical wave.

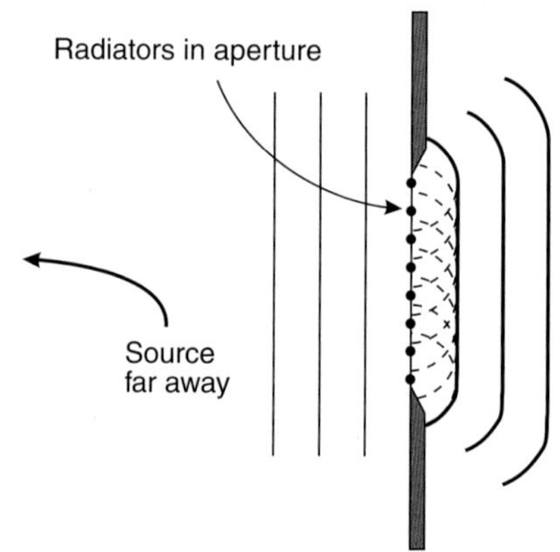

Fig. 4.5 Diffraction of a plane wave at an aperture.

from a distant source encounters an aperture, a sudden disturbance occurs in the beautiful symmetry. Points inside the aperture emit Huygens wavelets up, down, or out, but no corresponding sources over in the shadow of the aperture provide canceling interference from that direction. In the center of the aperture, the wave propagates much as it did in the free-field, but off at the edge, some of the energy leaks over into the geometric shadow. The little points emitting wavelets are called *elemental radiators* or *secondary sources*. These radiators aren't envisioned as truly hanging like beads at the aperture. They are only an infinite number of mathematical points.

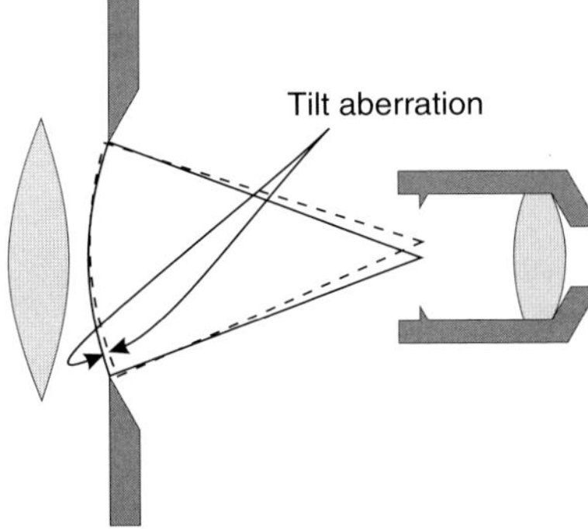

Fig. 4.6 A lens, aperture, and eyepiece showing tilt aberration.

4.3.1 Diffraction and Focusing

Diffraction in astronomical telescopes is caused by the restriction of light to an entrance aperture. Focusing is a convenience imposed on the optical system. This is implied in Figure 4.6, where the aperture and the focusing power have been divided into two independent functions. One function is the windowing of light by an iris or a rectangular cover—the diffracting aperture. Diffraction occurs whether the light is focused or not; it happens even in pinhole cameras.

All optical work is done on the nondiffractive focusing parts of the optical system. Thus, the optician's task is not to make an existing problem worse by adding additional aberrations to a lens or mirror. An optician cannot correct the window through which the telescope is looking.

The following discussion assumes that the focusing is already done when the wavefront encounters the aperture. This convention does not change the general principles involved.

4.3.2 Fresnel Zones

At focus, all of the elemental radiators should oscillate with the same phase. Every point of a perfect aperture appears to be cycling together with every other point. A sensor at such a focus is awash in light. However, at a slight angle with the optical axis (as in Figure 4.6), the sensor sees some of the points oscillating a little behind the other points. They aren't really doing anything different, but part of the aperture is farther away. Because time passes as the light goes the extra distance, those radiators are perceived at the offset point's location to be lagging behind the rest. Partial wave cancellation results in the total intensity of the light being less at this lateral distance from focus.

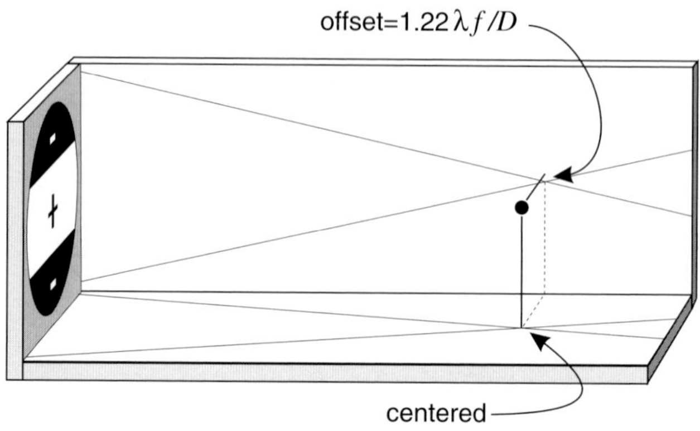

Fig. 4.7 Fresnel zones as seen from a point offset from a star. Notice the projections of the light cone on the bottom and far wall of the telescope. This offset is above the focused image. The positive and negative areas cancel so that this offset is the first dark ring.

If the sensor is far enough off axis, some distant portions of the aperture are seen to be once again in phase. Actually, they are a full wavelength behind the near portions of the aperture, but since the oscillation is sinusoidal, we can't tell the difference. The sources of those waves are atomic transitions that occur quickly, but the burst of light can be a single wave packet many meters and millions of wavelengths long. In the course of a few wavelengths, we really can't see any difference from one wave crest to another.

Different delays are indicated in Figure 4.7 by *Fresnel zones,*[2] dark and light regions on the aperture. Zero phase occurs only at the interface between two zones. Elemental radiators that are ahead in phase are denoted as "+" regions. Points that are behind in phase are indicated as "−" regions. The dark or light color is only an identifier of each region; it denotes a phase change of a half wavelength. It does not represent the actual appearance, nor does it represent the illumination. In reality, these regions gradually fade into each other. The actual phase varies smoothly, but these regions jerk from one sign to the other.

The detection position in Figure 4.7 is indicated by a tiny dot, which is in the focal plane of an eyepiece. The pattern on the aperture indicates the Fresnel zones as seen from that dot, rather than from our outside perspective.

This pattern changes rapidly. In only 9.2×10^{-16} seconds, the pattern is reversed from positive to negative (remember, the waves are flying into

---

[2] Named after physicist Augustin J. Fresnel, 1788–1827.

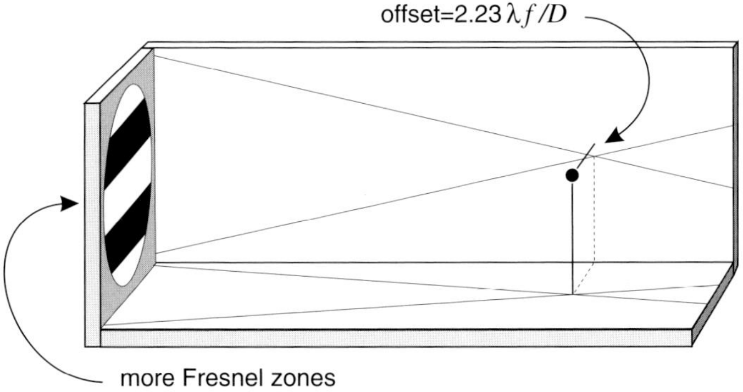

Fig. 4.8 Second minimum of diffraction pattern at 2.23$f\lambda$/D.

the aperture). If we follow the motion of the bars incrementally over small time scales, they are rushing down. If the detector is on the lower side of focus, they rush upward. The actual received wave at any given location is the average over time of the summed phase values of all these tiny radiators.

The Fresnel zone description of phases on the aperture should not be confused with the appearance of the image. Each location on the image has its own set of Fresnel zones. The Fresnel zone patterns are not observable directly since they refer to the wave field and we are capable of measuring only intensity at the sensor. This Fresnel zone picture is a convenient model, nothing else. Nevertheless, it succeeds in demonstrating many of the phenomena of diffraction and is the basis of the more accurate calculations described in Appendix B.

At risk of being too simple-minded, let's think of that aperture as being a large disk of construction paper. To determine the net wave intensity, we cut out all the like-colored areas of the disk and throw them into two separate piles, one positive and the other negative. We then weigh each pile and come up with two values, say 5 grams negative and 4 grams positive. Four grams cancel each other, leaving us with 1 extra gram of negative weight—call that value the "wave sum." Thus, the average wave sum is $-1g/9g = -\frac{1}{9}$. To calculate a number proportional to the intensity of the light, which is always a positive quantity, we need to square this sum to yield an intensity of $\frac{1}{81}$.

The intensity is weak, so our offset must be somewhere between the rings. Figure 4.7 is actually at a balance point. It is precisely between the rings, and so the dark area should cancel the light area, depending on the quality of the drawing and our skill with the imaginary scissors. Figure 4.8 shows the next dark ring, which occurs at a distance farther from the axis.

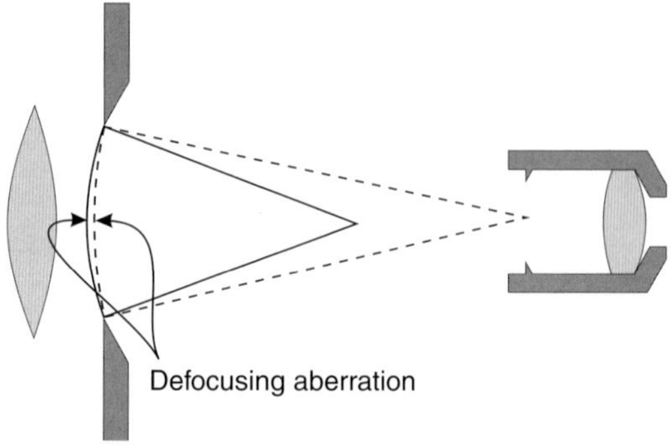

Fig. 4.9 Defocusing aberration. The secondary sources are vibrating on the solid arc, yet the eyepiece's focal plane is at the center of the dashed arc. The aberration resembles a smoothly pushed-in deformation. The center of the aperture appears delayed relative to the outer portions.

Here the tilt is worse, so more zones show, and two more bars are visible. Between the offsets of Figure 4.7 and Figure 4.8 is the first diffraction ring, where the positive and negative sections do the worst job of canceling each other. Five bars show there, but the two outside bars are very thin.

The important thing to learn from Fresnel patterns is that the wave sum is inefficient everywhere except precisely at perfect focus. Think about the construction paper again. At focus the aperture everywhere has the same sign—nothing cancels. The wave sum is $9g/9g$, and the intensity is 1. On even the brightest portion of the first ring, you would get an uncanceled wave sum of about $1.2g/9g$ and a light intensity only about $\frac{1}{57}$ the brightest value. The farther you go sideways, the worse the situation gets. At some point you realize that all you're cutting out are almost equal narrow ribbons of construction paper.

4.3.3 Fresnel Zones with Defocus

The wavefront changes character when the eyepiece is defocused. Figure 4.9 shows the induced aberration at the aperture if the eyepiece is drawn back. The resulting Fresnel zones are circular, as in Figure 4.10. With the passage of half a wavelength, the pattern reverses, producing a negative spot in the center. However, the square of the wave sum is the same, so the intensity has the same positive value. For sensor locations beyond focus, the passing of time would show a collapsing pattern with new zones appearing at the edge of the aperture. These new zones move inward to disappear at the center. For locations inside focus, we'd see identical snapshot patterns, but new zones would appear exploding from the center and moving out.

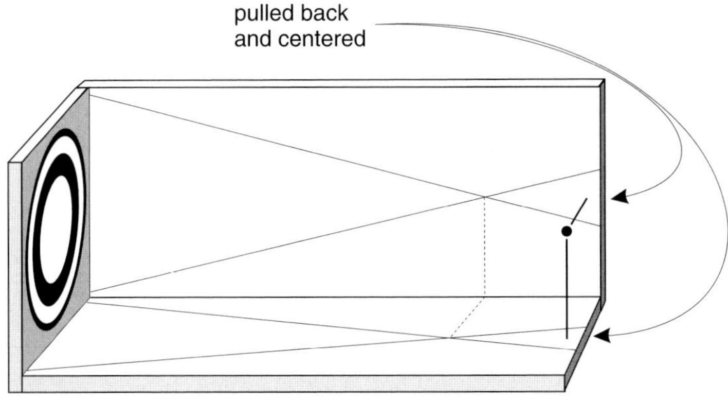

Fig. 4.10 Fresnel zones at a defocusing aberration of 1.7 wavelengths.

Even though the pattern is collapsing and we could have picked any snapshot, the wavefront phase is chosen so that the central spot is biggest. The total amount of defocusing aberration in Figure 4.10 is about 1.7 wavelengths. This value may be determined by counting colors from the center and moving out; the edge is about 3.4 half-wavelength zones out.

We deliberately choose the wavefront phase so that the area of the central positive zone is exactly equal to the areas of all other zones ringing it. The wave sum should be zero when the number of zones is 2, 4, 6, etc. The more elaborate theory predicts that the on-axis image intensity goes to zero when the defocusing aberration is 1, 2, 3... wavelengths, just as anticipated by this simple Fresnel zone model.

Moving upwards, as in Figure 4.11, shows the effect of mixing tilt

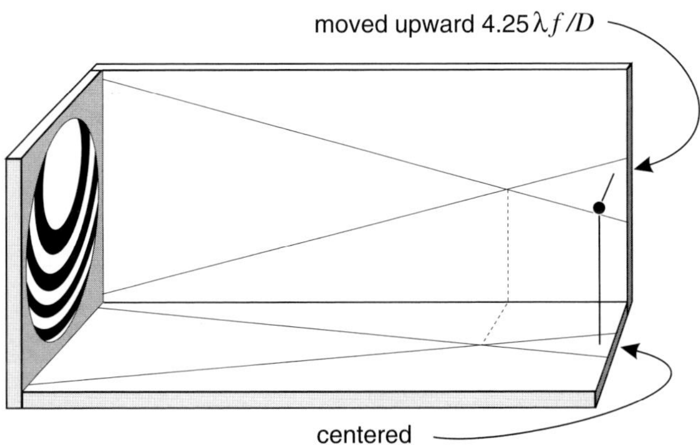

Fig. 4.11 An off-center bull's-eye Fresnel zone pattern showing two aberrations (defocus and tilt) mixed together.

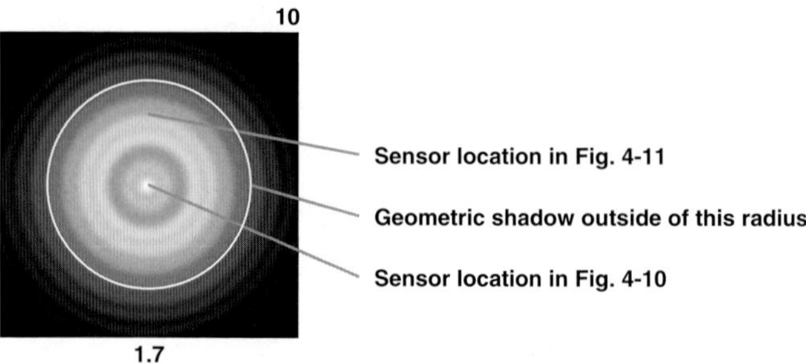

Fig. 4.12 The indicated location is where the intensity is calculated as the Fresnel zone sum of Figures 4.10 and 4.11. The bright ring indicates the boundary of geometric shadow.

and defocusing aberrations. Figure 4.12 is the same image pattern with an arrow pointing out the sensor location on the image. The edge of geometric shadow is shown as a light ring.

The strong, outer ring seen in most defocused patterns corresponds to the location where the central Fresnel spot begins sliding off the aperture. (In fact, it is half gone at the beginning of geometric shadow.) Thus, the slightly brighter edge ring in defocused images represents the last flourish of the central Fresnel zone before it disappears and darkness closes in. Notice how the brightest parts of the disks do not fill the circle of geometric shadow and how some of the light has escaped the disk to occupy the shadow. If that pattern were described by ray optics, the disk would be an absolutely featureless circle with perfect darkness outside the radius of geometric shadow.

This image shows another intriguing behavior. Soft rings terrace the pattern, even in the shadow zone. As the sensor is offset further, fresh Fresnel zones come into view, each causing an oscillation in the intensity.

4.4 Nodes and Antinodes

Certain locations in the volume around the brightest image point appear to be dark compared with their immediate surroundings. Figures 4.7 and 4.8 showed two such points on the first and second dark rings. In defocusing, on-axis darkness is found when the number of Fresnel zones is an even number. These locations appear to be quiet while the tumult rages around them. They are nulls, more commonly called *nodes*. The opposite of a node is an *antinode*, where the wave action is strongest and brightness is at a local maximum. Good examples of antinodes are the peaks of the diffraction rings, as well as the highest point on the central image spot itself.

Everyone has seen nodes, even though many aren't aware of it. If you

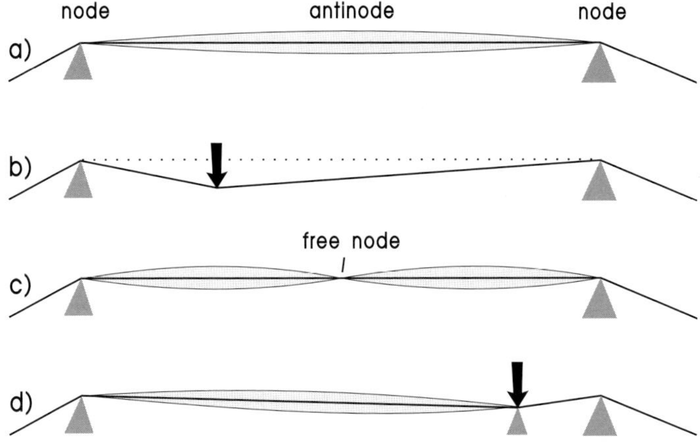

Fig. 4.13 Nodes and antinodes of a guitar string.

look carefully at a guitar string you can see the simplest type of vibration, as shown schematically in Figure 4.13a. There are two nodes at the bridge and fret, and a single antinode at the center. This situation is called a *standing wave*. Of course, the presence of the two nodes at the edges of the string is hardly a surprise. These positions are mechanically constrained and can hardly be expected to move much.

But nodes can hang freely in space, seemingly held by no physical restraints. If you pluck the guitar string as in Figure 4.13b, you will see the pattern of Figure 4.13c shortly after you release the string. After more time passes, it will decay to the situation in a). When you plucked the string, the stroke contained a multitude of frequencies (remember Fourier and the impulse response function). The strongest frequencies are the "fundamental" in a) and the "first harmonic" in c). No modes of vibration exist other than multiples of the fundamental frequency.

Situations a) and c) are an octave apart. What if you want a tone somewhere in between? You can't find it with a fixed distance separating the fret and bridge. To obtain an intermediate tone you must depress the string at another fret (Figure 4.13d), which implies a new fundamental and new harmonic sequence.

Now, if nodes and antinodes were only found in stringed instruments, they would hardly be useful for a discussion of diffraction. Standing waves are everywhere, however. For example, they are visible on the surfaces of rapidly vibrating liquids. These standing waves are sometimes seen in a coffee cup or other container shaking on the same tabletop as an appliance or tool with a fast electric motor. The surface appears almost stationary, even though you know intellectually that the speed of water waves is much faster than the almost leisurely drift of the waves. An equivalent to the bridge and fret in this case is the edge of the container, and the standing wave appears in the two-dimensional surface of the fluid. If the container is round, the most likely vibration is radially symmetric. If it is square, you see a sort of checkerboard figure.

A three-dimensional case is occasionally found in an old microwave oven. Microwave cooking patterns can be stirred by a hidden metal fan placed over the exit portal of the microwave transmitter. These fans can fail, and the only noticeable change is that cooking becomes very uneven. A stationary standing wave pattern is set up that overheats at the antinodes and leaves the food almost raw at the nodes.

Even with a working fan, holes and bright regions can interfere with even cooking. For this reason, the food must be moved at least once during cooking, and most modern microwaves feature turntables. Everyone has experienced the tiny hot spots that scorch overheated popcorn in micro-wave ovens. These antinodal regions are roughly 1–2 cm across.

What are the bridge and frets here? The microwave oven is a metal cavity whose walls are nodes because electrical conductors cannot have electric fields deep within them. Since any form of electromagnetic radia-tion has both electric and magnetic fields propagating in tandem, the metal cavity reflects microwaves efficiently. Even in the window, you are look-ing through a metal mesh that behaves as a solid metal wall to microwaves. The food is always elevated on a low tray because it won't heat right against a node.

Here's the connection to telescope optics. A telescope aperture is oscillating at a given unchanging frequency (like the microwave transmit-ter). Bright places in the image are the equivalent of hot spots in the oven. Dark places are similar to cold spots that leave food uncooked.

We can thus interpret the behavior of the diffraction structure near focus as a standing wave. A slice through the focal region of an image that suffers from a small amount of spherical aberration is depicted in Figure 4.14. This figure shows how the light collapses to focus and expands again beyond it, with the region closer to the objective on the left above the label –2 and the region farther from the objective above the label 2. It is printed at very low contrast to emphasize the low-level structure. The nodes and antinodes are easy to identify. Because the aperture is a rigid geometrical structure, there are locations where the wave doesn't fit correctly and places where the vibration is favored.

This situation is very much like a brook trickling over a rocky stream-bed. Over one particular pebble, the surface of the water is elevated a small amount. In other places the surface seems undisturbed. The overall look is deceptively frozen, and we forget that in only a few moments all the water

Fig. 4.14 Longitudinal slice near focus of an aperture suffering from slight spherical aberration. (Numbers explained in Appendix D.)

has moved downstream to be replaced by fresh water occupying the same configuration.

Similarly, we interpret the standing wave of the image as some sort of fixed artifact, but energy is rushing through the instrument at the speed of light. The nodal points that can be seen as dark rings in the focused image and corrugations in the defocused image are induced by the geometry of the situation. They are more like the pebble than the water. Aberrations disturb the geometry and in doing so affect the standing waves.

4.5 Other Aberrations—The Pupil Function

The tilt and defocusing aberrations are correctable errors. True, you can't see much if the image is defocused, but you can regain a crisp image with only a turn of a knob. Tilt error merely requires you to redirect your gaze to the center of the stellar image. The observer can't get rid of the smearing of the image caused by diffraction, but if the magnification is small enough, that spread is acceptably small.

Up to now, the discussion has been mostly about perfect apertures. What happens when some imperfection causes the wavefront to buckle as it passes through the aperture? Our little secondary sources, or radiators, are then no longer on the surface of a sphere. They are confined to some sort of surface shape for which no unique focus position can be found. Asymmetries form in the node patterns (as in Figure 4.14).

The best focus position no longer sees these radiators at a single phase. The wave sum can never equal the whole area of the aperture, no matter where we move the focal point. Of course, for most of the realistic

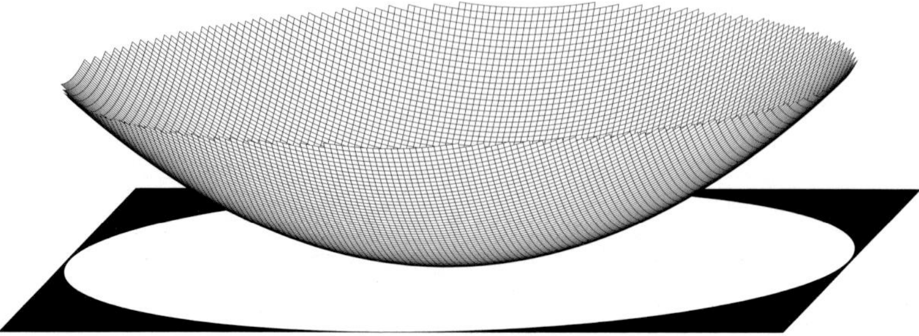

Fig. 4.15 The aberration function of defocusing, with the aperture indicated in black below the wave-front surface. The direction of focus is up.

aberrations considered in this book, the errors at best focus will never get so large as to demand more than one coarse Fresnel zone to show them. Not until aberration amounts to ½ wavelength peak-to-valley will more than one Fresnel zone appear on the aperture at best focus. A more accurate way of doing the wave sum will show that the peak intensity is always less than unity for aberrated apertures.

These Fresnel zones are a good educational device, but they don't allow sufficient gradation to permit detailed examination of what is happening with radiators at the aperture. For that, we need to introduce the *pupil function*. For now, we can define the pupil function as a surface containing information about both the phase and the transmission of the aberration. It is defined more precisely in Appendix B.

Wavefront phase for simple defocusing appears in Figure 4.15. The aperture is shown below the wavefront. Fresnel zones are contours of such a surface. The transmission portion of the pupil function looks like a flat tabletop because this aperture has uniform transmission to the edge. Most of the pupil functions appearing later will have similar on-off transmission behavior. The aberration or phase function is shown most often because it is generally the most interesting component of the complete pupil function.

Figure 4.15 shows the defocusing aberration function outside focus. The aberration function inside focus is a low mound. For a perfect aperture, the aberration function would be a flat plate. All of the optical difficulties appearing in this book will have an associated pupil function. Far from being just a conceptual device, the pupil function contains the essence of image formation in the telescope. Mathematical operations on the pupil function lead to the diffraction pattern and the modulation transfer function.

Chapter 5
Conducting the Star Test

Astronomers have long used the star test as a comforting touch with reality. A quick turn of the focuser is often enough to confirm that the telescope is aligned, cooled off, and ready for operation. Making such a determination requires no reference to numbers, but occasionally you'll want to use the star test to estimate the strength of aberrations on the glass. You must then know exactly how far to defocus the telescope in terms of quantities that can be compared with wavefront deformations.

Also, convenient use of the star test often has nothing to do with "stars." In some cases, atmospheric turbulence, apparent stellar motion, or waiting for a clear night makes using true stars too difficult. To conduct daytime or earthbound tests, you can no longer rely on the great distances and small angular extent of actual stars. When using an artificial source, you must figure the minimum distance at which it can be placed and the maximum size of the pinhole. If you are sloppy about these points, you could unfairly decide that a good telescope is bad.

This chapter deals primarily with the following three topics:

1. Translating defocusing aberration to the more familiar but less general topic of how far to move the eyepiece in the focuser.

2. Sizing, making, and placing artificial sources.

3. Setting up and doing an "official" star test, as opposed to the usual check of the operating conditions of the telescope.

Important results of this chapter are listed in tables. For completeness, the derivations are sketched either here or in the Appendices. Those who are interested can see where some of the concepts in this chapter originated, but performing an effective star test requires no more than careful use of the tables.

5.1 Defocusing and Sensitivity

Figures appearing in this book avoid any reference to how far you must turn the focuser forward or backward to obtain a certain amount of defocus. Instead, any defocus amount refers to *defocusing aberration* measured in wavelengths on the aperture pupil (see Section 4.3.1). Why use the defocus parameter in such a peculiar manner, when the straightforward method of eyepiece motion can be easily understood by everyone?

The answer is simple; telescopes are too different from one another. The amount the focuser must travel to show a given pattern varies with the focal ratio of the instrument. Telescopes with equivalent aberrations and obstructions will show identical patterns, but they all do so for different eyepiece motions. If you specify focuser motion, you must also give the aperture ratio. The result is a muddy picture of a truly simple concept.

Defocusing aberration, as measured on the pupil wavefront, is a sort of universal coordinate system that classifies identical behavior in many different telescopes. In an effort to reduce the multiplicity of patterns, defocusing aberration is used as a generic variable. It has features that transcend telescope type. More important, it is easy to go the other way and calculate how far you must move the eyepiece to yield a given defocus aberration.

5.1.1 Focuser Motion Related to Defocusing Aberration

The defocusing aberration was mentioned briefly at the end of Chapter 4, but no derivation of it was made. It is a simple expression that describes the differences in the sagittae of two different wavefront spheres.[1] The difference between two eyepiece positions for a given number of wavelengths of defocusing aberration is derived in Appendix E. The result is

$$f - f' = \Delta f = 8F^2 \Delta n \lambda \qquad \qquad \textbf{5.1}$$

where F is the focal ratio, λ is the wavelength, and Δn is the change in defocusing aberration in wavelengths.

If n is allowed to go from $+\frac{1}{4}$ to $-\frac{1}{4}$, this quantity becomes $\Delta f = 4\lambda F^2$. This is an expression for the *depth of focus*, or the maximum range for improperly setting the focuser. Since the diameter of the diffraction image is $2.44\lambda F$, the image spot is much longer than it is wide. In fact, the length-to-width ratio

$$\frac{\text{length}}{\text{width}} = \frac{4F^2\lambda}{2.44\lambda F} = 1.64F \qquad \qquad \textbf{5.2}$$

is nearly 25 for an *f*/15 instrument.

The length of this pencil-shaped region is very helpful. It permits small amounts of error in the setting of focus in accessory instruments (such as cameras). It makes less difference in adjusting visual focus

---

[1] The *sagitta* is the amount by which the sphere protrudes through the aperture. Because the radii to the edges of this partial sphere look like stretched bowstrings, it was natural to name this quantity *sagitta*, or *arrow*. (See Figure E.1 in Appendix E.)

because focus is usually fine-tuned by the eye itself. Only those observers who have had cataract surgery or who have little flexibility in their eyes may be forced to rely exclusively on the focusing action of a telescope.

Sidgwick gave another formula for depth of focus: $\Delta f = 4(1.22\lambda F^2)$. This factor is 1.22 larger than the one derived here (Sidgwick 1980, p. 425). The seeming discrepancy comes from the methods used to derive the expressions. Neither formula is meant to be a crisp limit, only the point where the image begins to degrade noticeably. Both expressions are proportional to the focal ratio squared. Thus, an $f/5$ telescope has only a fourth of the focusing tolerance of an $f/10$ telescope, and a ninth that of a classical $f/15$ refractor.

Tables 5.1a and 5.1b list eyepiece motions for varying focal ratios and defocusing aberrations. For example, if we defocus the image of an $f/6$ Newtonian by 8 wavelengths, we can see from the tables that we must change the focus by 0.050 inch (or 1.27 mm). In the convention used here, one must focus *outward* when the defocus aberration is given as a positive number and focus *inward* if it is given as negative.

Much can be learned by carefully examining these tables. They show that defocus distances are vanishingly small for fast focal ratios. The first tabulated column shows values for 0.5 wavelength of defocusing aberration, or about the depth of focus mentioned above. Yet to achieve focusing within $\pm\frac{1}{4}$ wavelength at $f/4$ (or $\Delta n = 0.5$), one must hold focus to within 0.0014 inch, or 0.035 mm. Clearly, if our eyes weren't somewhat internally adjustable, we would struggle to focus fast instruments. Such vanishingly small depths show why it is so difficult to focus a camera in fast instruments. Slow-motion helical or motorized focusers would seem to be justified for these low focal ratio telescopes.

At the other end of the chart are extremely long focal ratios over $f/20$, which would describe such odd instruments as two-mirror Kutter schiefspieglers. To induce 12 wavelengths of defocusing aberration in such instruments, we would have to move the eyepiece over an inch. It is apparent that on such slow instruments, we will scrutinize only small defocusing aberrations before running out of focuser travel. Long focal length telescopes, however, are usually lunar-planetary instruments. They are deliberately tested to higher standards, so small amounts of defocus are the most interesting. A focal ratio of 50 is included because you might mask down your instrument to see a supposedly perfect image.

Please note that for long aperture ratios and high values of defocus, these tables are only approximate. The defocus in these cases becomes a significant portion of the focal length and the effective f-number of the instru-

Table 5.1a

Defocus distances for different
focal ratios and defocusing aberrations
(distances in inches)
Wavelength is 2.165×10^{-5} in

Focal ratio	Defocusing Aberration (wavelengths)					
	0.5	2	5	10	12	16
4	0.0014	0.0055	0.0139	0.0277	0.0333	0.0443
4.5	0.0018	0.0070	0.0175	0.0351	0.0421	0.0561
5	0.0022	0.0087	0.0217	0.0433	0.0520	0.0693
6	0.0031	0.0125	0.0312	0.0624	0.0748	0.0998
7	0.0042	0.0170	0.0424	0.0849	0.1019	0.1358
8	0.0055	0.0222	0.0554	0.1109	0.1330	0.1774
10	0.0087	0.0346	0.0866	0.1732	0.2079	0.2772
12	0.0125	0.0499	0.1247	0.2494	0.2993	0.3991
15	0.0195	0.0780	0.1949	0.3898	0.4677	0.6236
20	0.0346	0.1386	0.3465	0.6929	0.8315	1.1087
25	0.0541	0.2165	0.5413	1.0827	1.2992	1.7323
50	0.2165	0.8661	2.1654	4.3307	5.1969	6.9291

Table 5.1b

Defocus distances for different
focal ratios and defocusing aberrations
(distances in millimeters)
Wavelength is 550 nm

Focal ratio	Defocusing Aberration (wavelengths)					
	0.5	2	5	10	12	16
4	0.035	0.141	0.352	0.704	0.845	1.126
4.5	0.045	0.178	0.446	0.891	1.069	1.426
5	0.055	0.220	0.550	1.100	1.320	1.760
6	0.079	0.317	0.792	1.548	1.901	2.534
7	0.108	0.431	1.078	2.156	2.587	3.450
8	0.141	0.563	1.408	2.816	3.379	4.506
10	0.220	0.880	2.200	4.400	5.280	7.040
12	0.317	1.267	3.168	6.336	7.603	10.14
15	0.495	1.980	4.950	9.900	11.88	15.84
20	0.880	3.520	8.800	17.60	21.12	28.16
25	1.375	5.500	13.75	27.50	33.00	44.00
50	5.500	22.00	55.00	110.0	132.0	176.0

ment is different for the new position. The amounts of defocus necessary are squeezed inside of focus and stretched outside of focus. A first-order correction may be applied:

$$x^* = [1 + (2x/f)] \qquad\qquad \textbf{5.3}$$

where x is the value from the table, x^* is the new calculated value, and f is the focal length. For 12 wavelengths outside focus on a 250-mm $f/4$ telescope, we calculate the correct focal length as 0.8464 mm, not much change at all. But if we have stopped an instrument down to $f/50$, the corrections differ quite a bit.

On fast instruments, the star-test image will probably be evaluated at high values of defocus, beyond even 12 wavelengths. This is not too much of a problem because the test is still sensitive to the relatively severe deformations that pester these instruments.

The best way of using Tables 5.1a and 5.1b is to look up the values corresponding to your focal ratio and write them down somewhere. It might even be convenient to calibrate your focuser knob. Rack it a full turn and see how much it advances focus. This procedure is easy on Newtonians and refractors. You just measure the change in the amount of protrusion in the focuser tube. For example, if one turn of the knob yields ¾ inch (19.05 mm) of focuser travel, a 30° twist gives about ¹⁄₁₆ inch (1.6 mm). This motion is equivalent to 10 wavelengths defocusing aberration for a telescope working at $f/6$.

On Cassegrain-type catadioptrics of the Schmidt or Maksutov type, it is less obvious how to tell which direction focus is tracking or how far it moves. These instruments often achieve focus not by physically transporting the eyepiece, but by internally moving the primary mirror toward the secondary. There is nothing to measure on the outside, so defocus measurement requires another trick. First, focus the telescope with an eyepiece sitting firmly in its socket. Then, loosen the eyepiece and draw it 10 mm or so outside of the socket. Tighten the eyepiece set screw. Find focus once again, being aware of the direction and angle that you have turned the focuser knob (it may help to stick a temporary pointer on the end of the knob). You have found the direction and amount of an effective 10 mm *inward* focus change. On one Schmidt-Cassegrain, this motion was counterclockwise, but telescopes may differ.

5.1.2 Sensitivity of the Star Test

In the equations governing diffraction (Appendix B), defocus is added as a uniformly interfering term to the summation. The point where errors show themselves with highest sensitivity is right at focus because no defocusing term has been added to dilute the image. However, the focused image has two problems:

1. You can't tell where on the surface the error originated.

2. The way the error is expressed is in the intensity of the diffraction rings.

A good example is the degradation caused by a secondary mirror. Light has been kicked from the center of the focused image into the rings, but you can't tell why by peering at the focused image. Furthermore, the human eye is not skilled in determining absolute intensities of the rings. Yet, by drawing the eyepiece out of focus, you see the shadow of the diagonal begin to appear. The shadow more-or-less follows its origins, and aberrations behave similarly. As you defocus the instrument, the light partially unmixes and you can at least guess at the difficulty causing the trouble.

Unfortunately, if you defocus any image far enough, it will look the same regardless of the extent of the error—a flat, uniform disk of light with obstructions clearly delineated. Even the spider shows itself if you defocus too far. The star test loses sensitivity with increased defocus. The light becomes completely unmixed.

Our goal is to find an intermediate focus where the light has become slightly less entangled but is still mixed enough to show optical errors. In my opinion, most optical errors are best shown and identified at around 8 to 12 wavelengths defocusing aberration (prominent exceptions are zones, which often are best identified when defocused further, and astigmatism, which is easiest to identify nearer focus). This 8 to 12 wavelength value is neither too far out of focus—where sensitivity is lost—nor in the region where sensitivity is still so high that good instruments are condemned. The best feature of the star test is that this region of appropriate sensitivity has a stellar image which is not nearly as sensitive to turbulence as a focused imaged. For example, if you are testing a 6-inch *f*/12 Maksutov-Cassegrain, the 10 wavelength defocused image is about 0.53 mm, or an arcminute, across. The structure of the image may shake from turbulence, but its shape is much easier to see than the about 1.5 arcsecond focused image. Even on a mediocre night, you may be able to provisionally judge a telescope.

What is the upper limit of sensitivity under ideal conditions? A photograph in Figure 1.5 depicts the actual defocused diffraction pattern of a nearly perfect circular aperture under nearly monochromatic illumination. In the theoretical pattern of Figure 1.4, the defocus was adjusted until the theory closely matched the photograph. Variations of only $\frac{1}{50}$ wavelength were enough to destroy the match. Even then, the photograph betrays slight differences caused by microscopic projections at the edge of the aperture stop, which had been made from punched metal. Under laboratory conditions, using monochromatic light, the star test can detect wavefront distortions of $\frac{1}{50}$ wavelength peak-to-valley, an outstandingly sensitive measure.

Welford (1960) stated that the star test was accurate to $\frac{1}{20}$ wavelength for broad deformations of the wavefront and $\frac{1}{60}$ wavelength for sharp variations. Welford's estimate matches the observations here. Later on, in Chapter 11 on zonal defects, it is noted that the star test is sensitive to sharp zones even in the presence of gross defocusing. Again, observation concurs with Welford's result.

Field conditions will lessen this sensitivity considerably. Even so, the star test is more than adequate. Using light of more than one color and testing under less than optimal conditions, the slight spherical aberration of telescopes having total wavefront error of less than $\frac{1}{10}$ wavelength is still easy to detect. Even though perusal of advertisements makes it seem that tenth wavelength is about the minimum acceptable, if you find a smooth wavefront truly contained within two concentric shells separated by a tenth wavelength, you possess an outstanding instrument.

One other point should be made about the visual star test. It is not equally sensitive at all focus positions. Welford's $\frac{1}{20}$ to $\frac{1}{60}$ wavelength sensitivities are good for very small values of defocus. As we see later (particularly for the chapters on spherical aberration) the sensitivity drops with increased defocus. This property, however, is not a disadvantage. It will allow testers to tune the sensitivity so good instruments are not condemned. Less than $\frac{1}{4}$ wavelength of low-order spherical aberration becomes difficult to see for defocus values beyond 10 wavelengths, so demanding that the aberration still be visible at plus or minus 10 wavelengths defocus is a way of gauging its severity.

5.2 Artificial Sources

The star test is often performed in the field on a real star. However, for critical testing, you will find that an artificial source of light is convenient and less variable. Artificial sources are preferred for a number of reasons:

1. You can control the brightness.

2. With bright sources, the color can be readily altered and adjusted using filters.

3. The close-in nature of the test allows less of the turbulent atmosphere to intervene.

4. Since the source is fixed with reference to the telescope, it requires

no tracking. If required, you can hold the telescope rigidly.

The beautifully steady pattern that results from use of an artificial source will spoil you. There are some disadvantages, however:

1. The use of an artificial star usually demands a nearly horizontal telescope position. This places maximum demands on the optical mounting cells. Unusual astigmatism, misalignment, or warping may originate solely from the vertical position of the cell.

2. Some of the recommended source placement distances listed in Table 5.2 demand that you look across horizontal spans of 1600 feet (about 0.5 km) although most entries in the table are much less.

3. You need to make a point-like source of light. You must know the diameter of the pinhole to ensure that it is smaller than the resolution of your instrument. Unlike real stars, the source is not always guaranteed to be small.

4. The instrument is designed for focus at infinity. Use at less than this distance often induces spherical aberration.

5.2.1 Distance of Artificial Sources

A paraboloidal reflector is the perfect single mirror for light originating at astronomical distances. However, if the source is at a distance of only twice the focal length, a sphere is the perfect surface. At an intermediate distance, the ideal mirror is a prolate spheroid. Thus, three different forms do the best imaging at three different distances. A telescope could perform adequately when star tested with a nearby source yet fail when directed at the distant sky. Even worse, a fine telescope could be unfairly misjudged by failing the test on a source that is much too close. How much is the test disturbed if the source is placed nearby?

The largest effect of using an artificial source close at hand is to induce spherical aberration in the system. When a perfect paraboloidal mirror is forced to peer myopically at a nearby source, it shows spherical overcorrection that is not found when directed skyward. W.T. Welford says the star test should be conducted with an artificial star placed more than 20 focal lengths away, but he also warns his readers to do accurate ray tracing of the optical system before it is used at questionably close source distances (Welford 1978). We can see below that Welford's suggestion of 20 times the focal length is a very good one for normal apertures and focal ratios, but it fails rather badly for fast paraboloidal mirrors.

In the May, 1991, issue of *Sky & Telescope*, Roger Sinnott traced rays through paraboloidal mirrors and determined how close the source could be placed before an unacceptable spherical overcorrection of ¼ wavelength was noticed. This empirically-derived formula, rewritten for the

Table 5.2

Telescope-to-pinhole distances resulting in
¼ wavelength overcorrection error in paraboloids
(multiples of focal length)

		Aperture Ratio					
		4	5	6	8	10	15
D(in)	D(mm)						
2.4	61	13	10*	10*	10*	10*	10*
3	76	16	10*	10*	10*	10*	10*
4.25	108	22	11	10*	10*	10*	10*
6	152	32	16	10*	10*	10*	10*
8	203	42	22	12	10*	10*	10*
10	254	53	27	16	10*	10*	10*
12.5	318	66	34	19	10*	10*	10*
14	356	74	38	22	10*	10*	10*
16	406	84	43	25	11	10*	10*
18	457	95	49	28	12	10*	10*
20	508	105	54	31	13	10*	10*
24	610	126	65	37	16	10*	10*

* See text on page 92 for explanation of asterisks.

notation used here, is

$$N[\text{ft}] = 28\left(\frac{D}{F}\right)^2 \quad [\text{D in inches}] \tag{5.4}$$

where F is the focal ratio and D, the aperture diameter. This equation is re-written to calculate the multiplier of the focal length:

$$\text{Mult} = \left(\frac{336}{FD}\right)\left(\frac{D}{F}\right)^2 = 336\frac{D}{F^3}[\text{D in inches}]. \tag{5.5}$$

The formula can be put in any unit system by pulling out the unit of wavelength (550 μm):

$$\text{Mult} = (336)(2.17 \times 10^{-5})\frac{D}{F^3\lambda} = \frac{1}{137}\frac{D}{F^3\lambda}. \tag{5.6}$$

By comparing a parabola and an ellipse that are tangent at the center and touching at the edge of the mirror and doubling the greatest difference, a similar result can be derived analytically. This results in the slightly different formula

$$\text{Mult} = \frac{1}{128}\frac{D}{F^3\lambda}. \tag{5.7}$$

The difference is insignificant, and it probably occurs because Sinnott's program automatically accounts for the different angle of exit from a mirror's surface while the analytic calculation ignores this angle. Table 5.2 uses Sinnott's result because it leads to smaller separations. Even though it is numerically derived, it probably is more accurate for the low focal ratios of greatest interest. An example calculation: a 6-inch *f/4* paraboloid has ¼ wavelength of spherical overcorrection when directed toward a source placed at 63 feet (multiple of focal length: 31.5). The distances in this table induce approximately ¼ wavelength of spherical aberration in a Newtonian reflector (Sinnott 1991).

Since the distances appearing in this table are only on the edge of optical respectability, readers are advised to double or even triple them for use in the star test. Wavelength enters in the denominator of Equations 5.6 or 5.7, meaning that a perfect paraboloidal mirror will appear less than ⅛ wavelength overcorrected with the source placed at twice such a distance.

Don't fret if you are forced to position the source a few steps closer than you would like. In my years of telescope testing experience, few instruments possessed less than ⅛ wavelength total error on the wavefront.

Distances greater than Welford's empirical rule are required only for low focal ratios, large apertures, or peculiar optical systems. Most slower optical systems demand less separation than Welford's generous limit. The telescope with the highest chance of breaking Welford's rule will probably be the Newtonian reflector for which this table is calculated, but there are some strange exceptions. Naturally, this discussion just deals with induced aberrations; whether the beam from a nearby target will succeed in exiting the telescope without hitting a baffle or hardware is another matter.

Values that will be less than 20 times the focal length (after doubling) are suppressed in the table and marked with asterisks. Such close source distances will stretch the focus outward more than ¹⁄₂₀ of the normal focal length of the instrument. As the focuser tube seldom allows such motions without running out of travel or causing the tube baffles to vignette the telescope, such close source distances are not recommended if they can be avoided.

In fact, a good policy before testing is to remove the eyepiece after focusing on the artificial source. Place your eye at the approximate focal plane and verify that the whole optical system is still in view (you may need to use a flashlight to illuminate the inside of the tube). If not, you've got to move the source farther away and try again.

One trouble zone occurs in this table. In fast and extremely large mirrors, the tolerable distance increases explosively. For a 24-inch *f/4* mirror, the source should be placed (after doubling) at a distance of at least 2016 feet (0.61 km). These huge instruments are comparatively rare, and more important, they are seldom tested for quality much beyond the diffraction limit. Large Newtonians are employed for light-gathering power, so using the ¼-wavelength distances in Table 5.2, undoubled, is better than not doing the test at all.

If you are testing more ordinary instruments, twice the distance in the table is easily achieved. For the more common 12.5-inch *f/5*, the source distance has contracted to a manageable 354 feet (0.11 km).

Even though Table 5.2 strictly applies only to Newtonian reflectors, we shall use it as a general guide for the star test. After many telescopes were modeled in ZEMAX, few optical system were found with higher multipliers. There is some variation in the sign of the correction error. Most refractors at finite conjugate distance (i.e., source positions nearer than infinity) have induced undercorrection instead of overcorrection. The same is true of classical Cassegrains, yet Schmidt-Cassegrains display overcorrection like Newtonians. Maksutov-Cassegrains can be either under- or overcorrected. Keep in mind that the multipliers of Table 5.2 are only a suggestion and if there is any question about minimum distance, a detailed ray trace will settle the question.

Where the focusing mechanism of the telescope itself affects the optical correction, Table 5.2 could be seriously in error. This situation is common for catadioptric telescopes that change the mirror-secondary separation to achieve focus. In one case, the minimum source separation to keep all the rays inside the diffraction spot has been estimated for Schmidt-Cassegrains by Rutten and van Venrooij. For a 200-mm (8-inch) *f/10* having surface shapes approximating those available in commercial units, the separation is approximately 48 meters, or 24 focal lengths (Rutten and van Venrooij 1988, p. 87). In the Schmidt-Cassegrain modeled in Appendix G.5, ¼-wavelength of overcorrection happens at 12.3 nominal focal lengths away, so doubling the distance for ⅛-wavelength optics occurs at a multiplier of 25. For the moving-mirror Maksutov-Cassegrain later in this chapter, the ¼-wavelength undercorrection multiplier happens at 15.7 nominal focal lengths, meaning the ⅛-wavelength focal-length multiplier is 31. Oddly enough, moving-mirror focusing mechanisms seem to make either variety of catadioptric more tolerant of close focus position.

Another telescope for which this table doesn't work is the refractor, either the apochromat or the simple achromat. Refractors accept remarkably close source positions. Indeed, some smaller and slower refractors will not generate ¼ wavelength of spherical aberration even used as close as the two focal lengths of equal conjugate positions (i.e., the target is as far in front of the objective as the image is behind it). However, it is useful to test them anyway at over 10 to 20 focal lengths. Otherwise, the lens would have to be removed from the tube to avoid the obstruction of the baffles.

One additional instruction for the reader: If you cannot measure the source distance because of intervening impediments, try to err on the long side. You cannot place the source too far away. If your telescope seems to have a smooth correction error, put the source at a greater distance and test again before you conclude that the error is on the glass.

5.2.2 Diameter of Artificial Sources

We must carefully select a pinhole size in the artificial source so that it is smaller than the resolution of the instrument. On the other hand, it must be large enough to allow sufficient illumination to fill a defocused image with light. To calculate such a diameter, we extend the Airy disk radius to the distance of the pinhole. If that radius is chosen as the *diameter* of the source pinhole, we ensure that the source is no more than half of the angular extent of a point image.

This calculation is done for distances twice those in Table 5.2. The results are listed in Tables 5.3a and 5.3b. A quick glance at these tables shows some pinholes that will be extremely difficult to make or measure. It is certainly no easy task to make accurate sizes of pinholes only 0.07 mm (or about 0.003 inch) across. But these pinholes pertain to unusual telescopes—a 3-inch *f*/5, for example. Twenty times the focal length of a 3-inch *f*/5 is only 300 inches (7.6 m). It is relatively simple to use a pinhole 4 times as large (0.28 mm) and place it 4 times more distant (100 ft.). The artificial source is still only a backyard away.

The pinhole can be punched in aluminum foil. To check its size, optically expand it in a microscope while a coarse Ronchi screen of known periodicity sits on it. If you do not possess a Ronchi screen, estimate the pinhole's size projected to a distance of 250 mm. For example, if you are viewing through a microscope with compound power of 100, and the virtual image of the pinhole appears to occupy the same angle as 25 mm on a ruler held 250 mm from your eye, then you have a pinhole 25/100 mm = 0.25 mm in diameter. Take the largest dimension as the maximum effective diameter. The hole need not be round although a round image allows it to be as bright as possible for the effective diameter.

You can effectively shrink a large pinhole to a tiny one using another technique. If you have access to a good microscope objective, you can image a large pinhole onto a small replica of itself. Here, place the large pinhole where the microscope's eyepiece normally sits (4–6 inches from

Table 5.3a
Maximum diameters in inches
for artificial sources

		Focal Ratio					
		4	5	6	8	10	15
D(in)	D(mm)						
2.4	61	0.003	0.003	0.003	0.004	0.005	0.008
3	76	0.003	0.003	0.003	0.004	0.005	0.008
4.25	108	0.005	0.003	0.003	0.004	0.005	0.008
6	152	0.007	0.004	0.003	0.004	0.005	0.008
8	203	0.009	0.006	0.004	0.004	0.005	0.008
10	254	0.011	0.007	0.005	0.004	0.005	0.008
12.5	318	0.014	0.009	0.006	0.004	0.005	0.008
14	356	0.016	0.010	0.007	0.004	0.005	0.008
16	406	0.018	0.011	0.008	0.004	0.005	0.008
18	457	0.020	0.013	0.009	0.005	0.005	0.008
20	508	0.022	0.014	0.010	0.006	0.005	0.008
24	610	0.027	0.017	0.012	0.007	0.005	0.008

Table 5.3b
Maximum diameters in millimeters
for artificial sources

		Focal Ratio					
		4	5	6	8	10	15
D(in)	D(mm)						
2.4	61	0.07	0.07	0.08	0.11	0.13	0.20
3	76	0.09	0.07	0.08	0.11	0.13	0.20
4.25	108	0.12	0.08	0.08	0.11	0.13	0.20
6	152	0.17	0.11	0.08	0.11	0.13	0.20
8	203	0.23	0.14	0.10	0.11	0.13	0.20
10	254	0.28	0.18	0.13	0.11	0.13	0.20
12.5	318	0.35	0.23	0.16	0.11	0.13	0.20
14	356	0.40	0.25	0.18	0.11	0.13	0.20
16	406	0.45	0.29	0.20	0.11	0.13	0.20
18	457	0.51	0.33	0.23	0.12	0.13	0.20
20	508	0.56	0.36	0.25	0.14	0.13	0.20
24	610	0.68	0.43	0.30	0.17	0.13	0.20

the threaded end of the objective). Shine light through the pinhole and then through the microscope objective, in a reverse direction to the way that microscopes customarily process light. The microscope objective is pointed at the distant telescope and the source appears to be floating a few millimeters in front of the objective. For example, if a 1-mm pinhole is placed 100 mm behind a 5-mm focal length microscope objective, the source appears to be diminished to about 0.05 mm. You can also use high-magnification eyepieces for this purpose (also used backwards), but micro-scope objectives are best.

5.2.3 Using a Reflective Sphere instead of a Pinhole

Even if you have success with the painstaking manufacture of the tiny source itself, you must then attach it to a lamp or flashlight and focus some light through the tiny opening. Making a true pinhole source can be laborious, but there are easier ways of fabricating a source that works every bit as well.

Since the star test is a supposedly easy technique for evaluating finished telescope optics, readers probably do not want elaborate and difficult plans for constructing auxiliary equipment. Optical devices cannot often be reproduced without having access to the same lenses and other minor gadgets used in the source document.

Yet the problem of adequate point-source construction must be solved somehow. Astronomy popularizer John Dobson figures his mirrors using the star test. Glints and reflections viewed at great distances serve as the point source. He even claims to have performed the final test of one telescope using the reflection in a bird's eye.

Using the glitter of the Sun in small spherical reflectors is an excellent way of achieving the requisite small size and dazzling intensity. Dobson very likely uses a variation of the device described below when a well-placed and patient bird is not available.[2]

Conveniently, every year huge numbers of ideal spherical reflectors are made in the form of blown-glass tree ornaments. These balls are silvered on the interior and commonly available in sizes of 2.5 to 7.5 cm (1 to 3 inches). If we calculate the size of a bright light source reflected in such a decoration, we can use these objects to make tiny virtual-image pinholes of accurately known size. The derivation of the approximate size of the glint involves some straightforward mathematics; it has been placed in Appendix F.

One important thing to notice from this derivation is that you don't have to do a trigonometric calculation every time you plan to use one of these spheres. You merely estimate that the glitter image will be less than about $\frac{1}{300}$ the diameter of the sphere for an average solar reflection. If the light is reflected back toward the Sun (i.e., a nearly centered reflection appears in the sphere), the half-degree solar disk shrinks to about $\frac{1}{450}$ of the sphere.

---

[2] I do not recommend the star test as the sole method of testing during fabrication. It is best used as an independent check of workshop tests. The star test can be quite confusing when two or more types of surface deformation are contributing to the errors.

Table 5.4

Approximate virtual pinhole
maximum dimensions
(light source is the Sun)

sphere (in)	pinhole (in)	sphere (mm)	pinhole (mm)
0.5	0.002	12.5	0.04
1.0	0.003	25	0.08
2.0	0.007	50	0.17
3.0	0.010	75	0.25
4.0	0.013	100	0.33
5.0	0.017	125	0.42
6.0	0.020	150	0.50
7.0	0.023	175	0.58
8.0	0.027	200	0.67
9.0	0.030	225	0.75
10.0	0.033	250	0.83

This $\frac{1}{300}$-diameter approximation has certain consequences when viewed in the context of Table 5.3. We see a range of needed pinhole diameters from 0.003 inches to 0.027 inches. Three hundred times these diameters would call for reflective spheres from about 1 inch (25 mm) to 9 inches (23 cm) diameter. This latter sphere was calculated for a 61-cm (24-inch) *f/4* reflector seen at a distance of 615 meters (0.38 miles).

Table 5.4 gives approximate maximum dimensions of virtual pinholes for various sizes of reflective sphere where the source of light subtends $\frac{1}{2}°$ (approximately the solar diameter). Let's say you need a sphere 5 inches in diameter.

Few ornament bulbs are as large as 5 inches across. However, you can use a smaller glitter point. The reason so much care is taken with source size is because you're going to have to place that sphere a long way from your fast reflector, and you want bright images.

The sphere need not be whole. A convex adhesive mirror, commonly available in automobile parts stores, is used for increasing the field angle of flat rear-view mirrors. If you draw a circle that is the diameter of the sphere you want, you can take it to the store and estimate which of these inexpensive reflectors most closely matches the circle.

Partial sphere reflectors must be aligned so that the reflection is visible in the direction of your instrument, but this task is easy and it will last long enough that you can do an unhurried test.

One last note: Other authors have recommended the use of black spheres because they obscure the reflection of the surroundings while still reflecting the Sun. You could get an equivalent effect by using a neutral density 1.0 to 2.0 filter at the eyepiece (dimming not just the sphere, but the whole field of view). Using a glossy black surface may be a good policy during inspection of the dazzling image close to focus, but it is not productive when defocusing farther. If you must diminish the intensity, go ahead and do so with an eyepiece filter, but start with the brightest image possible. Alternatively, it is often convenient to have two spheres in close proximity, one smaller than the other (such as a shiny ball bearing next to the tree ornament). A check for small amounts of astigmatism, for example, is best done by rocking the focus very slightly on either side of the most compact image. The large sphere's image is often too bright and the smaller sphere serves well as a point source for small defocus.

5.2.4 Setting Up a Nighttime Artificial Source

You can arrange the same light reflection at night by providing your own illumination. Light no longer comes from a constant angle of the Sun, so care is required in arranging a lamp. Attach a flashlight on a photographic tripod or other adjustable support and direct it towards your sphere. The flashlight performs best when stopped down to about 1 cm and placed about 1 meter away. This procedure will approximate the $\frac{1}{2}°$ angular diameter of the Sun, so the tables above will work the same. You may achieve a tighter beam if you stop the flashlight down off-axis, so that the filament of the lamp is not directly visible. Use a flat sheet of aluminum foil to make the mask. The exit portal need not be perfectly round. The flashlight should be placed as near to the line of sight as possible.

You might also wish to color balance the light more closely to the output of real stars. Coloring the source lamp is particularly useful when testing refractors. Most incandescent filaments have a black body color temperature of around 2500°, or as red as Betelgeuse. The spectrum should be filtered either with a pale blue (Wratten 80A) eyepiece filter or by filtering the output of the flashlight with a similar light blue filter (Berry 1992). For most testing, though, you will prefer a single color, and the reddish tinge of the flashlight is welcome. In fact, you may prefer to test in green- or gold-colored spheres, which are also available as tree ornaments.

Choose a site without stray sources of light. A daytime test can be conducted nearly anywhere because the Sun is the brightest light source, but a nighttime test can be compromised by the presence of street lights nearby. Your flashlight is still likely to be the most prominent source reflected in the sphere. After all, it's close and is directed right at the sphere. However, nearby interfering street lights provide secondary glitter points, and other glints may make interpretation difficult.

Excluding stray sources of light is easy. For example, you could construct an "accordion" box to shade the sphere from all but lights in the direction of the telescope. This is a little cardboard box blackened on the inside and having a black poster board accordion-folded at the back. This folding acts as a non-glossy beam dump. Most often, though, all that is really required is a carefully-placed hood or the open end of a cardboard box. Easier yet is to go where there are no spurious lights. Your usual observing site supposedly is a fairly dark locale. Just take the test equipment along.

You have one luxury that daytime testers lack: you can move the source closer to the sphere until the pinhole begins to exceed the Airy disk. In fact, this process is made easier if you form a square hole in the flashlight mask. Stop and draw the flashlight back a small distance when this squareness appears.

5.3 Performing the Test

Because people wanting to test telescopes don't usually have access to towers or convenient topography to elevate the sphere, they must test through ground turbulence. Daytime testing is probably best done during the early morning and over a grassy field, but different locations and times have their own behavior. Often, a quiet time of non-turbulent behavior can be found to briefly persist near sunset. Go ahead and try the test anytime and anywhere; you may be pleasantly surprised. Also, try to place the Sun at your back to shade the eyepiece and ensure an approximately round reflection in the sphere.

You can mount your sphere in a stiff piece of poster board, poking the hanger through a hole cut in the board. If you don't use the board and wish to hang the sphere on a bush or a tree, then be sure to paint, tape, or otherwise obscure the hanger region of the ornament. It has nonspherical curvature that may present a second interfering point of light. If you are using a curved rear-view mirror, be sure to obscure the edge if it is shiny. (See Figure 5.1.)

One might think that a black poster board would work best, but a uniform color of dark green also works well. The uniformity is more important than the color, although bright colors should not be used.

Most telescopes require moderate distances, but large Newtonians of low focal ratio demand long, clear testing fields. Long distances that meet other requirements are sometimes hard to find. You may be able to locate straight stretches of country road, perhaps straddling a convenient valley. Be sure to set up your testing range over grass on the upwind side of the

Fig. 5.1 Spheres are illuminated by a large spotlight to exaggerate the glitter reflection. A convex rear-view mirror is shown connected to the tripod head. The stray reflection points have yet to be obscured by tape.

road. You don't want the heat from asphalt to disturb the test. Also, avoid optical paths that cross over building roofs. Public parks are ideal, since they feature large expanses of grass and might even be relatively deserted early in the morning.

If your telescope has a thin mirror, be prepared for astigmatism. Think of the thin mirror as flexible. Upended, it sags. If your telescope has a heavy mirror (thick or thin), also anticipate some warping. This is particularly true if you aren't supporting the mirror gently in a sling, although it appears occasionally even then.

Such patterns will not resemble the neatly symmetrical diagrams appearing in Chapter 8 because only two supports are likely to squeeze. If you can't eliminate this effect by careful mounting, you will have to test at elevated angles on real stars.

You may find the alignment of the telescope is changed when the instrument is pointed at the horizon. To avoid strains, optics are mounted loosely. Try to set up a situation with a slight upward tilt of the optical axis. With luck, the optics will lean back to rest against their natural supports. Either choose a testing range with a natural upward slope or mount the source higher. At worst, you may need to do temporary fine-alignment for the test. One suspects that glass power-line insulators are so popular as curved reflectors only because they are conveniently mounted on towers.

Now try a "snap test." Rock the focuser on either side of the sharpest image and see how difficult it is to set the focus. In the most desirable situation, the focus seems to snap into place, and no matter where you halt, you are always convinced that the limiting factor in focusing is your inability to stop your hand from turning precisely at the crispest image. (Much of this depends on the focal ratio and how steadily the telescope is mounted.) The least-desirable situation is one where the focus looks about equally good over a range of focuser travel. You drift through the region of best focus, unable to decide. Hand-eye coordination is far from being the limiting factor (Suiter 1990).

If your visual power of accommodation is strong, then you must provide a dominant field object to hold the eye's focus while you vary the focuser. A reticle on the field plane of the eyepiece provides such an object. If you have an illuminated reticle eyepiece of 12 mm or below for guiding, you can use it (possibly with a good Barlow). If you don't have a guiding eyepiece, stretch a scrap of black electrical tape halfway across the field stop of a high magnification eyepiece (the stop is the hole on the underside). If you have placed the tape close to the best focus of the ocular, you will see half of the eyepiece's field occluded by a sharp-edged shadow.

Place the point source image close to a straight edge of the artificial pattern. Your eye will naturally focus on the large, high-contrast edge. Then you can vary the image focus at will while the focus of your eye is held as if it were in a vise. This psychological trick was once common in darkroom enlarger focusing aids.

Look for individual aberrations following the instructions in the chapters that deal with them. In testing refractors for optical errors having nothing to do with color correction, you may find it helpful to use a green filter on the eyepiece. In fact, using a colored filter is a good policy for all telescopes, whether reflector or refractor. Even though reflectors have no overt color errors, the star test still suffers some confusion from the finite bandwidth of the eye. For example, red light of wavelength 630 nm may be 10 wavelengths out of focus, while deep blue light of wavelength 420 nm is 15 wavelengths out of focus. A colored filter reduces the range of contributing wavelengths in the image.

As an example of how to use these tables from beginning to end, as well as a road map of the procedures and pitfalls you may encounter in star testing telescopes, the rest of the chapter describes the testing procedure for imaginary instruments.

5.3.1 8-Inch *f/6* Newtonian Reflector

We find in Table 5.2 that the source distance for ⅛-wavelength error is 2×12*f* or 24 times the focal length of 4 feet. Multiplication yields a source distance of 96 feet (29 m). Table 5.3a or 5.3b says the source diameter is about 0.10 millimeter, or about 0.004 inch.

This pinhole diameter—multiplied by 300—indicates a solar reflection in a 30-mm sphere. Since this is reasonably close to the 1-inch (25-mm) ornament bulb, we will use that. Don't do a lot of extra work matching these parameters precisely. You also don't need to measure a 29-m testing range carefully. The sensitivity of the star test is not helped or hindered by this sort of exaggerated precision. Pace the distance off; thirty-five steps should be enough. If the only available spherical reflector is 2.5 inches in diameter, you can easily go twice as far to about 200 feet (60 m). *You* are in control, not the test.

The range is set up over a grassy lawn. It is about 8 A.M. with the sphere placed to the south-southwest. The Sun is over the tester's left shoulder.

We note from Tables 5.1a and 5.1b that the focusing travel necessary to achieve a defocusing aberration of 12 wavelengths is 1.9 mm (0.075 inches). The focuser in this telescope withdraws ¾ inch per rotation, so we will give it less than about ±⅒ of a turn for most aberration checks.

The first thing you notice when viewing the defocused image is the seeming appearance of severe misalignment. When you set the telescope up, you checked the coarse alignment, and it was fine. You replace the sighthole eyepiece. The mirror's dot is still on the center. What is happening?

Then you notice that the focuser is racked back 50 mm farther than usual. In fact, you had to dig in your eyepiece box to find the focus extender tube. You look in the sighthole again and this time see the problem. At this focus setting, the edge of the diagonal mirror cuts off the outer portion of the mirror. The vignetting is a little worse on one side than the other, which explains the off-center diagonal shadow. The out-of-focus disk isn't really that far misaligned, it just isn't completely illuminated.

You walk out to the sphere, pick it up, and take it another 30 steps farther back. Returning to the telescope, you put in an eyepiece and focus; it is now about an inch closer to the tube. Pulling the eyepiece out and putting the sighthole back in, you can now see the whole mirror reflected in the diagonal. Offhand, you realize that you could have used a bigger sphere at this new distance. You decide to give this one a try anyway; it seems bright enough.

This time you notice that the secondary shadow leans slightly to the left side of the defocused image. This appearance indicates real misalignment. You twist the appropriate screw on the primary mirror cell (see Chapter 6) and then check the image again. It's worse. Returning to the adjustment screw and undoing the previous adjustment, you give it a slight turn in the opposite direction.

Checking again, you see much better alignment. A few more minor adjustments and alignment is finished. Collimation will probably have to be redone before the telescope is used on elevated objects because the tube assembly is unusually strained for this horizontal angle. At this point, you look for pinching or astigmatism. You see no such effects, but then this mirror is small, full-thickness, and gently mounted.

Performing the snap test, you see that the image goes through focus rapidly. That's good news. You stare at the defocused image and try to detect a stationary pattern indicating surface roughness in the slight turbulence. You don't see any, but you will test for this condition later in the dark. It's difficult to detect roughness in any turbulence, even the slight amount troubling the instrument now. One good sign is the smooth appearance of the diffraction rings; they would be coarse or broken if roughness were severe. You put a yellow filter on the eyepiece. It seems as if the filter isn't deep enough because many colors can still be seen. You put on a green filter. Now the minima are more visible.

Roll the focus back and forth equal distances on either side of focus to look for spherical aberration. Recall that this is exhibited by a hollow center on one side of focus and a bright center that diminishes toward the edge on the other side. Some tendency in this direction is visible, but it is not severe.

This is a subjective judgement, so you replace the green filter with a neutral density filter to change conditions somewhat and try it again. Using a red filter, you obtain a fresh estimate. Putting the green filter back on and refocusing, you use the calibrated knob (remember, 1 turn equals ¾ inch) to defocus ¹⁄₁₆ inch by turning the focuser outward about 8.3% of a full turn. You are now 10 wavelengths out of focus. Remember the distribution of light. Turn it the same distance inside focus and look at the distribution again. If you are not struck by an obvious difference in the two distributions, then your instrument is probably safely below the specified cutoff. Look forward at Figure 10.12 in the 33-percent 10-wavelength defocused star-test diagrams for low-order spherical aberration. Notice how this aberration makes the transition from subtle to obvious in moving from ⅛ to ¼ wavelength. Unfortunately, 10-wavelengths defocus is not enough to completely hide the undercorrection, but it is by no means as bad as ⅓ wavelength.

Taking the secondary mask off, you next defocus the instrument from far inside focus to far outside focus, all the while looking for dips or extra bright rings that would indicate zones. You see no tight circular structure.

The last test is for turned edge. The way you normally look for this is to put on a deep-colored filter (red is good) and inspect the visibility of diffraction minima inside of focus compared with outside of focus. If turned-down edge is the only aberration, the rings are strong and crisp on the outside and weak or washed out on the inside. However, the effects of turned edge are competing with the effect of undercorrection, which also tends to wash out the diffraction rings on the outside of focus. You peer through the filtered eyepiece and can't really decide this point. One aberration fogs another.

Assessment: This telescope should perform passably on the planets, but it could do better. It is right at the edge of specifications, so you shouldn't complain to the maker. The optics don't seem to be severely rough, but a test for roughness will have to wait for a dark field of view and less turbulence. The telescope was acquired for general-purpose use, a task it should perform well.

5.3.2 16-Inch *f*/4 Dobson-mounted Newtonian

This telescope will require a larger separation from the source. Table 5.2 indicates that a distance of 84 focal lengths will cause ¼ wavelength of spherical overcorrection, so you double it and go to 168 focal lengths; 16 inches $\times\, 4 \times 168 = 10{,}752$ inches, or 896 feet (273 m). The required pinhole size is 0.018 inches, or 0.45 mm. To use a reflection of the Sun will require a reflective sphere over 5 inches in diameter.

You cannot find a suitable sphere in the tree ornament box, so you go to the auto parts store and examine the adhesive wide-angle rear view mirrors. A small mirror found there would seem to be about 7 inches in diameter if it were a complete sphere. It's a little too large, but close enough. Remember, you have a factor of two in the pinhole diameter table before your source exceeds the Airy disk size.

You tape over the outer few millimeters of the mirror because of a secondary reflection in its shiny bezel. You use masking tape because the color really doesn't make any difference. Looking around for an adjustable mounting, you think of a cheap camera tripod with an adjustable ball head. A scrap of wood attached to that tripod head will hold the curved mirror.

Then the search for a test site begins. You finally find a long upslope to the north. It is crossed by a road, which could give some trouble with turbulence, but you're willing to take a chance. You set up the source tripod at 9 A.M. on the high end of the slope. Going a few steps in the direction of the telescope and dropping your head to the sight line, you make sure the Sun is centered in the curved mirror. Also, you check the tripod for spurious reflections and tape over any prominent gleams.

Go downslope about ⅕ mile (or 0.3 km) and set up the 16-inch. A quick look in the 6-mm eyepiece confirms that the optical path is seriously disturbed by turbulence (it's too late in the morning). You give up and decide to attempt the test during a night session at your observing site. This way, the telescope is likely to be operating in a less turbulent environment.

About 9 P.M. the very next night, high cirrus clouds move in and ruin the view. You decide to test the cooled-down mirror and hope, in the meantime, that the clouds go away. Driving along the gravel access road that leads to the site, you again place the sphere about 0.3 km away. This time, however, the tripod is carrying the flashlight, so the sphere is hung from a pasture fence. The flashlight has a 1-cm hole in the mask, and it is placed about 1 m from the partial sphere. You locate it just off the optical path between the sphere and the telescope, so you will obtain a small, round reflection.

Since the path is closer to level, the telescope might lose alignment as it is heeled over. Looking at the sphere, you find that indeed it has lost collimation. As you raise the tube toward zenith, you shake it gently, and ease it back down. That cures the problem.

You then inspect the image for pinching and astigmatism. The image does tend to form a cross at focus, but it's not so bad that it will interfere with the important test for spherical aberration.

In the snap test, the image does not focus as rapidly as you would like. However, you can readily focus your eyes from 200 mm to infinity, and suspect that your accommodation makes the snap test unreliable. You replace the 6-mm eyepiece with a 12-mm illuminated reticle eyepiece on a high-quality Barlow lens and repeat the snap test, all the time being watchful that the crossed wires of the reticle stay sharply in focus. The focus is still indeterminate over a significant region. That's not good.

Attaching a 33-percent mask to the back of the spider, you check the correction. The shadow appears almost at once on the inside and remains big and dark until the telescope is grossly defocused. It doesn't emerge on the other side until you twist the focuser 6 to 10 times farther, after which it emerges from a bright core. Filters alter this situation only slightly. The mirror seems profoundly undercorrected. In the unlikely event the test is being skewed by a bad eyepiece, you choose another from the eyepiece box and again very carefully center the image in it. The undercorrection is still there.

You have a suspicion that the large sphere size is perhaps dazzling your vision. Returning to the source, you increase the distance separating the flashlight and the sphere to 2 m. Even with the darker image, the severe undercorrection still shows. The focus is set at 10-wavelength defocus on either side of best focus and this confirms the undercorrection. It appears somewhat between the ⅓ and ½ wavelength diagrams of Chapter 10.

The rings aren't visible on either side—possibly because of roughness or turbulence. Rings don't appear with a green filter either. One test does not decide roughness, however, especially with as large an instrument as a 16-inch, so you reserve judgement about that point.

Undercorrection is so bad that you don't even look for turned edge or zonal aberration.

Assessment: The telescope fails. It needs to be refigured. If it were even a little worse than a marginal ¼ wavelength, it would be disappointing but acceptable. After all, you don't often expect diffraction-limited optics at *f*/4. But this instrument is far beyond the limits of acceptability.

5.3.3 6-Inch *f*/12 Apochromatic Refractor

Because this telescope is expected to perform well under the most difficult circumstances, it will be tested in a comparatively harsh manner. Twenty times the focal length is 120 feet, or 37 meters. To avoid vignetting from the baffling, you will push this distance to 80 m. Interpolating Table 5.3, you see that a pinhole diameter of about 0.16 mm would be correct at 37 meters, but you're going twice that far, so you want one twice as large. Three hundred pinhole diameters of 0.32 mm yields 96 mm, or about 4 inches.

You only have a 50-mm tree ornament, but since you will be doing this test at night it's easy to move the 1-cm masked flashlight to about 60 cm from the sphere instead of the usual 1 m. The 60-cm distance means that the hole will subtend a little less than a 1° angle as viewed from the sphere. You back the flashlight mask with a light blue camera filter in order to achieve better color balance for the chromatic aberration tests.

Taking the telescope to your usual observing site, you hang the sphere about 250 feet from it. The flashlight is pointed at the sphere from a couple of feet away on the near side. Because the fully-assembled instrument is inconveniently high when directed toward the horizon, you place it between the seats of two sturdy "movie director's" folding chairs. You will sit on the ground.

You attempt to point the telescope by moving the rear chair. The tube is directed at the feet of the tripod, so you elevate the front by slipping in a magazine. The sphere is now slightly too low.

It seems easier to move the target than the telescope, so you walk to the source and move the sphere higher. You rearrange the flashlight, verifying that the brightest reflection is back toward the telescope.

The image needs only a jiggle to center it. You slip in a higher magnification eyepiece. The first thing to examine is color correction. The disk has a slight magenta or reddish fringe inside focus and a green fringe outside focus. In focus, there is no apparent color haze. Rainbow smearing is not apparent in any direction, which indicates that decentering or wedge error is absent. A brighter image would be helpful, so you move the flashlight to about 30 cm from the sphere.

Now, the Airy disk is noticeably bloated, but no color haze is visible. You return to the flashlight and move it back.

Putting a green filter on the eyepiece, you look for astigmatism or stretching as an indicator of misalignment. None can be seen. Defocusing either way, no apparent difficulty with correction appears. The telescope snaps well. You defocus a long distance and look for zones. None are seen. Turned edge doesn't show, but this is a refractor. The lens cell obscures the far edge.

You are disturbed by the lack of contrast in the diffraction rings. This could indicate a problem with roughness. Then again, your eye just may be unaccustomed to the delicacy of the rings. You put in a deep-red filter, but that cuts out too much light, so you turn once again to the green filter.

The in-focus image seems to have several asymmetric thickenings in the rings, but that could be caused by slow-moving air currents between the image and you. You watch long enough to decide that the pattern is fixed, but tomorrow you will do the test from a ladder. You could be looking through cooling artifacts next to the ground.

Spherical aberration is tested at 5 wavelengths defocused either way, and the aberration is below the ⅛-wavelength diagram of Figure 10.11.

Assessment: This telescope might suffer a slight medium-scale roughness, which would compromise the images on perfect nights, and then again, it might not. Such a small amount of aberration would have gone unnoticed in the other instruments. Nevertheless, it is worrisome in a lunar-planetary refractor. However, you decide to do the formal star test again and evaluate it a number of nights on planetary images. Roughness is a difficult aberration to unambiguously separate from turbulence, and you could have misdiagnosed it.

5.3.4 8-Inch *f/*10 Schmidt-Cassegrain Catadioptric

To test this Schmidt-Cassegrain, recall that the telescope diverts rays beyond the edge of the Airy disk at about 48 m (157 ft), or about 24 times the focal length. (Table 5.2 results in an incorrect distance because the internal focusing mechanism compromises the optical correction.) You decide to increase separation distance to at least 100 m (328 ft). A source at this range is 2.5 times the 20 focal lengths recommended in Table 5.2, so it requires 2.5 times the 0.134-mm pinhole size of Table 5.3, or 0.335 mm. The reflective sphere should thus be 100 mm in diameter. You don't have a 4-inch sphere, but you can find a convex mirror that would be 7 inches in diameter if it were a complete sphere.

The pinhole can expand a factor of 2 before it exceeds the Airy disk size. Seven inches, though, is a bit close to the limit. Then you recall that if you test with the Sun directly to your back, with the glitter point centered in the sphere, the divisor is closer to 450 than 300. This condition would permit a sphere at least 6 inches across, and 7 inches is not too much more.

You set up the test early one morning with the Sun low on the eastern horizon and the sighting range to the west. A quick look in the eyepiece confirms that the image is too bright. After hurrying inside to get a 2-inch tree ornament, you set the new sphere at a range of 70 meters and point the instrument again. A peculiar tube current elongates the secondary shadow on one side of focus, but it goes away after a few minutes.

First comes alignment, a relatively straightforward operation since it involves only adjusting one element. Slipping on a green filter, you do the snap test, but you can't say for sure whether the instrument snaps adequately. Focus seems soft, but not seriously defective.

You detect a small amount of spherical aberration, but you can't tell if it is undercorrection or overcorrection. The telescope is refocused with the regular high magnification eyepiece and green filter, the set screw loosened, and the eyepiece withdrawn 4.4 mm. Checking Tables 5.1a and 5.1b, you see that this amount corresponds to about 10 wavelengths outside focus. Repeat inside focus, only this time focus the instrument first, and after doing so, push the eyepiece all the way in to the stop. The edge of the pattern is stronger outside of focus, so the system is overcorrected. By the different size and overall prominence, an estimate of about ¼ wavelength is made.

Having a built-in 33% obstruction, you check the shadow break-out points and estimate a 2.5:1 ratio (see Section 10.6.1). Since this instrument has an induced spherical overcorrection of 0.12 wavelengths at 100 m, the infinity-focus residual could be well within tolerance.

In looking for roughness, you find an unusual amount of turbulence even this early in the morning. You'll have to test again when it is quieter.

Assessment: The instrument shows signs of being excellent, but ground turbulence is too severe. Observation of the planet Saturn that night showed Cassini's division looking sharp and black between bouts of bad seeing. You will try the test again at night using a flashlight source, just as a matter of curiosity, but the telescope is a keeper.

5.3.5 6-Inch *f/*12 Maksutov-Cassegrain Catadioptric

A Maksutov is a special case in that higher-order spherical aberration (also called "zonal spherical aberration") cannot be removed from the system. In properly made Maks, this error is not present to such a degree that it is a significant flaw, but it is something that you have to learn to test for and recognize so that you can properly ignore it. The key procedure here is to not run the star test at too high a sensitivity. On the other hand, you should not test at such a low sensitivity that you miss other aberrations. In essence, you split the test, looking first for simple figuring errors and second, verifying that the expected low level of higher-order spherical aberration is present. Again, you would be safest by using a considerably bigger multiplier than that indicated in Table 5.2, because this is a strange optical system and you do not want to underestimate. Using 30, you multiply the 6 ft focal length to 180 ft. You happen to know that a fence across and down the street is 220 feet away, so you place the target next to the fence.

At that distance, it is apparent that a 3-inch sphere is allowed, but you do not have one, so you will try to make do with a 2-inch sphere. You set up in the early evening over grass to catch the solar-heating crossover. You look at the roundness of the defocused disk, and are pleased to see that there seems to be no misalignment. Misalignment is somewhat problematical in Maksutov-Cassegrains because most of them are finished so that

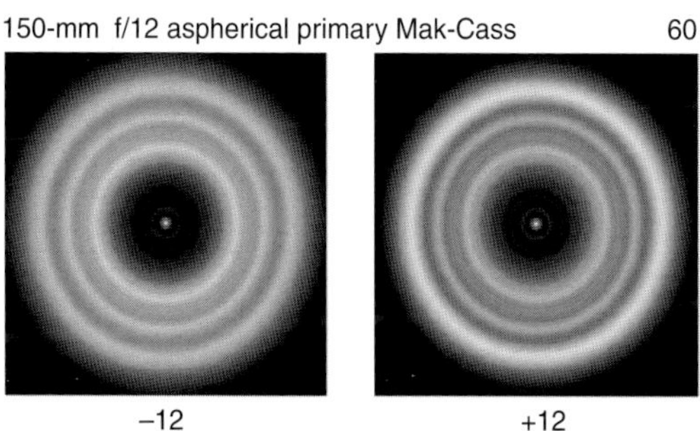

Fig. 5.2 Typical residual high-order spherical aberration characteristic of the Maksutov-Cassegrain telescope.

there is little you can do about alignment other than return it to the factory. Even high-end Maksutov-Cassegrains tend to be non-adjustable, and the makers do not want you changing them.

The first thing you look for is chromatic aberration, any unusual color that appears around a focused image. I have never seen a Maksutov with curves ground so improperly that it displayed chromatic aberration, so you are unlikely to see it. In passage through a focused image during the "snap" test, you are also unable to detect any low-order spherical aberration error. Evidence for this would be if either side of focus had a more prominent hole than explained by the diagonal shadow. It does not, and both sides appear fairly flat. Nevertheless, you defocus by about ±10 wavelengths, and inspect the outer edge of the secondary shadow. An improper amount of low-order correction error shows itself by the apparent size of the shadow. Actually, there is a small deliberate low-order correction applied as a counterbalance to the higher-order spherical aberration, but if it has the right value, it produces a nicely balanced pattern. You look and can see no easily apparent difference between the brightness of the rings just outside the secondary shadow.

Finally, you look for the higher-order spherical aberration. You don't really expect great amounts of this aberration, since it is designed right into the instrument. If they miss on the fabrication of the design, they generally tend to make errors on lower orders of aberration. It is a general rule that unskillful hands make coarser errors. You should defocus ±12 wavelengths and look for appearances not much worse than Figure 5.2. Even though there are trifling differences between the two images in the figure, they are difficult to detect if viewed serially and not side by side. The design of Figure 5.2 has a Strehl ratio greater than about 0.95, neglecting the secondary obstruction of course. The slightly brighter outside ring in the outside-focus image is caused by the roughly ⅛ wavelength rolled edge characteristic of high-order spherical aberration.

Assessment: The instrument is excellent.

Chapter 6
Misalignment

A straightforward way of greatly improving a telescope's image is to align the optics. One of the most ignored optical problems, misalignment is also one of the most curable. The improvements derived by aligning previously neglected instruments are always noticeable, and sometimes the enhancement is startling. Refractors have a larger diffraction-limited field and so are (usually) permanently aligned at the factory, but most reflective optics must be maintained by owners. Users tend to be reluctant to touch the alignment for fear of making it worse, but this chapter will teach you to overcome this fear. The benefits are great and the risks are few.

The star test cannot only be used to diagnose misalignment, it is also useful to achieve fine alignment. Once you become familiar with the star-test method of making that last small adjustment on the collimation, you will make it a standard procedure during every observing session. With practice, star-test alignment becomes easy.

6.1 Kinematic View of Alignment

An optical surface is a three-dimensional object of finite size. To locate such an object in space, one must first find some point (usually its geometric center) and move that point's three linear coordinates. Once its center is placed, two remaining orientation angles must be fixed to specify its position. Actually, mathematicians describe three such rotation angles, but symmetry usually allows one to be ignored. For alignment, this hidden angle is a final meaningless rotation around a symmetrical optical axis.

The three coordinates of the center as well as these three angles make up the six degrees of freedom necessary to know the position and orientation of a solid object precisely. The mounting points are well designed if they are sufficient to fix the object's location, but not to overconstrain it. Three-legged stools are stable on the roughest floors because their three-point design is both necessary and sufficient. Stiff four-legged stools usually rock because the design has too many supports. The fourth leg cannot be precisely made to occupy the plane of the floor defined by the other three legs. This difficulty also accounts for the reason that optics mounting cells are so often multiples of three points.

Alignment is usually separated into two independent tasks. First, the centers of all optical elements are placed on a single axis line and properly spaced. Second, the tilts of each optical element are oriented so that each becomes a circle of revolution around the axis. Thus, if one twirls the optical layout around the axis, it looks the same in all positions. The placement of the center on-axis is called *centering*, and obtaining the correct tilt orientation is called *squaring-on* by some authors.[1]

6.2 Effects of Misalignment

If alignment isn't precise, any number of bad effects can result. Although we will calculate patterns mostly for misaligned Newtonian reflectors, the behavior depicted here qualitatively describes a wide variety of systems and is generally useful. For example, the image of a misaligned commercial Schmidt-Cassegrain is similar to that of a Newtonian reflector, although the amounts differ versus angular field radius. Likewise, Maksutov-Cassegrains of the type employing an aluminized spot on the rear of the corrector for the secondary will display off-axis coma, or a fan-like stretching of the image in a 60-degree angle.

Off-axis, a Newtonian reflector exhibits a mixture of two pure aberrations, coma and astigmatism. Astigmatism is a cylindrical deformation of the wavefront that results in two separate low-quality focal distances, with focus at each appearing as a linear stretching of the image. In misalignment, astigmatism isn't ground into the glass permanently. It is caused by viewing the mirror at an oblique angle. Similarly, coma is induced merely by the tilt of the Newtonian mirror.

As misalignment worsens in Newtonians, astigmatism overtakes coma to become the dominant aberration. Coma worsens linearly with collimation error, while astigmatism increases as the square of the distance to the optical axis. This behavior is illustrated in Figure 6.1. For all practical cases of misalignment, the crossover point is always beyond the level of even coarse alignment. Coma is always stronger than induced astigmatism for Newtonian telescopes.

The bright parts of a comatic image develop wing-shaped appendages, and some of the light is smeared away from the optical axis. Coma effects are the same inside or outside focus. For Newtonians, the aberration stretches the image away from the optical axis. For more complex optical

---

[1] Tilted-component telescopes don't follow such a simple sequence. They require a procedure matched to the particular instrument.

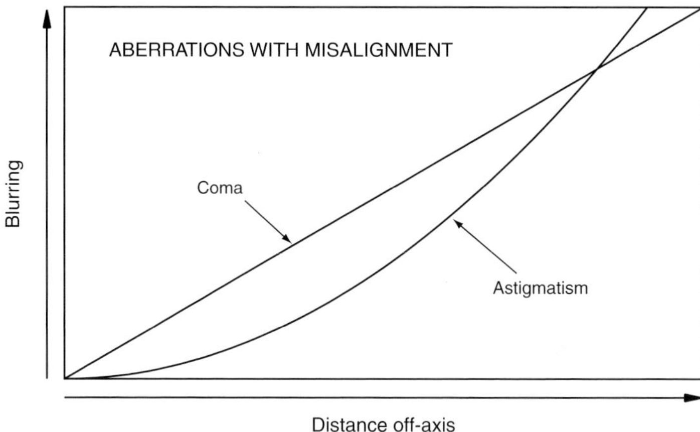

Fig. 6.1 Astigmatism and coma aberration as misalignment becomes worse in Newtonian reflectors.

systems, coma can pull light either toward or away from the axis. Spot Maksutov-Cassegrains, if they don't have enough degrees of freedom to eliminate coma, often appear to stretch the light in the opposite manner to a Newtonian.

Misalignment of the main mirror in fast Newtonians seems easier to perceive outside of focus. The skewed optics no longer have a centered obstruction, and outside focus the off-center shadow of the diagonal adds to the stretching caused by coma. The offset shadow makes coma seem even worse. Inside focus, the off-center shadow leans in a manner that partially counterbalances the coma stretch. Alignment on either side of focus is possible, but the preferred method is to pull the eyepiece back.

Astronomical telescopes have such a constricted field of view that coma caused by a tilted objective is generally not visible unless the image is viewed at high magnification. Any off-axis smearing of the image in low-power eyepieces is primarily the fault of the ocular and cannot be adjusted out.

Unfortunately, coma and astigmatism in composite form are common errors in poorly collimated Newtonian reflectors and catadioptrics. Well-designed refractors and advanced reflectors use the multiplicity of surfaces to reduce or eliminate the coma error. Good refractors chiefly reveal misalignment through astigmatism.

Poor alignment is probably the single greatest contributor to the undeserved shabby reputation of Newtonian reflectors. These instruments are commonly made at very low focal ratios. Their owners don't realize their telescopes must be kept in razor-sharp alignment. A fast Newtonian usually spends its entire existence in a wretched state of collimation. Of course, if low magnifications are used, the zone of excellent alignment will usually be found somewhere within the field stop of the eyepiece. However, at high magnifications (where optical quality must be at its best), the focal-plane region of good quality can be at one edge of the field stop or possibly outside it entirely.

As an example, let's look closely at the common 10-inch (250-mm) *f*/4.5 Newtonian reflector. Schroeder (1987, p. 96) says that *f*/4.5 paraboloids can tolerate misalignments of about 1.8 arcminutes. Superb images are thus confined to a region at the focal plane only 0.05 inch (1.2 mm) in diameter. This tolerance is a bit tight, however. Refocused imaging would allow a misalignment of about 3 arcminutes, a circle on the focal plane of 0.08-inch (2 mm) diameter. The passable region is a circle ⅕ the diameter of the full moon. This alignment is ruined with only about ⅒ turn of a main mirror adjustment screw.

6.3 The Aberration Function of the Misaligned Newtonian

If focused near the smallest image blur circle, the aberration function of misalignment is similar to the complicated form that appears in Figure 6.2. That figure scarcely does justice to the deformation. The phase surface is like a drumhead that has just taken an enormous off-center blow. One half-circle is deformed downward, and the other half-circle has a balancing bulge upward. Added to this form is a small amount of the saddle-point deformation characteristic of astigmatism. The aggregate result is a wonderfully interesting surface to investigate by calculations, but it is bad news if it describes the wavefront of your telescope.

The functional form of the coma aberration is

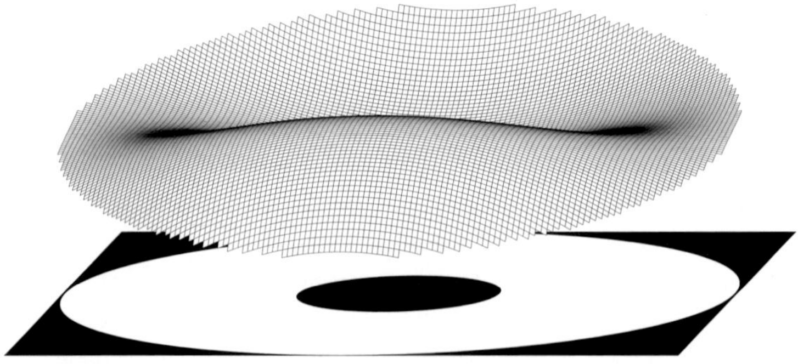

Fig. 6.2 Misaligned Newtonian wavefront just after it has passed through the aperture.

$$W_{\text{coma}}(\rho, \theta) \;=\; \frac{A_3^{\;\text{coma}}}{2}(3\rho^3 - 2\rho)\cos(\theta)\,, \qquad\qquad \textbf{6.1}$$

where ρ is the distance from the optical axis (normalized to reach 1 at the edge of the aperture), $A_3^{\;\text{coma}}$ is the total coma aberration, and θ is the angle from the axis of misalignment (Born and Wolf 1980). The slight amount of astigmatism mixed with this coma has the form appearing in Chapter 14, and the total deformation is $W_{\text{misalign}} \;=\; W_{\text{coma}} + W_{\text{astig}}$. The amounts of each aberration depend on the aperture and the focal ratio.

6.4 Filtering of a Misaligned Newtonian

Since the phase surface is not axially symmetric, its MTF depends on which way the stripes of the MTF target are oriented. If we calculate the MTF at an average orientation, we can see how the behavior improves with better collimation. Five curves are depicted in Figure 6.3. The first is the perfect circular aperture. The next is a 30% obstructed but otherwise perfect aperture. Third is the filtering caused by a shift of the 10-inch $f/4.5$ focal plane sideways by a generous 2-arcminute tolerance. If this shift is doubled, the contrast sags considerably in the fourth curve. The last and bottom curve is the transfer expected from a severely misaligned mirror. Any alignment effort whatsoever results in better collimation than the last curve, but some Newtonian owners are so frightened of producing worse performance that they refuse to touch the adjustment screws.

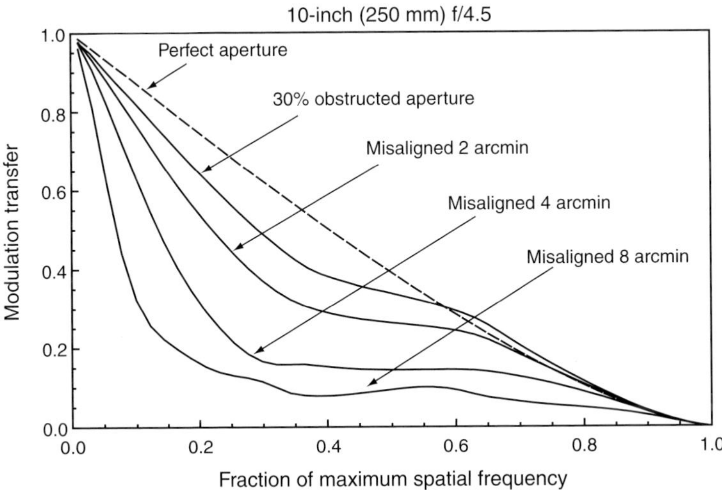

Fig. 6.3 Modulation transfer function of steadily worsening Newtonian alignment (45° orientation of MTF target bars with respect to the direction of misalignment)

The worst MTF curve dips quickly. The interesting behavior is found low on the spatial frequency axis. The initial drop indicates that the telescope is performing as well as a perfect aperture of about ¼ its full diameter. Severe misalignment reduces the low spatial frequency contrast to that of a 2.5-inch telescope!

Even correcting the misalignment until the axis is tilted by twice the tolerance only improves the contrast so this mirror acts like one of half its aperture. This MTF curve is about the same as would occur for a 50% or 60% obstructed aperture. No one would stand for such huge obstructions, but many telescope owners casually accept misalignments of this magnitude.

The filtration of even a slightly misaligned telescope is enough to severely degrade the images. Be fearless in attempting to collimate your telescope. You can hardly do worse than an unaligned instrument, and the potential improvements obtained with only a small effort can be profound.

6.5 Aligning Telescopes

An individual alignment procedure for every commercial telescope cannot possibly be given here because each maker has tiny variations in the manufacture of cells, mirror holders, and adjustment hardware. For this reason, we will examine the generic features of the process used in aligning only three common telescopes. In spite of this restriction, the descriptions which follow represent a large number of optical systems. You will not be given detailed instructions in the form of a recipe to be learned by rote ("then turn thumbscrew A"). That job is best served by the instructions that came with the instrument. Instead, you will be taught procedures that can be applied to nearly every telescope.

Other collimation procedures use specialized methods or tools. Fussy readers are encouraged to pursue them. However, the instructions below produce reasonably straightforward collimation without cutting too much into valuable observing time. The justification and ordering of the steps will be the primary objective, with side comments concerning tricks and pitfalls.

1. *Establish the axis line.* The axis line is defined by two points, usually the center of the eyepiece and the center of the objective lens or mirror. It is customarily directed along the tube's center.

2. *Center the optical components on this axis line.* If the telescope has more elements than an objective and eyepiece, at least one non-trivial centering must take place. Sometimes, centering is set at the factory and is not adjustable by the telescope user.

3. *Establish the tilts of the elements.* Generally, the tilt adjustments must be made in a certain order from one of the two points that define the axis to

the other point. Poor ordering makes alignment much more difficult.

4. *Repeat steps 1, 2, and 3 as an iterative procedure.* Because the adjustments are seldom completely independent, one alignment step can disturb previously correct settings. Coarse alignment is like raking leaves. The first pass doesn't gather every leaf; it takes a few swipes.

5. *Adjust only one element in fine alignment.* Fine alignment takes place on a real image. Normally, a tiny adjustment must be done on the element for which adjustment is most convenient.

Although it is not strictly necessary, you can save time and effort by using a helper to align. On all but the most compact telescopes, collimation is a job best done by two people, one at the eyepiece and one making adjustments.

6.5.1 The Newtonian Reflector

The presence of the diagonal mirror and the many confusing reflections make this adjustment the most difficult of the three discussed here.

Before the alignment begins, you have to make preparations. The first item needed is a centered sighting hole. This blank or lensless "eyepiece" has a small opening at the back instead of a lens element. The purpose of such a device is to allow you to unambiguously place your eye on the axis of the focuser tube. A less obvious purpose is to allow you to sight from a location right at the focal plane. It obscures the view unless the eye is placed close. Thus, the eye can be located not only at the center of the focuser tube but at the correct distance from the diagonal. You want to place this sighting hole near the location of the focal plane when the telescope is focused on distant objects.

An inexpensive source of these sighting holes is a *clear* plastic 35-mm film can with it bottom cut out and a 3-mm hole drilled into the lid. Wind it with a few turns of tape so it will fit snugly. With the digital imaging revolution of recent years, these have become less ubiquitous, so you may be forced to search for and use an alternative like a plastic bottle.

The second preparation is to place a marker dot at the geometric center of the mirror. Taking care not to touch the mirror's surface accidentally, lay a stiff ruler across its diameter. Along the ruler, you can determine the center very precisely, but at right angles to it, you can only estimate. At the center, draw a short (3 mm) line at a right angle to the straight edge. The center will be at some location along the direction of this stubby line. Do the same after a quarter turn of the mirror. Look closely at the center. What you should see is two short lines slightly offset as in Figure 6.4. The true center is at the intersection of these two lines if they are extended until they meet. This is the place to put the dot.

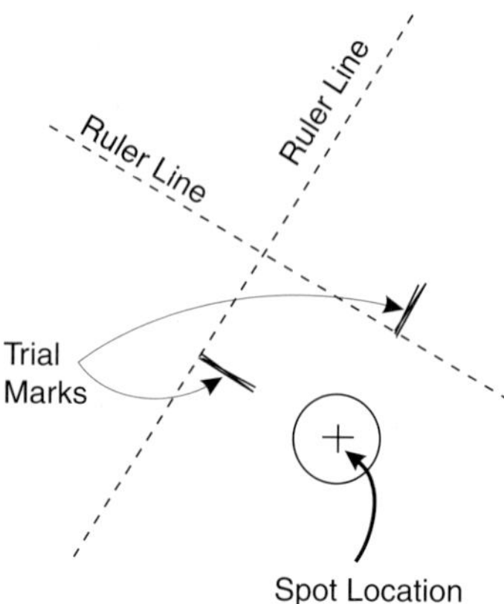

Fig. 6.4 The procedure for finding and marking the true center of a circular disk.

Don't commit yourself to making a large dot at first. It's best to make a barely perceptible mark at this location and measure it both ways to verify it is the center. When you make a bigger mark, you will probably want to extend it out on one side to correct for the inevitable error. This dot can be something as simple as permanent black marker ink or as elaborate as white paint. (Some people claim a white dot or annulus is easier to see when you are attempting alignment in the dark with a flashlight.) Make the dot large enough to see easily (6 mm or ¼ inch). Do not worry about the dot or short lines interfering with optical quality. The marks are in the shadow of the diagonal mirror.

Steps 1 through 3 are best done in a brightly lit room illuminated by diffuse light. (Fluorescent bulbs are ideal.) The inside of a telescope tube is not a well-lit place. During coarse adjustment, tape a sheet of white paper inside the tube across from the focuser. The paper will provide a lighter background to the outline of the diagonal holder than the dark interior of the tube.

Step 1: Establish the axis line.

The axis line will be from the center of the eyepiece's field of view to the center of the mirror. This step would be trivially easy were it not dependent on the right angle break that happens at the diagonal. Once the diagonal is aligned, it will be simple to check, but it is impossible to verify now. We instead pay attention to one crucial fact, the need for the focuser to slide the eyepiece linearly along the optical axis, thus making sure that both your eyepiece and Barlow lens will be on this axis. A misaligned Barlow is a disaster for those who must wear eyeglasses, since the Barlow and medium-power eyepiece are used in preference to extremely short-focus eyepieces.

We will assume that the optical axis will be along the center of the tube and that the diagonal will be set precisely at 45°. The problem of focuser motion reduces to making sure the focus tube is pointed at the tube's center line and that it doesn't lean toward either end of the tube. If you have an accurately round tube and the focuser fits snugly against it, you probably needn't worry, but such instruments are rare.

It has been mentioned that the diagonal need not be pointed at 45° and that the focuser can lean to the front or back of the tube, but such a procedure tosses out the convenience of having the optical path be semi-aligned just by assembling the telescope. Since the focuser must be aligned so that the optical path strikes the axis of the diagonal anyway, the adjustment at right angles is easy enough to do as well. What is theoretically possible may be interesting to speculate about, but minimizing the time we waste in alignment is paramount.

I can't give instructions for measuring the lean of the focuser because every situation is different. I can only point out some tricks for magnifying the lean direction. The first is to put something besides the eyepiece in the focuser. A good choice is a long tube the same diameter as an eyepiece. We will call it the "long eyepiece." If you extend such a tube on the outside of the focuser, it will become easily apparent if it leans toward either end of the tube. Make a right angle template out of a manila file folder or other thin cardboard sheet. Cut out enough of the corner so that it clears the focuser hardware, and lay it against the protruding tube. Any lean of the focuser along the telescope tube becomes immediately apparent.

Remove the diagonal, and extend the long eyepiece to the center of the telescope tube, or as near as you can get. See if the long eyepiece is pointed at a skewed position to either side of the telescope axis. This step is especially easy if your Newtonian has a spider. Just place a pointer or screw in the hole from which you removed the diagonal holder (if it has one) and look through the other end of the long eyepiece tube. Of course, center the spider first. If the focuser axis is skewed, shim the focuser until it points squarely at the center of the diagonal axis.

Once you've verified that your focuser fits squarely on the telescope and advances and retracts eyepieces toward the diagonal axis, you are ready to continue. This measurement usually need not be done again.

Step 2: Center the optical components along the axis line.

Since the axis will go through the center of the mirror and the center of the image plane wherever they might lie, "centering the components" means centering the diagonal along that line. Remember, diagonal placement is

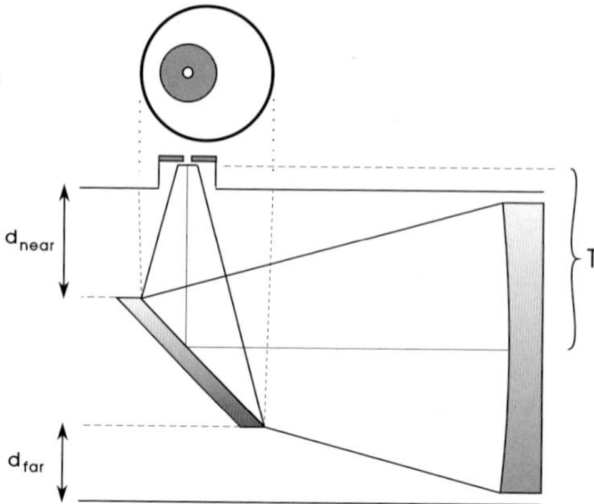

Fig. 6.5 A perfectly aligned Newtonian telescope with focal ratio about 1.75. It displays centered and non-centered features.

achieved without any reference to reflections. At this stage, the main mirror need not be inside the instrument. The diagonal could well be a block of cement. In fact, it is helpful to think of it as nonreflective. You must suppress any urge to center the reflections you see in the diagonal. That doesn't concern you now.

Consult Figure 6.5. This diagram represents a perfectly aligned Newtonian. (Focal ratio has been exaggerated to make the anomalous behavior readily visible.) Note that the distance d_{near} is greater than d_{far}. Alignment is perfect by definition, yet it seems as if the diagonal mirror is not centered. Already we seem to violate general condition 2.

Most important, the components must be *optically* centered, not physically centered. When you look through the little sighting hole replacing the eyepiece, what do you see? The diagonal should appear as a perfect circle neatly centered at the bottom of the focuser tube. The offset is caused by perspective foreshortening of the far side of the diagonal. The far side is so much further away that it actually appears smaller. You need to slide the diagonal sideways and down to center it on the converging cone of light. The view through the sighting hole is sketched in Figure 6.6.

The amount of forward offset is determined by trial-and-error. Move the diagonal forward and back until it *appears* centered. Please note that the word "appears" is emphasized in this statement. Because judgement may be difficult, you may be tempted to remove the sight hole and peer along the focuser tube at a grazing angle. Then you would incorrectly compare the front and back sides and incorrectly set the diagonal. The correct method is to view it from the center and move the diagonal until it only *looks* centered.

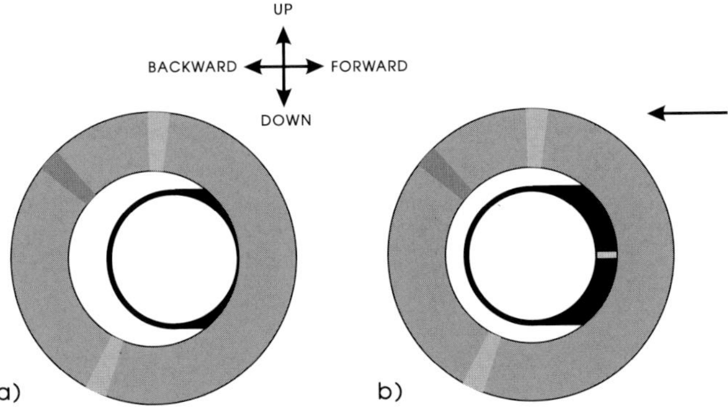

Fig. 6.6 Centering of the diagonal at the bottom of the focuser tube. (This tube is depicted as the dark annulus at the outside with gray radial streaks in it.) a) An improperly placed diagonal. b) The correct position as a result of a backward slide of the diagonal holder.

You may detect an error in the "up-down" direction of Figure 6.6. That error is probably caused by a previously undetected tilt in the focuser or a side shift of the spider after it was actually installed. Correct it either by adjusting the shim of the focuser or by moving the spider before continuing. Figure 6.6b shows a properly centered diagonal.

The amount of offset of the diagonal away from the eyepiece is impossible to judge just by looking. It must be calculated and measured until the quantities d_{near} and d_{far} are correct. This offset can be straightforwardly calculated using analytic geometry. (Such a derivation appears in Appendix C.) Here are the answers and some easy approximations:

If D is the diameter of the mirror, L the diameter of the field of 100% illumination, T the distance from the center of the tube to the focal plane, f the focal length, and s the sagitta of the mirror surface, then the offset is $(s \cong D^2 / 16f)$,

$$\text{Offset} = \frac{T + nL/2}{1 - n^2} - T \text{ where } n = \frac{D - L}{2(f - s)}. \qquad \textbf{6.2}$$

For a small field of 100% illumination and a shallow mirror, this is approximately

$$\text{Offset} \cong \frac{T}{4F^2}, \qquad \textbf{6.3}$$

where F is the focal ratio.

Say you have two equal stacks of coins. If you remove one coin and place it on the other stack, the stacks differ in height by two coins. For the same reason d_{near} is larger than d_{far} by twice the offset. A number of esti-

Table 6.1a

The difference between the two measurements from the tube
to the edge of the diagonal, $d_{near} - d_{far} = 2$ Offset

Field of 100% illumination is zero

		Focal Ratio						
		4	4.5	5	6	7	8	10
Diameter (in)	T			2 Offset (in)				
3	3.5	0.11	0.09	0.07	0.05	0.04	0.03	0.02
4.25	4.4	0.14	0.11	0.09	0.06	0.04	0.03	0.02
6	5.6	0.18	0.14	0.11	0.08	0.06	0.04	0.03
8	7.0	0.22	0.17	0.14	0.10	0.07	0.05	0.04
10	8.4	0.26	0.21	0.17	0.12	0.09	0.07	0.04
12.5	10.2	0.32	0.25	0.20	0.14	0.10	0.08	0.05
14.25	11.4	0.36	0.28	0.23	0.16	0.12	0.09	0.06
16	12.6	0.39	0.31	0.25	0.18	0.13	0.10	0.06
17.5	13.6	0.43	0.34	0.27	0.19	0.14	0.11	0.07
20	15.4	0.48	0.38	0.31	0.21	0.16	0.12	0.08
24	18.2	0.57	0.45	0.36	0.25	0.19	0.14	0.09

Table 6.1b

The difference between the two measurements from the tube
to the edge of the diagonal, $d_{near} - d_{far} = 2$ Offset

Field of 100% illumination is zero

		Focal Ratio						
		4	4.5	5	6	7	8	10
Diameter (mm)				2 Offset (mm)				
75	90	2.8	2.2	1.8	1.2	0.9	0.7	0.4
110	110	3.5	2.7	2.2	1.5	1.1	0.9	0.6
150	140	4.4	3.5	2.8	2.0	1.5	1.1	0.7
200	180	5.6	4.4	3.6	2.5	1.8	1.4	0.9
250	210	6.7	5.3	4.3	3.0	2.2	1.7	1.1
320	260	8.1	6.4	5.2	3.6	2.6	2.0	1.3
360	290	9.0	7.1	5.8	4.0	2.9	2.3	1.4
400	320	10.0	7.9	6.4	4.4	3.3	2.5	1.6
450	350	10.8	8.6	6.9	4.8	3.5	2.7	1.7
500	390	12.2	9.7	7.8	5.4	4.0	3.1	2.0
600	460	14.4	11.4	9.2	6.4	4.7	3.6	2.3

mates for $d_{near} - d_{far}$ are listed in Table 6.1. These typical offsets are generated by using Equation 6.2 with reasonable estimates for T.

At long focal ratios, the offset is very small, so use of these tables should only be critical in their lower left corners—for fast and big mirrors. For that reason, the approximation in Equation 6.3 is more than adequate. Optimally, fast Newtonians should be designed with very small regions of full illumination at the focal plane (Peters and Pike 1977).

Most spiders can be adjusted to allow the diagonal to be offset deliberately merely by changing their mounting-screw adjustments. Do not worry if the vanes on opposite sides of the diagonal are not precisely lined up. This condition only modifies the spider diffraction pattern. The amount of light scattered by the spider is the same as before. In a way, the spider diffraction pattern with 8 less-bright spikes may be better than 4 strong ones in some observing circumstances.

Step 3: Establish the tilt of the elements.

Once you have the diagonal visually centered in the focuser tube and offset the proper distance, you are ready for the tilt adjustment. If you were to carefully set the tilt of the primary mirror first, the setting would go amiss when the diagonal is finally adjusted. Thus, the tilt of the diagonal must be fixed before the main mirror is touched.

At this point, visualize the main mirror as painted white. All you can see at the primary are mirror clips jutting onto its surface. Ignore any reflections. A useful trick is to become painfully aware of dust on the mirror. Concentrate on the dust and the mirror clips. Look *at* the mirror, not *through* it. Figure 6.7a shows the diagonal with incorrect tilt.

Rotate the diagonal holder in the spider until it looks like Figure 6.7b. Then adjust the screw in the holder base that tilts the diagonal either toward or away from your eye to center the main mirror clips as in Figure 6.7c. Most diagonal holder bases don't contain springs. If you loosen one screw, you must tighten the other two.

Oddly, the two most common adjustments are rotating the whole diagonal on its axis and adjusting only one screw at the base. You may well ask why diagonal bases have three hard-to-reach screws. A spring-loaded hinge and one adjustment screw make more sense. One explanation is that telescope makers are extremely conservative, and diagonals have been mounted with three screws for a long time.

The above description had all of the wrong turns and frustrations removed. At first, it seems impossible to make these adjustments to simul-

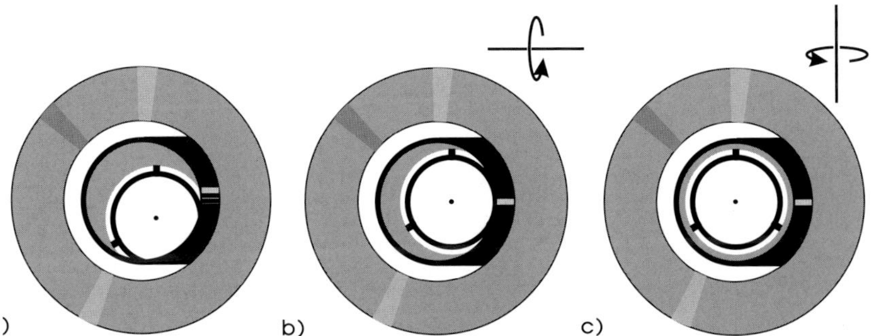

Fig. 6.7 Setting the angle of the diagonal: a) A misaligned diagonal. b) After spinning the diagonal in the spider. c) After tilting the diagonal using screws at the base.

taneously achieve diagonal alignment and keep the connectors tight. If the wrench is turned on the spindle, diagonal rotation is especially hard to prevent. Fifteen minutes of careful work can be lost while you try to torque the diagonal securely into place. What you eventually learn is how to predict the effect of the wrench. Ultimately, you will know how far to offset the finger-tight alignment to achieve wrench-tight alignment. Also the above description assumed you have the very common tube-centered spider mounted on a spindle. Small Newtonians may have other mounting schemes, including a stalk extending from the focuser. The general principles don't change however, and you have to adapt them to your mounting hardware.

This diagonal tilt adjustment is a lot easier if you own a laser collimator. In this case, you place the laser in the focuser as if it were an eyepiece and adjust the tilt until the laser beam hits the center dot. Be sure to check whether the mirror still seems centered below the focuser and neatly contains a reflection of the main mirror, though. This adjustment just orients the diagonal at the correct angle; it does not place it in the correct position below the focuser. It is easy to imagine a case where the diagonal catches and deflects the laser beam toward the center of the mirror, but intercepts the laser near its edge.

The last job of coarse collimation is setting the tilt of the main mirror. Finally, you may now consider the reflection in the primary. The little hole through which you are looking is reflected somewhere near the mirror dot. By adjusting the screws on the back of the main mirror, move the hole reflection until it is behind the dot.

If you use a laser, orient the main mirror so it returns through the hole it came from. A laser alignment is equally dependent on the accuracy of the placement of the alignment spot, the centering of the downgoing beam and the centering of the return beam. An easy mistake to make is to center the return beam without attention to the downward alignment, incorrectly reasoning that since the beam returns, it is approximately centered. If you replace the laser by a sighting hole, you'll see that the alignment is poor.

Figure 6.8a is an example of a misaligned main mirror. Figure 6.8b shows correct coarse alignment. The stubby arrow shows the direction to the reflection of the focuser's base.

Step 4: Repeat steps 1, 2, and 3 as an iterative procedure.

You are certainly allowed to start this process over again to correct any difficulties. Remember to order the steps properly. Don't adjust randomly, even though you are sorely tempted. A good example of one such tempta-

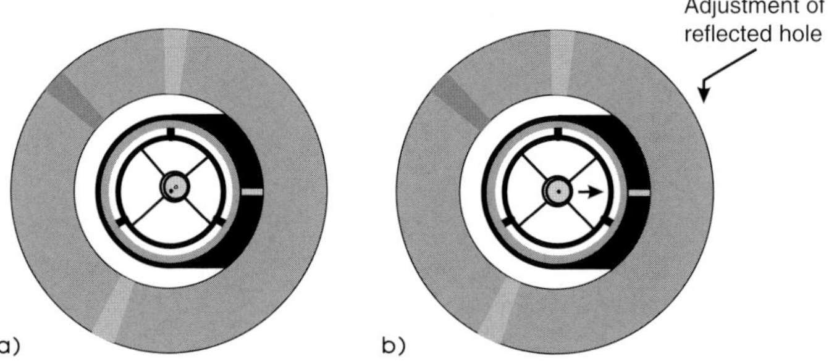

Fig. 6.8 With the diagonal aligned, the screws on the back of the main mirror are adjusted until the dot is superimposed on the sighting hole: a) the starting appearance, b) correct coarse alignment after moving the reflection of the hole in the indicated manner.

tion is the decentered appearance of the second reflection in the diagonals of Figures 6.5 and 6.8b. Remember, these systems are correctly aligned, but the reflection of the bottom of the focuser is still decentered, an especially large effect in fast mirrors with big diagonals.

When they first notice this decentering, many first-time telescope collimators start tugging at the diagonal, and end up destroying their hard-won coarse alignment. People are unusually disturbed by this decentering, and they often skew the alignment until these circles are approximately concentric as well.

Unfortunately, this mistake has the same effect as tilting the focuser over, and by implication, tilting the focal plane. The eyepiece is no longer transported along the optical axis. Barlow lens performance may suffer, and the edges of photographic film frames become indistinct and blurry. Because of the natural curvature of field in Newtonian reflectors, one side of the frame is severely out-of-focus while the other is not. (Indeed, such photographic behavior can be used to diagnose a tilted focal plane.)

Supposedly, Figure 6.5 is perfectly aligned. Why then, should a telescope in perfect collimation ever show any form of decentering? The problem is the three reflections that occur when you look though the telescope in this manner. The usual path of light is one reflection off the primary and one reflection off the diagonal. When we stare backwards through the telescope, the path is one reflection off the diagonal, a second reflection off the main mirror (okay so far), and *a third reflection off the diagonal*. We should expect centering only as far as the main mirror, as in Figure 6.7b. Neither the diagonal as reflected in the main mirror nor the bottom of the focuser tube as reflected twice in the diagonal needs to be centered.

While it is true that a valid alignment can be made with a centered secondary in double reflection (i.e., with no downward offset), the centering of Figure 6.7 does not happen in precisely the described manner. For this alternate form of alignment, we must have the discipline to center the main mirror at the bottom of the focuser tube and have the diagonal edge appear eccentric in single reflection. This alignment has the effect of offsetting the region of 100% illumination to one side. Most people do not have this discipline, and fruitlessly try to center both the single and double reflections with the resultant tilted field. I prefer to deliberately make the offset and learn to anticipate the appearance of Figure 6.8.

One modification should be made when you are satisfied with the coarse alignment of the instrument. Untwist the vanes of the spider so that they are once again perpendicular to the main mirror. This step is best done gently with a narrow-jawed crescent wrench. Cradle the vane near its middle or somewhat nearer the tube. Look through the sighting hole and twist the spider until it appears narrowest. This ensures that the spider obstruction causes as little blockage as possible.

Step 5: Adjust only one element in fine alignment.

Unless you are aligning a slower Newtonian (*f/*6 and up) or have learned to trust your geometric alignment procedures implicitly (which may well work for you), you must fine-align on a star. Use your accustomed configuration for high magnification. If you normally use a Barlow lens for high powers, include it in the optical train. You should be more concerned with hitting the axis of that Barlow than with attaining a pretty alignment of the dots. Pick out a moderately bright star of high elevation, and center it in the field. Rack the focuser out or in until defocusing aberration is about 10 wavelengths. If the telescope is severely misaligned, you see something like Figure 6.9d. This pattern is unlikely if coarse alignment was done carefully. A misalignment of 8 arcminutes in our example means that the reflection of the sighting hole is decentered by 3 mm, or at the edge of the dot. (Consider how bad the alignment must have been before you started!)

The model used to generate these patterns is somewhat inadequate. The obstruction is still centered in the calculations, but it isn't centered in a real misaligned reflector. This effect diminishes, however, as defocusing is lessened for critical alignment. At 5 wavelengths defocus, this offset is buried in the overall blur of the pattern.

More likely, you see less distorted patterns as in Figures 6.9c or 6.9b, which are examples of milder misalignment. For these cases, the pattern Figure 6.9b is difficult to distinguish from a perfect pattern when it is defocused as far as 10 wavelengths. Choose a dimmer star and defocus a lesser amount. Something like the images in Figure 6.10 should appear. If seeing

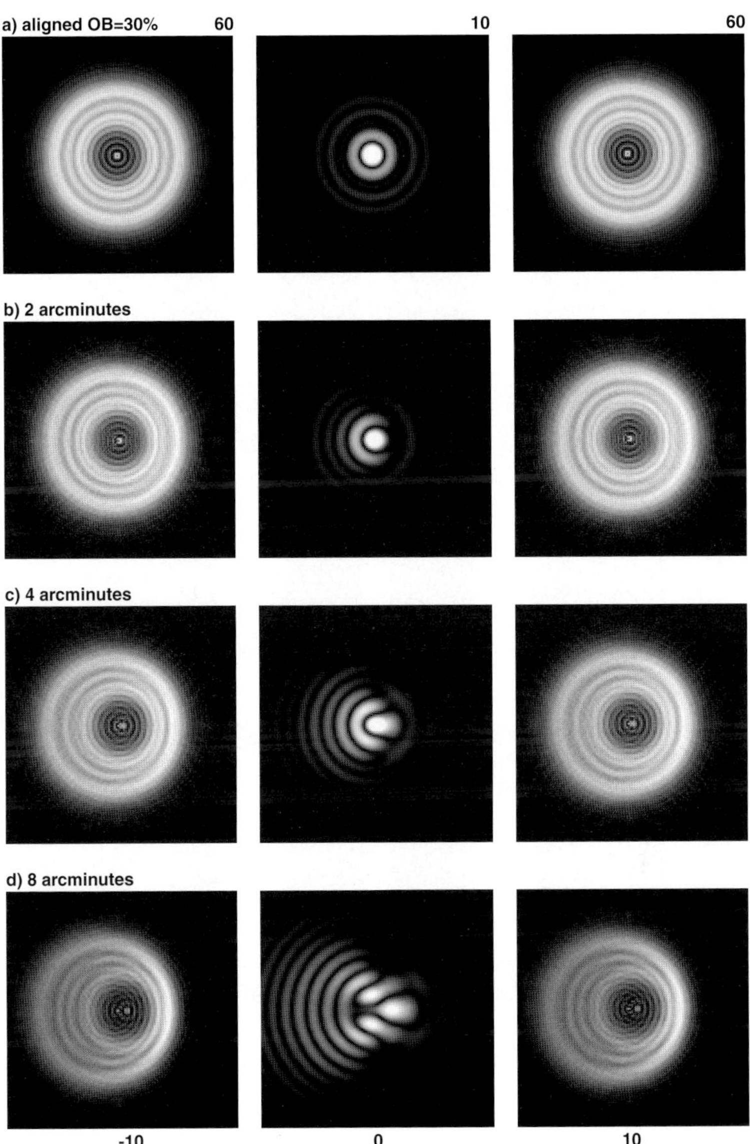

Fig. 6.9 Star test patterns showing increasingly bad misalignment of a 10-inch, (250-mm) f/4.5 Newtonian reflector: a) the expected pattern if the telescope is perfectly aligned, b) misaligned by 2 minutes of arc (the worst misalignment that delivers a good image), c) misaligned by 4 arcminutes, and d) misaligned by 8 arcminutes. The focused patterns are magnified 6 times compared to the unfocused patterns. See Appendix D for labeling information.

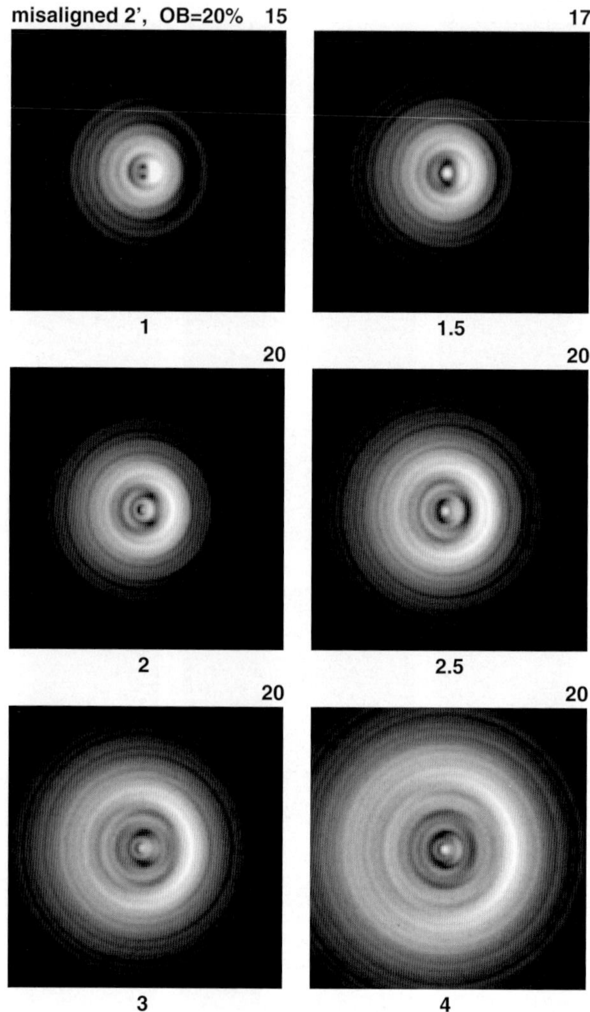

Fig. 6.10 The 2-arcminute misalignment of Figure 6.9 as defocusing is adjusted from 1 wavelength to 4 wavelengths. The misalignment is clearly shown.

is excellent, you can perform the last critical adjustment on a focused star. Unfortunately, the seeing is seldom good enough to do so. Take comfort in knowing that if turbulence is bad enough to make star-test alignment difficult, seeing is probably the limiting factor in your wobbly stack of filters.

The optical axis in all of the figures is far to the right of the pattern. You need to move that axis back to the center of the field, so you want to move the image to the *left*. The real situation will not be so cooperative. The coma flare can be pointed at any angle. You must be able, somehow, to correlate the pattern you are seeing with a cell adjustment screw.

A Newtonian has two mirror reflections, including one at a right angle. Perhaps some people can visualize the three-dimensional situation well enough to determine which screw to turn purely by logic, but they are few. The best technique to identify this screw is to decide toward which clock angle you want to move the image—say, 7 o'clock. Then rack the eyepiece a long distance out of focus. Placing your hand in the optical path of the telescope and noting where the shadow intrudes on the defocused disk, you can decide what part of the mirror corresponds to a certain clock angle. You can bring the hand shadow either to 7 o'clock or 1 o'clock. Then follow this orientation back along the tube to the mirror cell.

When you trace this line back to the mirror, you will emerge either close to a screw or across the tube from one. Give that screw a small adjustment in either direction. It is not important which way you turn it, just that you remember the modification. Recenter the star and see if the situation is worse or better.

If it is worse, undo the damage and turn it the other way. Look at the image again. It should now improve. Pick a new adjustment direction, and repeat the whole process. Always recenter the star before deciding on the next step. If the cell adjustment bottoms out and you cannot tighten a screw, remember that loosening the other two is equivalent.

As you home in on a fine-aligned mirror, you should be following a path like that in the example of Figure 6.11. You can only tilt the mirror along the edges of the adjustment triangle. Please remember, what you are really changing is the position of the optical axis, not the image, so you want to move the image in a reverse direction to all of these adjustments.

I personally find image movement very confusing, so I don't even watch the image shift in the field of view. I just decide the clock angle and use the trial-and-error method described above. Again, it is more important to move slowly and methodically than to understand every twist and turn. The path of Figure 6.11 is not the only solution to this initial condition or even the most efficient, but it succeeds.

6.5.2 The Refractor

Most smaller, less-expensive refractors cannot be collimated because the makers prefer locking down the factory adjustment. Reasoning that incompetent users cannot botch the alignment, they don't fit the telescopes with adjustments. Fortunately, small refractors use a long-focus design that can tolerate large misalignments, so the absence of adjustments doesn't often affect the image.

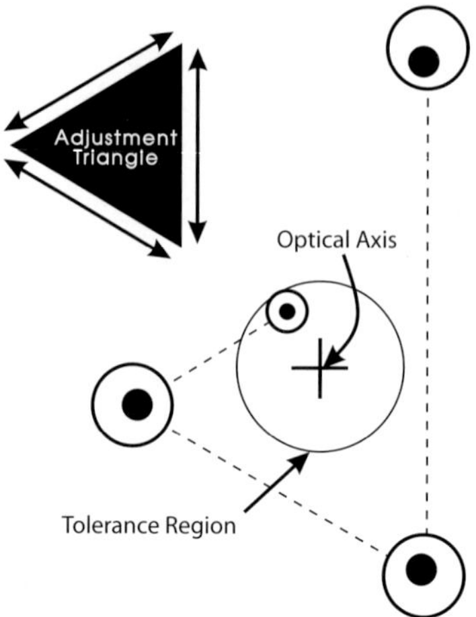

Fig. 6.11 The adjustment triangle of a three-sided cell, and the first few steps in fine alignment on a star-test image.

In recent years, partly because of the resurgence of interest in big refractors and the advent of new apochromatic and advanced designs used at low focal ratio, refractors are being supplied with adjustable cells once again. If your refractor isn't adjustable, you can still check it using the method described here. Unfortunately, telescopes lacking adjustments might have to be returned to the maker for collimation. So-called "telephoto" designs or Petzval designs may also need to be returned. These instruments are permanently fitted with a hard-to-reach corrector/field flattener as the last lens group, far down the tube.

Refractors are not difficult to align. Geometric alignment is usually sufficient because their advanced optical designs yield wide, well-corrected fields. The problem is one of technique and equipment. A refractor is aligned sitting on a table with the lens cap on.

The usual device used to inject light into the darkened tube is called a Cheshire eyepiece (Sidgwick 1980, p. 185), a modified version of the sighting hole used to align the Newtonian. In that case, ordinary room light on the translucent film-can cap was sufficient to provide enough backlight to see the dot. For refractors, a great deal more light is needed. Ordinary glass only reflects about 5% of the energy that strikes it, transmitting the rest. Coated lenses reflect even less. The Cheshire eyepiece is designed to provide a target bright enough to see, even after inefficient reflections. You can obtain one commercially or make it yourself.

Figure 6.12 shows one such alignment tool. It has a long tube with a porthole drilled in the side to allow light to enter. A dowel cut at 45° is

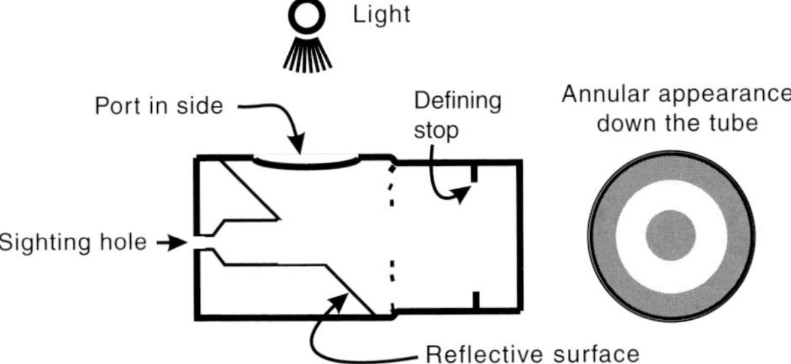

Fig. 6.12 The Cheshire "eyepiece." The Cheshire is just an illuminated sighting hole; it contains no curved or high-quality optical surfaces. The reflective target pattern is shown at right.

inserted in the end of the eyepiece and drilled so that you can see through it. The oblique side of the dowel is at least painted glossy white, and if its surface is polished metal, so much the better. The inside of the sighting hole is carefully blackened. The purpose of the defining stop is to put a crisp edge on the target. Such a refinement is not really necessary if the porthole is carefully placed.[2]

Taylor advocates the use of a white card tilted at 45° into which a sighting hole is placed, a "Cheshire" without the "eyepiece" (Taylor 1983). This suggestion, however, presumes that a great deal of care is taken with mounting the card and centering the sighting hole. I do not advise this procedure but—if you must use it—blacken a small elliptical area around the hole so you will see a circle at 45°.

Most adjustable lens cells use some variant of the push-pull system depicted in Figure 6.13. This lens cell, when precisely adjusted and locked down, is extremely stable. A telescope flange is fixed on the tube, and the lens cell floats on three adjustable "push" screws threaded into that flange. Because the push screws are not sufficient to prevent the cell from dropping off, a matching pull screw associated with each push screw is added, making 6 screws in three groups around the tube. Since no springs are used, as one screw of each pair is loosened, its partner must always be tightened. Figure 6.13 shows a wide separation of the lens cell and the telescope flange to demonstrate the function of a push-pull pair. In fact, the starting configuration should always have the lens cell fitted nearly against the telescope flange. The adjustment will be longer-lived if the gap is small. Also, most lens cells carefully tuck the adjustment screws out of the

---

[2] If you use a Cheshire to align Newtonians, make sure you have sufficient in-focus travel to allow the eye hole to be placed close to the actual focal plane, and with homemade Cheshires make certain that the long sighting hole does not obscure the outside of the optics.

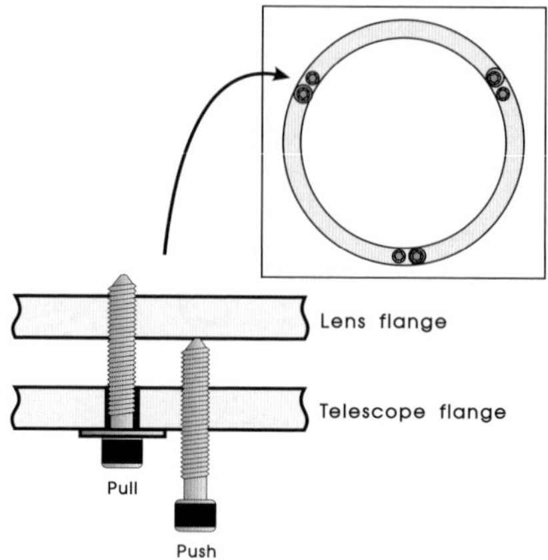

Fig. 6.13 A "push-pull" adjustment screw pair. Inset shows three pairs making an adjustable cell.

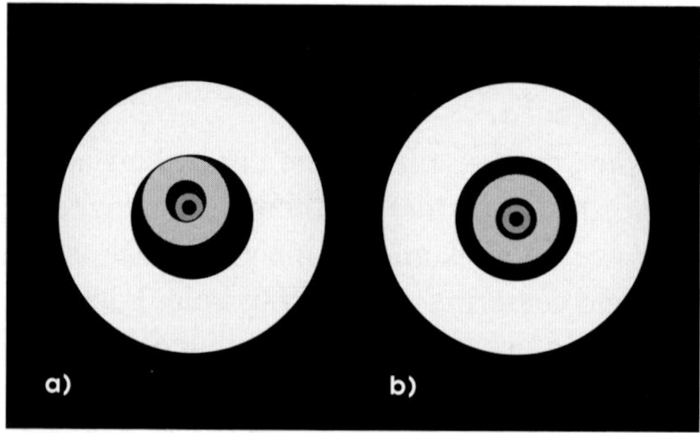

Fig. 6.14 The Cheshire reticle reflection patterns from a 152-mm $f/12$ air-spaced apochromatic refractor: a) before collimation, b) after collimation. Neither alignment gave noticeably different images at the eyepiece.

way so they will not disturb the clean lines of the telescope. Some designers are so clever at hiding these screws that the instrument may not at first even appear adjustable.

If you can make this adjustment with a helper, you are strongly advised to do so. A couple minutes of alignment with two people quickly expands to an hour when you do it yourself. Put the Cheshire in the focuser and shine light from the side, carefully shielding your eyes from the light source. If possible, use a low-magnification, close-focusing telescope looking directly through the rear hole. This ploy removes your eye from the bright sidelighting and expands the reflections so you can easily see them.

In one apochromatic refractor, the pattern of Figure 6.14a was visible through a close-focusing finder telescope behind the sighthole of the Cheshire. The big annulus was a bright steely gray and was just about the size of the Cheshire. It must have been reflected from an air-to-glass surface of low curvature. The next annulus inward was pastel blue and somewhat dimmer. It was smaller, so it must have originated at a more sharply curved surface. The smallest ring was a very dull red or magenta and was not even observable with unmagnified vision. It perhaps relied on two or more of the inefficient reflections at coated interfaces.

This apochromat had six air-to-glass surfaces, and only three reticle reflections were seen. Fewer reflections are anticipated in a doublet. In fact, the Cheshire tool may not be very useful with some cemented doublets or refractors which use optical couplants (oils or gels) between the lenses. Only one reflection might be bright enough to see, and you wouldn't be able to compare it to others.

By adjusting the push-pull pairs, you can quickly make the reflection look like Figure 6.14b. Turn the telescope over and check the pattern again. You will probably discover that the lens rattle designed into the cell compromises the centering. This condition is nothing to worry about; simply adjust it until it is about equally misaligned at each orientation. Even with the situation in Figure 6.14a, misalignment was not noticed in the image.

After you achieve this grade of alignment, you are usually done. You can check the out-of-focus image but you probably won't be able to detect any astigmatism caused by misalignment (although other causes are still possibilities). The well-corrected field of a typical refractor is enormous.

If you do notice some astigmatism, you can certainly try to adjust it out using the star test. The direction of adjustment is less clear than it was with more coma present, since the direction of the optical axis can either be along the short dimension of the out-of-focus stellar disk or along the long dimension. For example, the optical axis may be found at 4 o'clock, 10 o'clock, 1 o'clock, or 7 o'clock, depending on whether you're inside or outside focus. With coma present, the angle was unique.

If the eyepiece is set inside focus, the optical axis can be found on either side of the *short* axis of the astigmatic oval. If the eyepiece is outside focus, the optical axis can be found along the *long* axis. For obvious reasons, you should decide on a certain side of focus and stick with it.

When the cell is adjusted at $90°$ to the proper direction, the stretch direction of astigmatism rotates rapidly. Also, undoing the adjustment and going an equal distance on the other side doesn't improve anything; it just reverses the rotation.

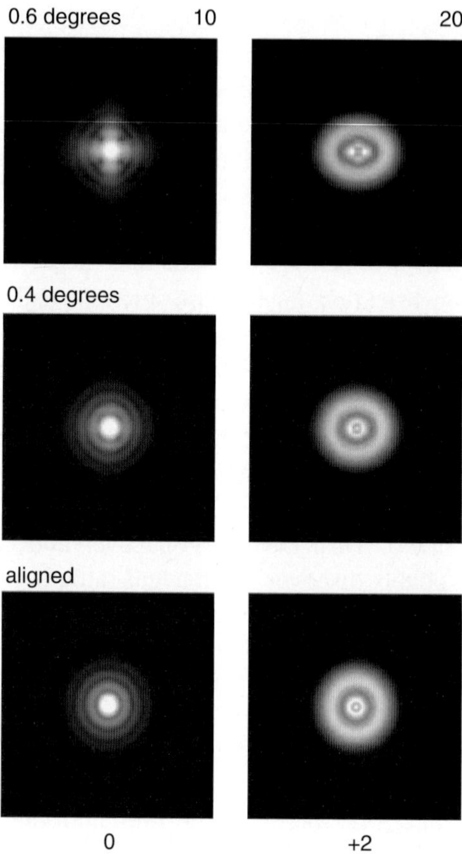

Fig. 6.15 Misalignment patterns calculated for the 152-mm f/8.4 apochromatic refractor of Appendix G.2. The defocus pattern is taken slightly outside of focus. The 0.4 degree pattern decreases the Strehl ratio by only 0.95 of what it had been on axis.

In star-test alignment, only a tweak of the push-pull screws will be enough. After all, the telescope should be very nearly collimated. Tiny changes at the screws mean enormous changes at the focal plane. If you are unable to remove the astigmatism by collimating it out, your telescope may be suffering from pinched optics or a true cylindrical deformation ground into the glass.

Let's review the general steps involved in alignment and see how they applied to refractors:

1. *Establish the axis line.* It was defined as the center of the tube.

2. *Center the optical components on this axis line.* Since most refractors have only one closely-spaced group of lens elements held in an accurately machined cell, this step was automatic. The focuser is assumed to transport the eyepiece along the axis. (In small, inexpensive refractors,

this condition is not always met.)

3. *Establish the tilts of the elements.* This step was accomplished by centering the reflection of the annular reticle pattern of the Cheshire eyepiece.

4. *Repeat steps 1, 2, and 3 as an iterative procedure.* Check the alignment with the refractor turned over, and adjust until the Cheshire reflection looks about equally misaligned at all orientations.

5. *Adjust only one element in fine alignment.* This step was probably not needed, but if it were, it would have taken place on the objective. In Figure 6.15, misalignment of 0.4 degree would be easily observed in geometric alignment.

6.5.3 The Schmidt-Cassegrain

Schmidt-Cassegrains of effective focal ratio $f/10$ have a primary mirror of about $f/2$ multiplied by a five-power convex secondary mirror. The center of curvature of the both the primary and secondary mirrors must be along the axis of the corrector. Since the main mirror is not adjustable by the user on most Schmidt-Cassegrains, that adjustment must be set properly at the factory or the telescope cannot be collimated.

An unacceptable main mirror adjustment is difficult to diagnose, but some clues exist. First, go through the rest of this collimation procedure to the best of your ability. Then, using a sighting hole (described on Section 6.5.1 under Newtonian alignment), look back through the optical system. If you don't see absolutely concentric circles, rings within rings, your best alignment may be a kind of compromise. You will be offsetting the secondary to partially compensate for the aberrations induced by a misaligned primary. Still, the main mirror misalignment must be fairly serious before you are really able to detect non-circularity in these tiny reflections.

The front side of the instrument is an easier location from which to detect misalignment of the primary mirror. For 200-mm Schmidt-Cassegrains, place your eye a couple of feet from the front (about ½ meter) and center the biggest reflection of the secondary outside the back of the secondary. By carefully adjusting the placement of the eye, you are able to see the reflection of the secondary as a thin annulus outside the true secondary. You are now near the alignment axis of the main mirror. If the mirror is seriously misaligned, it should be obvious that this axis does not coincide with the axis of the tube because the inside of the telescope will look tilted.

Another clue is derived from the way these primaries are mounted. The focusing action actually transports the mirror forward. The mirror's center is glued to a plate on the front of this axial focuser. Often, these mirrors get out of adjustment because of some sort of mechanical fault in the focuser. (Perhaps it has taken an enormous jolt during shipping.) As you

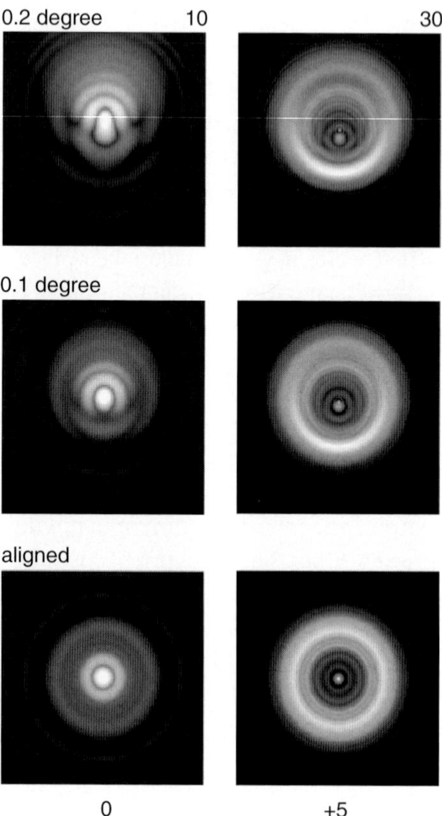

Fig. 6.16 Computed star test with photopic weighting and actual parameters likely to be used in a 200-mm Schmidt-Cassegrain telescope. The middle misalignment reduces the Strehl ratio to about 0.75 of what it was on-axis. The secondary mirror is tilted only 0.1 degree.

focus the instrument, the image is not seen to defocus in a fixed location but reels or loops across the field. In any case, such anomalous focusing behavior, if severe enough, demands factory service.

For now, let's assume you have an adequately aligned primary mirror. The only free adjustment is the tilt of the secondary mirror. If you have a severely misaligned Schmidt-Cassegrain, you may need to coarsely align by looking through it with a sighting hole. Center the reflection of the primary mirror in the secondary. Usually, this step will be unnecessary.

The final step is fine alignment with the star test. You can align the telescope in the daytime on an artificial star or at night on a real one. The Schmidt-Cassegrain is the most convenient of the example telescopes to align because of its compactness. If your arms are long, you can actually reach the adjustment screws while your head is behind the eyepiece. Of course, collimation is still easiest with two people, one calling out instructions and the other trying to obey them.

The secondary cell of a Schmidt-Cassegrain mount is a variation of the diagonal mount of the Newtonian. In both cases, loosening one screw is counteracted by tightening the other two. In changing the tilt of the secondary mirror, you must achieve alignment while keeping the mirror cell screwed tight.

It may seem expedient to overly tighten one screw as collimation is approached. Avoid such a shortcut. The secondary mirror is mounted in glass, and you might break the corrector. Also, the secondary mirror is held on a stiff plate, but this plate can be bent and the mirror strained. Finally, you might jerk the wrench out of the socket when forcing it and end up scratching the corrector. Tighten it snugly, but don't force it. If you have to move it a bit more, loosen the other two screws before tightening it further.

A misaligned Schmidt-Cassegrain will generate the same sort of star test behavior as depicted earlier for a Newtonian. Perform the star test without a 45° diagonal. You will have one less potential source of aberration, and you will be able to observe the angle of the optical axis. The correct screw to turn is straightforwardly determined. Soon, you can center the shadow of the secondary in the image. (Use the same method as was used in the Newtonian.) Refine the alignment by turning to a dimmer star and defocusing it less or, if seeing is excellent, leave the telescope in focus and adjust the image for symmetry.

To review, the steps involved in aligning a Schmidt-Cassegrain were as follows:

1. *Establish the axis line.* That line, by definition, is coaxial with the tube.

2. *Center the optical components on the axis line.* Centerings are factory-set and hence are not adjustable.

3. *Establish the tilts of the elements.* The tilt of the corrector is, to first order, unimportant. The tilt of the primary is a factory setting and depends strongly on the condition of the focusing mechanism. Only the tilt of the secondary may be adjusted. Check the orientation of the primary from the front of the tube.

4. *Repeat steps 1, 2, and 3 as an iterative procedure.* Because so many of the coarse alignment adjustments are out of the owner's hands, iteration is impossible.

5. *Adjust only one element in fine alignment.* This step is combined with step 3. It is done on a star or artificial pinhole source placed at around 50

meters or farther; only the secondary is adjusted.

If your telescope still displays asymmetric images at the end of these steps, it will have to be returned to the manufacturer. The primary mirror is probably tilted. This lack of adjustments on the main mirror is perhaps the weakest feature of commercial Schmidt-Cassegrain designs.

Chapter 7
Air Turbulence and Tube Currents

Some sources of aberration have nothing to do with the telescope itself. They come from the necessary immersion of the instrument in a changing optical medium. The light of astronomical objects must traverse a turbulent column of air that extends many kilometers from the top of the atmosphere to the focal plane of the instrument.

Little can be done by the amateur observer about turbulence high in the atmosphere, but it is easy to recognize in the star test. Many of the problems described in this chapter cure themselves after a time, particularly those having to do with the cooling of the telescope or of its immediate environment. The aim here is to teach you how to detect aberrations that originate in the motion of air, prevent them as much as possible, and recognize when they subside. Learning the star-test behavior of the atmosphere at your location will also allow you to identify those rare bouts of unusually steady seeing favorable to high magnifications.

7.1 Air As a Refractive Medium

The index of refraction of air is very near that of a vacuum, but it is measurably different. At $0°$ Celsius, dry air at sea-level pressure has an index of refraction of 1.00029, while vacuum has an index defined exactly as unity (CRC 1973). Such a small difference doesn't seem worth worrying about, but a wave going through the air inside a 1.5-meter focal length telescope is slowed, relative to passage through vacuum, by about 791 wavelengths.

Assuming that air is approximately an ideal gas and that the fractional part of the index of refraction changes linearly with temperature, one can easily figure out the lag for a small temperature difference. Recall that $0°$ Celsius is $273°$ Kelvin in units of absolute temperature. Therefore, a $1°$ K difference in temperature over a distance of 1.5 meters results in a delay of $791/273$ wavelengths per degree Kelvin, or about 2.9 wavelengths/$°$K (1.6

wavelengths/°F). Now stretch the tube of air up through the atmosphere, many kilometers overhead, with each layer at a different temperature. Propagation delays can be profound.

However, it doesn't matter if the pencil of light is delayed uniformly. After all, the light has been on a long journey. Who cares if it arrives a little late? We are only interested in the variations in direction or time of arrival as the light enters different portions of the long, skinny cylinder of air in front of our instruments. We detect those differences when we see the image degraded by bad seeing. When we look through Earth's atmosphere, we hope for a pressure and temperature uniformity that does not often exist. Some parts of the wavefront are a small distance behind other portions of the wavefront. Such aberrations can cause differences in intensity and apparent location (i.e., "twinkling"), but their most common effect on large apertures is to blur the image.

In defense of the sky, perhaps we are expecting too much when we peer upward through all of that material and demand perfect images. After all, the total pressure of the atmosphere is the same as the pressure of over 30 feet of water. Sub-arcsecond resolution would not be expected from the bottom of a diving pool. Yet such resolutions actually are achieved on images seen through the atmosphere. On exceptional nights, the atmosphere surprises us by becoming beautifully tranquil.

Still, a mechanism producing small-scale differences in refractive index must exist before atmospheric effects become troublesome. Air, by its gaseous nature, doesn't tend to maintain differences in pressure or temperature—except for wide layering caused by the force of gravity. Air mixes together, averaging differences until layered uniformity prevails. Statistical inhomogeneity won't persist without mixing. The two mechanisms of most interest to telescope testers are atmospheric turbulence and tube currents.

7.2 Turbulence

If the atmosphere changed slowly, turbulence would never start. However, the atmosphere often is forced to move quickly. As sunlight deposits energy on the ground, it heats the air immediately above it. That air expands, becoming lighter than the mass of air immediately above it. The situation becomes unstable or "active," and the denser air falls to replace the warmer air below it. It moves quickly enough to generate turbulence.

This type of fluid motion is called a "Rayleigh-Taylor" instability. Completely fill an empty soda bottle with water and invert it with a playing card over the opening. Carefully remove your hand. The water does not fall out of the container (Walker 1977). If you were to suddenly remove the card, you would have a bottle full of water poised over a space of air.

All the water can't fall immediately. Air pressure holds the water up in the same way that a column of mercury is held up in a barometer. All of the water molecules are holding on to each other so they won't fall away individually.

Faster than humans can perceive, the random jiggles of the surface will cause one portion to deform slightly upward and another part slightly downward. That's all it takes. Once this process starts, it drives itself. A bubble rises and finally breaks off. The bottle empties, but it doesn't do so uniformly. The instability must form again and again. The bottle drains noisily, gently kicking in your hand.

Incidentally, the instability doesn't take place if a sufficient external force holds the surface level. With the card, the surface was constrained by the structural strength of the paper. By the time the opening is reduced to the size of a soda straw, surface tension alone provides enough force to overcome the instability. You can lift narrow columns of water by merely plugging the top of a straw.

For the atmosphere, no bottle circumscribes the unstable region, but essentially the same process occurs. As the cold air falls, inefficiencies ensure that the edges of the falling region aren't smooth. The fall is quick enough that tiny vortices form, but smaller swirls don't move with the speed of the main convective cell. Even tinier vortices form at the edges of these eddies. Finally the swirls become lost in complexity. At some small scale, the model of collective motion breaks down and the energy is expressed in heat. All realistic cases of fluid flow, such as streams and large convective cells, move macroscopically only on the average. When considered microscopically, they are turbulent.

7.2.1 The Aberration Function

Statistical variations of the wavefront are usually handled with semi-analytic procedures. These procedures, which commonly assume a Gaussian form for the random variations, are quite useful for calculating the long-exposure MTFs and other features associated with rough surfaces (Schroeder 1987, p. 315). However, they do not allow us to calculate the appearance of a single example image. They are the behavior averaged over many such roughened apertures.

The method used here to simulate the wavefront is called the midpoint displacement fractal algorithm. It is used to generate wonderfully realistic fractal landscapes (Peitgen and Saupe 1988, p. 96; Mandelbrot 1983; Har-

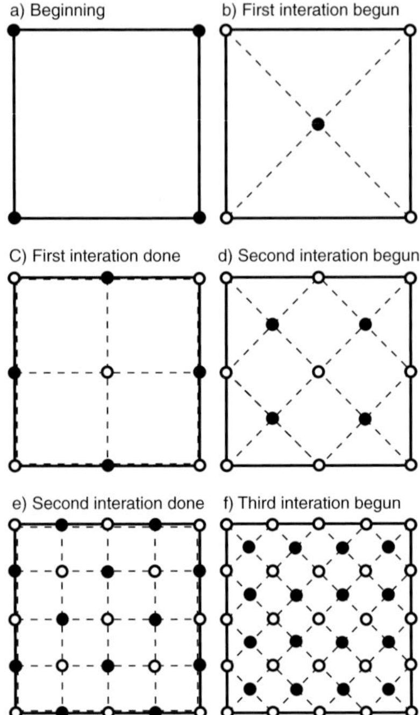

Fig. 7.1 The ordering of point assignments of the midpoint deviation algorithm. Here the roughness is out of the paper.

rington 1987). It can also render an artificial, pseudo-random wavefront just after it has passed through the aperture. Fractals were first applied to diffraction problems by M.V. Berry in the seminal article "Diffractals," in which a fractally-derived *phase screen* first appeared (Berry 1979, 1981). The method used here is an adaptation of an earlier article, where a one-dimensional variation of the midpoint deviation algorithm was used to calculate slit-type diffraction patterns (Suiter 1986a).

Because of the way the fractal must fill in the area of a two-dimensional grid, we may conveniently proceed by a two-step iterative procedure. The algorithm starts with the four corners of 129×129 point grid assigned arbitrarily to be of height zero (Figure 7.1a). Only the locations of the points are shown in Figure 7.1; the deviation is perpendicular to the paper.

For the first half of the first iteration (Figure 7.1b), the offset of the next finer division is calculated as the average of the previous four set points at the ends of the dotted lines (for the first iteration, this average is zero) plus or minus some random deviation $(\pm \Delta z)$. The center point is assigned this value. The points set only during each half-iteration are black; all previously assigned points are white. Points of the 129×129 grid that have yet to be assigned values aren't shown.

During the second half of the first iteration, the edges are filled in. The four nearest points are averaged again, assuming that the points off the edge of the area are zero, and to this number is added another plus or minus deviation. This time the maximum allowable deviation is divided by $\sqrt{2}$ before it is assigned. Notice that each complete iteration cycle averages the positions of the points first oriented at the diagonals to the point to be set, and then in the second half, averages at rectilinear angles. For obvious reasons, this algorithm can also be called the "×+" method.

At the start of the second iteration, the maximum allowable offset is divided once again by $\sqrt{2}$ to make it $\pm \Delta z / 2$. At the beginning of the third iteration, it is $\pm \Delta z / 4$, and so forth. Thus, the cell sizes are decreasing at precisely the same rate as the maximum deviation, setting up conditions so that the apparent random slopes average out to be the same at all scales. This is called *statistical self-similarity*, or an independence of scale for averaged behavior.

Finally, when the seventh iteration is completed (it would be the 15th frame of Figure 7.1, if the figure were allowed to go that far), the entire 129×129 grid is assigned. At this point, with the fractal algorithm finished, the area is conditioned to resemble a circular aperture. The aberrations and transmissions of points farther than 64 locations away from the center point are set to zero, and if a secondary is simulated, the same is done for all points within a certain radius. The outer parts of the 129×129 grid are cut off like the extra dough of a pie crust. The statistics of the clear aperture are then calculated, and the RMS deviation is scaled to the value demanded by the individual computation.

An easy misunderstanding of Figure 7.1 could occur here. Because of its superficial resemblance to the later image frames, readers could incorrectly assume that this non-physical mathematical procedure is used to calculate an image. This method merely simulates the random aberration function on the *pupil*, not an image. The image is calculated with the same Huygens-Fresnel theory used in other chapters (see Appendix B).

Physical reasons may determine that random variations do not persist equally over all scales. In fact, such is the case for turbulence. The characteristic width scale of turbulence, or about the distance between the "bumps," is on the order of 2–20 cm, with a good one-number estimate of 10 cm (Roddier 1981, p. 302; Schroeder 1987, p. 314). Abrupt deformations of the wavefront for slight lateral motions are not anticipated. The roughening of the wavefront is slightly rounded. This feature results in the

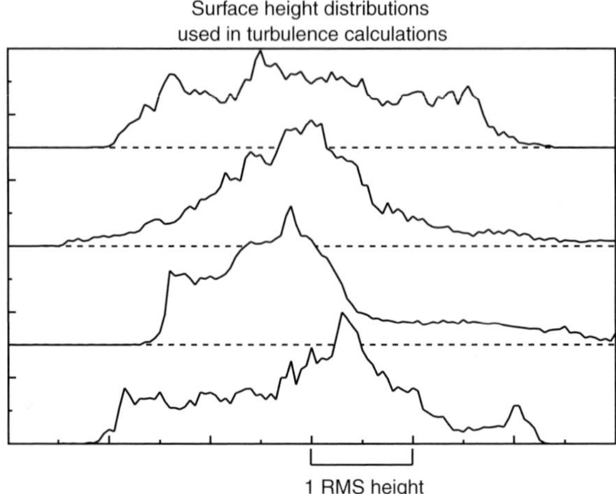

Fig. 7.2 Example wavefront-height distributions of the turbulent wavefronts, with the x-axis measured in units of the roughness RMS value.

well-known "small telescope" effect, where tiny instruments seem to give better images than large ones. Apertures smaller that 100 mm are looking through portions of the wavefront that are closer to a plane. Turbulence causes the image to jump around, but it appears to be well delineated from moment to moment.

For this reason, the algorithm is slightly modified. A *quenching factor* is applied to the $\sqrt{2}$ diminishing of each half iteration. The divisor in this case becomes $\sqrt{2q}$, where q is the quenching factor. For q greater than 1, the surface softens to make the gross variations more noticeable, with fine-scale roughness relatively suppressed. The surface resembles crumpled smooth paper instead of sandpaper. The quenching factor used here was between 1.1 and 1.6. No attempt was made to physically justify this model, but the lessening of finer scale eddies demands some sort of smoothing. The images generated with such quenching factors look the most authentic, and that realism justified their use in the images that appear here.[1] For quenching factors different from 1, the surface is no longer self-similar.

The other modification is a way of controlling the largest scale of the roughness. We would expect turbulence to show itself differently in a 24-inch behemoth than in a tiny refractor. In large telescopes, the image does less jumping around and momentary clearing of the blur happens less often. For the purposes of this chapter, the largest scale of the roughness was set at a significant fraction of the aperture. Thus, the calculations below are good for small telescopes with aperture of about 200 mm for most cases of turbulent air.

---

[1] Readers interested in the more physical time-averaged models can find a survey and a good bibliography in the review article by Hufnagel (1993).

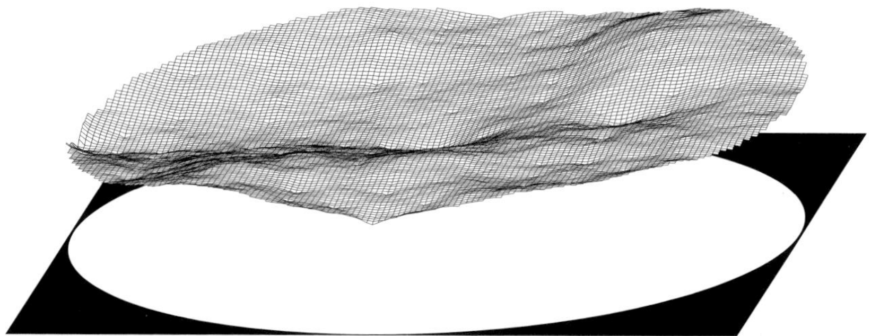

Fig. 7.3 An example modeled aberration function of air turbulence.

Some important characteristics of this fractal aberration will affect the quality of our simulation. First, the surface distribution is not Gaussian (bell-curve shaped). The surface distribution (shown in Figure 7.2) is roughly Gaussian in that it is more-or-less peaked, but under no circumstance is it Gaussian nor would it ever be Gaussian, even if the iteration were allowed to continue forever. More commonly, multiple peaks are found in the surface distribution. Such unusual action allows us to simulate detail that would be washed out of Gaussian models.

Second, the surface is locally correlated. A surface with memory will allow our images to contain realistic streaks and bumps. The MTFs appearing below are calculated with the modeled "snapshot" surfaces and will show statistical variations. This procedure gives us an important view that could not be attained if we figured the MTFs from long-exposure averages.

One feature compromises the algorithm described above. The fictional exterior points are assumed to be zero, so one might expect some unusual distortions to occur near the edges. Because the surface was pie-trimmed, the worst effects are toward the four compass directions of the pattern nearest those edges. Distortions were indeed seen, but the problem areas were small and the effects did not readily appear in diffraction patterns.

An example turbulent wavefront is shown in Figure 7.3, with the wavefront conveniently elevated a small distance through the aperture. The most pleasing aspects of this pattern are the pseudo-random creases running through it. Creases should model the shadow-band behavior of real turbulent air currents that cause speckles and temporary spikes in the image. The quenching factor has acted to smooth the fine-scale variation, which will become more apparent when compared to the aberration function for primary ripple in Chapter 13 on roughness. Because earlier iterations were capable of much more movement, occasional dimples appear in the aberration function.

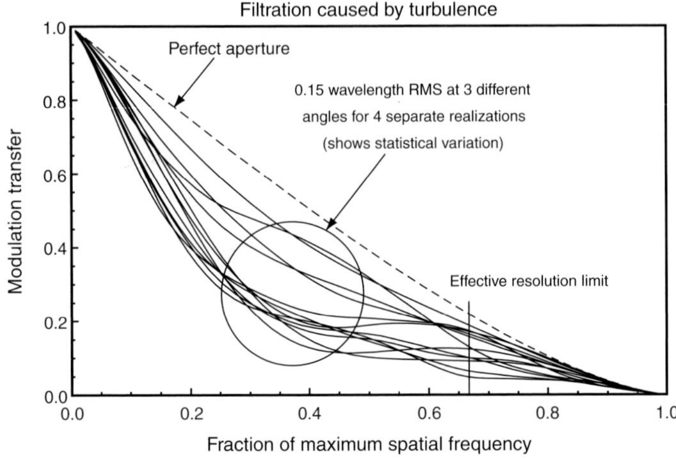

Fig. 7.4 Twelve MTF curves associated with 0.15 wavelength RMS air turbulence. Aberration functions are calculated with the fractal model described in the text.

7.2.2 Filtering Caused by Turbulence

These apertures are not circularly symmetric. Their ability to preserve contrast depends on the orientation of the bar pattern of an MTF target. Therefore, the MTF for each of the four generated surfaces was calculated along three axes. All 12 such MTF curves are shown in Figure 7.4. The amount of RMS aberration used was about twice the amount that leads to a Strehl ratio of 0.8. If one looks at Figure 7.2 and replaces the RMS height with 0.15 wavelength, it is apparent that the total wavefront aberration is about 4 times that value, or 0.6 wavelength. This aberration is about twice as bad as can be tolerated for high-resolution observing, but such aberration is by no means uncommon for air turbulence. Often, seeing is much worse.

Also notice the extreme fluctuation at the high-frequency end of the chart. Because the curve flutters around rapidly there, resolution is limited to about ½ to ⅔ of the theoretical maximum for the aperture. With a 200-mm aperture, the central blur circle has a radius of about 1 to 1.5 arcseconds.

7.2.3 Observing Turbulence

In the focused image plots of Figure 7.5, the modeled turbulence corresponds to a 5 on Pickering's 1–10 seeing scale, since the focused disk is always visible but arcs aren't often seen. This number corresponds to a "poor" seeing rating (Muirden 1974). Sometimes turbulence is much more severe. Good lunar-planetary viewing requires better.

Figure 7.6 tracks the focused image as turbulence aberration becomes

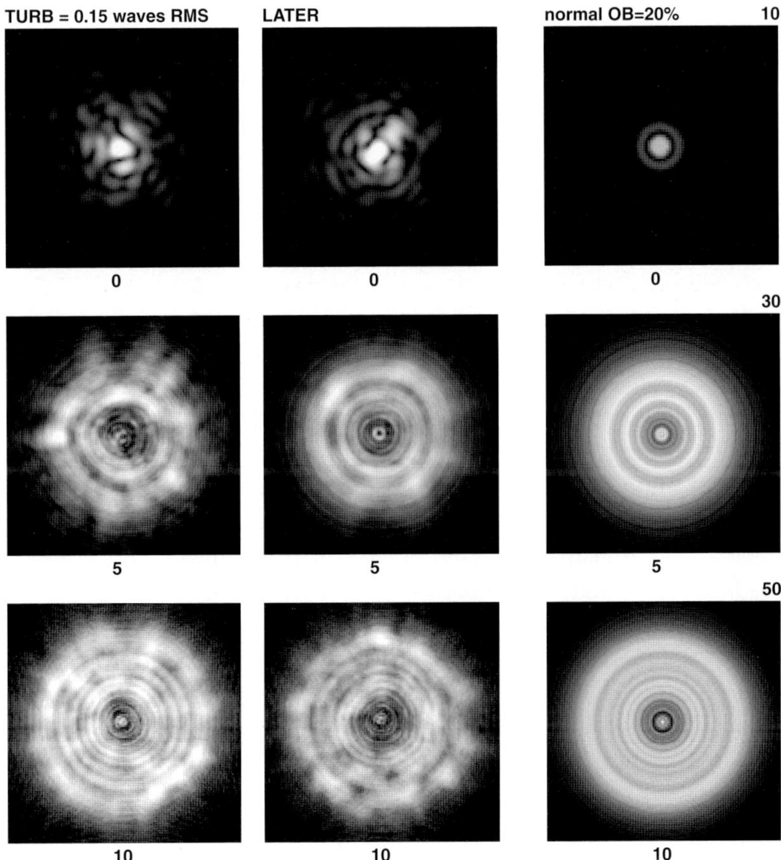

Fig. 7.5 Image patterns calculated for 0.15 wavelength turbulence. Perfect patterns are to the right. Central obstruction is arbitrarily set at 20% of the aperture. (For a description of the labeling of image figures, see Appendix D.)

less objectionable. Figure 7.6a shows long arcs and probably fluctuates between Pickering ratings of 6 and low 7. Figure 7.6b has a rating of high 8, since the rings are complete but are always moving. Figure 7.6c is about a high 9 or low 10, since the rings are stationary and the disk is crisply defined, but the weak ring still breaks up. With as much as $\frac{1}{20}$ wavelength of turbulence wavefront deformation, seeing still has a 10 out of 10 rating. The turbulence aberration is easily distinguishable from other aberrations. It moves quickly. In less than a second you see an entirely different pattern.

7.2.4 Corrective Action

You can do little about high altitude turbulence, since it is beyond reach. High turbulence is more a function of the climate rather than a local phe-

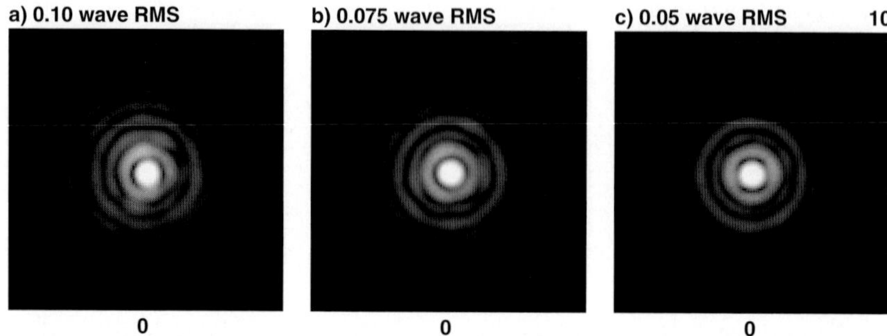

Fig. 7.6 Focused images as turbulence is lessened. The ring structure reappears.

nomenon. However, you can start a log of weather and seeing conditions and see if you can come up with correlations. In general, the presence of clouds and high wind indicates that surplus energy is being transported around in the atmosphere and that seeing is bad. Good seeing is not always associated with transparent nights and may in fact be negatively correlated. Tranquil nights tend to be a little hazy.

Local seeing, or turbulence that occurs within a few meters of the ground, is another matter. Local air turbulence may be caused by thermal currents from buildings or structures that have yet to cool from daytime heating. House shingles are notorious for their long cool-down times. Paving also retains heat and gives it up slowly. For this reason, observing over grass or trees is much preferable to observing over houses or roads.

Some authors have remarked on the interference caused by ground seeing (Muirden 1974), which is an effect located very near the telescope. Personally, I have never had any trouble with turbulence very near the telescope that was not caused by setting the telescope directly on asphalt or concrete, or cooling in the telescope itself. One exception to this general situation, however, is that the nearby observer is a very good furnace. Body heat can waft across an open tube quite easily. This problem is not too great in the summer, when temperature differences are lower, but in the winter it can do serious harm to an image. A cloth drapery for an open tube framework often helps here.

One thing to watch when several people are observing together is that those who are waiting to look don't congregate near the optical path of a telescope, or upwind of it. Heated air from their bodies or breath can intercept the incoming light beam. If possible, when hosting a public observing session, arrange the line on the down-breeze side of the optical path. Finally, if you have to transport the telescope to the site, be sure to park the vehicle so that air rising from its hot engine cannot interfere with anticipated high-resolution observation.

7.3 Tube Currents

Air at different temperatures is affected by gravity because cooler air weighs more. When unconstrained by exterior structures, it forms the convection cells discussed above. Air inside a tilted tube tends to follow the wall—hot air on the high side, cool air on the low side. The tube resembles a tilted stovepipe. As air is heated at the bottom and becomes less dense, cool air falls down the tube and forces the hot air upward. It rises to hug the tube on its upper side and eventually exhausts to the outside.

With a telescope at or near ambient temperature, temperature differentials aren't bad enough to cause tube currents. However, when a telescope is first taken outside, thermal inertia causes problems until the instrument reaches the environment's temperature. The thick glass of the objective is particularly prone to slow cooling.

7.3.1 The Aberration Function

Clearly, every telescope cools off differently. Some have other problems that may obscure or modify the tube currents, like a hot mounting or an observing pad that retains heat from the day. Some have peculiarly shaped tubes or partial tubes that would change the patterns modeled here. All cooling telescopes have an unavoidable amount of locally induced turbulence. Some telescopes have only stubby mirror boxes and no real tubes, and these generate different patterns than shown below.

Schmidt-Cassegrains and refractors must cool off only through their rear exit portals or directly through the tube by radiative and conductive cooling. One Schmidt-Cassegrain I examined displayed an extension of the secondary shadow on one side of focus and streaks parallel to the edge of the secondary on the other. At first, I thought it had a cracked corrector plate near the secondary mounting hole, but when the tube orientation changed, the pattern always followed an up-down direction. Presumably, a great deal of the cooling was taking place in or around the mirror or corrector plate perforations. An uneven thermal effect of the Cassegrain baffle tube may also have caused the problem.

Even though it is not matched by every telescope, the behavior modeled below is common enough in small to medium-sized Newtonians with round tubes. The model assumes that the warm air is confined to the upper side of the tube and that its effect is to advance the wavefront only along that upper side, leaving the rest of the optical path comparatively untouched. This behavior was described in Chapter 2 as resembling the turning of a page. Such behavior has been experimentally confirmed by Schlieren imaging (Greer 2000, 2004).

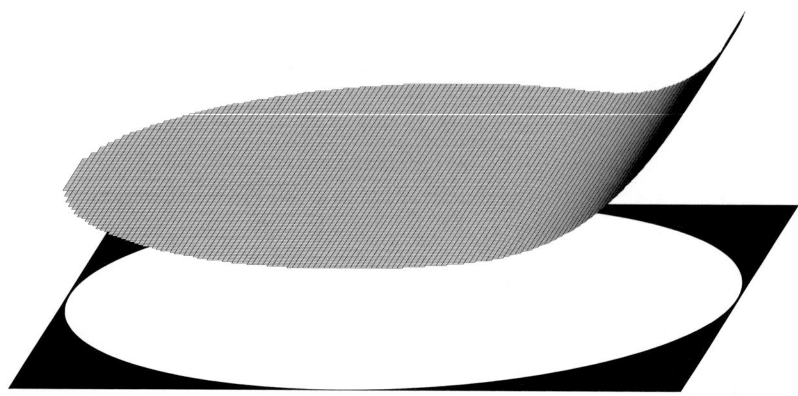

Fig. 7.7 The tube-current aberration function modeled over the aperture pupil.

The pupil function (W) model used here is

$$W_{tube}(x) = \frac{A_{tube}(x - 0.3)^3}{(0.7)^3} \quad (x > 0.3)$$

$$W_{tube}(x) = 0 \qquad\qquad (x \le 0.3)$$

7.1

where x is the linear coordinate in one direction across the surface with the origin at the center of the aperture, and A_{tube} is the coefficient. The value of x reaches 1 at the edge of the aperture.

The modeling does not allow variation in any other direction than the up-down coordinate. No roughness is superimposed on the effect of the tube current, even though it would surely be present.

The aberration plotted over the pupil is shown in Figure 7.7, with the up direction toward the right.

7.3.2 Filtering of Tube Currents

Again, the value of the modulation transfer function depends on the orientation of the bar pattern. The function was calculated for three angles: up-down, left-right, and a 45° tilt. Figure 7.8 shows two cases. The ½ wavelength example is as bad as an observer should tolerate. The 1 wavelength case is severe, but not unusual for telescopes that have just been moved from warm surroundings.

This transfer function plot possesses a number of interesting features. The first is the sudden drop of both the 45° and the horizontal MTFs. The sharp fall is caused by the localized nature of the aberration. The steep slope of the aberration kicks a lot of light out of the diffraction spot, affecting spatial frequencies even ¹⁄₁₀ of the maximum. For example, if you had

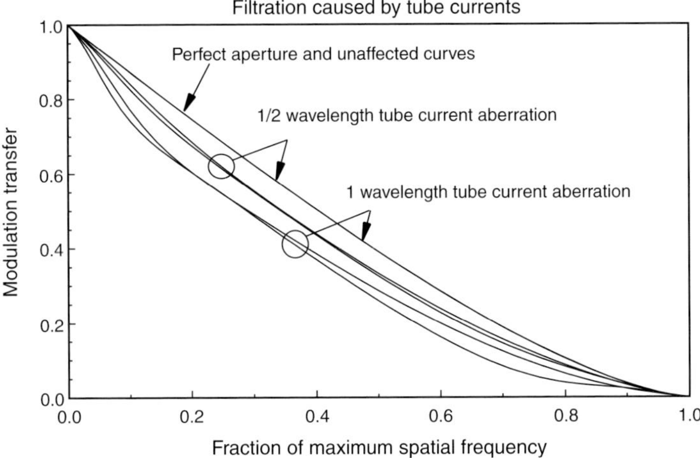

Fig. 7.8 The MTF for the modeled tube current aberration. Two different aberrations are shown. Each aberration has three curves, but the up-down bar orientation results in no degradation. The up-down MTFs of both aberration values are plotted identically on top of the perfect pattern.

a 200-mm aperture, resolution of 5-arcsecond details would be noticeably degraded.

Second, for one curve in each aberration amount, the contrast is unaffected. The smearing of the image in the vertical direction doesn't affect resolution of MTF targets with the bars oriented up-down. Of course, the unmodeled roughness would tend to break up this symmetry a little.

7.3.3 Observing Tube Currents

Tube currents are easy to see. The problem lies in determining whether or not the aberration is in fact caused by a tube current or is present in the glass. These currents can be remarkably stable. You would think that they would dance and sway like high-altitude turbulence. The patterns do change, but they do so slowly, like candle flames. Even though you may know intellectually that a candle is a dynamic process, when you look at the fire, you easily slide into the comfortable viewpoint that it is motionless. A candle flame seems perched atop the wick.

Because of the sharply-varying aberration functions, tube currents are often visible at values of defocus much greater than other aberration. In fact, they can be diagnosed more easily this way. Try racking 20 or 30 wavelengths of defocus and look for heated air wafting up from its source, whether it is the mirror or some other sort of conductive path into the optical air column.

Truly hot telescopes boil with turbulence, but small ones don't show this effect very long. As the telescope cools, the tiny temperature differences do not support the formation of massively turbulent air masses. The air moves slowly and smoothly upward. It drifts side-to-side languidly, but at any given time it is relatively quiet. Tube currents are always influenced by gravity and hence are oriented in an up-down direction. Tube currents arising in a larger telescope mirror will diminish over the course of a night, but they will not go away unless the ambient temperature stops falling.

You may easily determine the orientation of a suspected tube current wavefront deformation for refractors and Cassegrain-style instruments, just don't use a right-angled bend in the optical path. For a Newtonian reflector, however, determining the angle of the image is not straightforward. The cause is, of course, the built-in diagonal reflection.

Two remedies are suggested. The first is to use a star on the north-south meridian. Direction can be determined from the western drift by turning off the clock drive (if one is used). This trick cannot be used for an artificial source test. For such a telescope, a simple expedient is to rack the eyepiece far out of focus and then insert a fist or other obstruction from a known angle in front of the aperture. Up or down can then be easily located.

The diffraction pattern calculated for a total aberration of 1 wavelength appears in Figure 7.9. The pattern is squeezed-in on one side of focus and stretched-out on the other.

Once you see a tube current, make certain that some other difficulty isn't the problem. First, change the tube orientation. Locate a feature of the tube along which the stretching is pointed. If no tube landmark exists, make a slight mark on the tube or attach a little curl of tape. Then, rotate the tube by some reasonable angle. Unfortunately, fork-mounted Schmidt-Cassegrains are impossible to rotate. Use a test source at a different location.

Tube currents will still point up and down with the new orientation, but they stretch toward a different tube feature. Other difficulties, such as warped or damaged optics, now show a non-vertical tilt in the eyepiece.

7.3.4 Corrective Actions for Tube Currents

Tube currents are not really that serious for very small telescopes. Just wait until the instrument cools down to the ambient temperature. If environmental temperature varies so much that the telescope never really catches up to the local temperature, atmospheric conditions are so unstable that seeing will be poor anyway.

Sealed telescopes, particularly refractors, do not display the same tube current effects that are common in open-tube reflectors. Refractors are typically made with metal tubes that leak heat readily, so they quickly

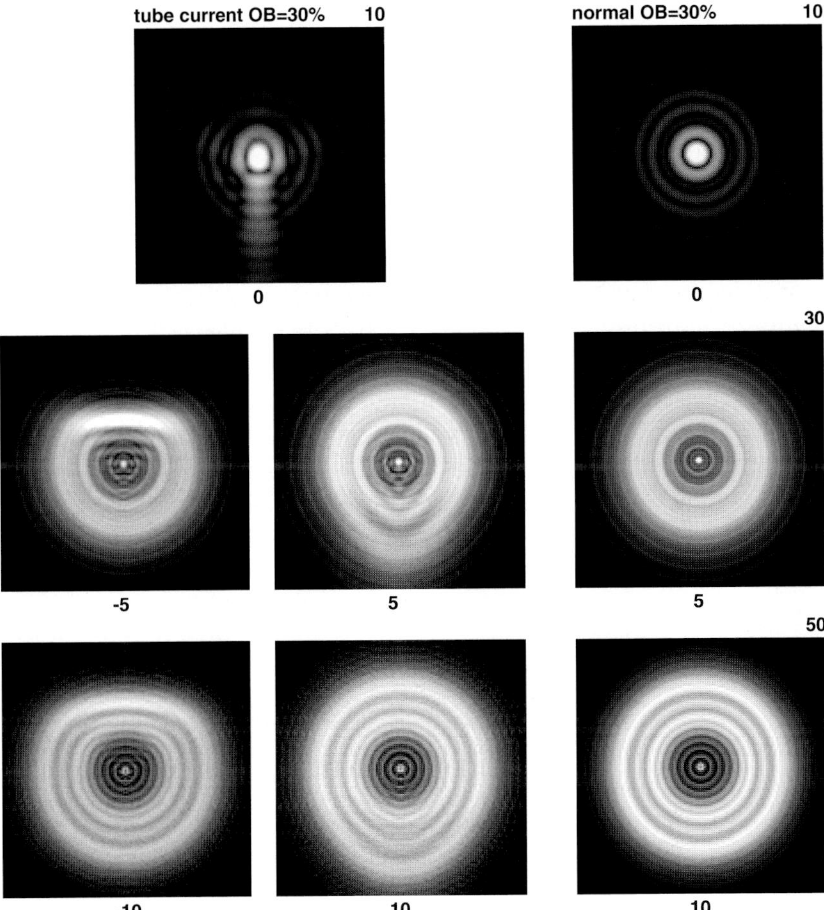

Fig. 7.9 The star-test patterns of 1 wavelength of tube current aberration. (The perfect aperture is in the column on the right.)

cool down. Setting up small telescopes in the late evening, with enough time to reach the ambient temperature, is often sufficient to eliminate this optical problem.

Observers who may find tube currents to be both objectionable and persistent are those using large or especially thick telescope objectives. I once worked on a 16-inch Newtonian mirror of 3-inch thickness (1:5 ratio). This mirror required half the night to cool down even under relatively benign conditions. Many nights it never cooled, but when it did, the mirror performed magnificently.

The best cure for both turbulence and tube currents is the use of a fan mounted to stir the air and cool the mirror (Greer 2004). Nevertheless, there are techniques to minimize atmospheric effects even in the absence of fans.

Telescopes should be prepared before they are required for observing. Transport in a warm automobile is one of the worst things that can be done to a thick mirror. Almost equally bad is storage in a sunlit shed or observatory. Even a garage can maintain a detrimental temperature difference. The optimum procedure is to open up the telescope or observatory in the early evening, long before the instrument is required. If the telescope is transported, try not to carry it in the heated passenger area of an automobile. Instead, haul it in a trailer or in the back of a truck.

Sometimes, the telescope must be set up on concrete or asphalt. Even with a properly stabilized instrument, the tube can catch external currents and act as a duct for them. In such situations, try sealing the bottom end of a Newtonian's tube with a plastic bag and see if the aberration decreases.

Chapter 8
Pinched and Deformed Optics

At the start of each observing session, you should star test for problems that may change with transport of the telescope or a remounting of the optics after maintenance. This chapter will discuss the star-test pattern characteristic of one such warping and will present ways to relieve unusual stresses on the optics. Beyond the image improvements derived by giving careful attention to the deformations, you may be able to avoid catastrophic damage. Edge fractures are common among mirrors with too-tight edge clips. I have seen one mirror that was so severely strained that it broke in half.

8.1 Causes

Imagine a mirror or lens as a thick circular chunk of gelatin. Because we don't ordinarily notice tiny deformations, we think of some materials as rigid, but all objects warp with pressure and temperature changes. An object that looks solid to unmeasured eyesight and touch becomes soft and pliable when we look at changes as small as a wavelength of light. Optics are rigid only in a macroscopic sense.

If we place the gelatin on three support points, the shape of the surface distortion becomes a complicated function of the thickness of the slab and the placement of those supports. The problem is even more complicated if we tilt the gelatin; different supports carry unequal portions of the weight. Clearly, the edges sag between the support points.

The supports underneath can be placed near the edges, causing the center to sag, or near the center, causing the edge to bend down. They can be placed at an intermediate radius, which causes the least deformation, but a more complex one. We can split each prop into 3 or 6, but what tradeoffs do we have to make?

That accounts only for the platform. What about the edges? If we screw on the edge supports too tightly, the surface pinches down around the three support positions. The clips could be properly adjusted, but when the telescope heels over, one clip may be forced to apply unusual pressure by the weight of the slab. Edge mounting has been solved a number of ways, ranging from the elegantly simple hanging-strap method used in Dobsonian mountings to a mercury edge-bag used in a few large equatorial

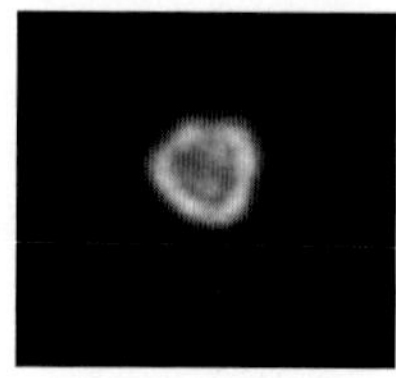

Fig. 8.1 Star-test pattern taken through a large refractor. Its three-sided symmetry betrays a pinching effect. When repairing this problem, the maker reported that the optical coupling agent was causing undue pressure on the lens.

reflectors.

Telescope mechanical design is far beyond the scope of this work, however. Only one example of surface deformation will be presented. Surface deformations cause an infinite variety of star-test patterns, none resembling the others. They may share some of the following characteristics, however:

1. They can be distinguished from turbulence because they are fixed patterns.

2. They often show 2, 3, or 6-fold symmetry.

3. They are usually weaker at high telescope elevations than low ones.

4. The pattern distortions may invert on passage through focus.

Although more common in reflectors, optics pinching is not unknown in refractors. One refractor I examined suffered a problem with a coupling agent used between the lens elements. The material hardened or bunched up in three locations, yielding a star-test pattern similar to the classic three-point pinched mirror cell. See Figure 8.1.

Also, lens cells have to be made a bit larger than lens elements to account for the more severe shrinkage of metal with cooler temperatures. Glass constricts, but it does not shrink as much as metal. If the cell is made too small, the optics can actually be squeezed at low temperatures, resulting in an astigmatic or otherwise deformed image. If temperatures drop too low, a tight cell can crack the lens like an egg.

8.2 The Aberration Function

There are as many forms of surface warping as there are optical surfaces, but only one will be modeled here. The arbitrary choice will be clip pinching or perhaps a thin mirror that has too few bottom supports. What characteristics must the model have to simulate the effects of tight mirror clips? First of all, it must be strongest near the edge. If the optics deform, they can do so most easily at the edge. Second, the aberration function should

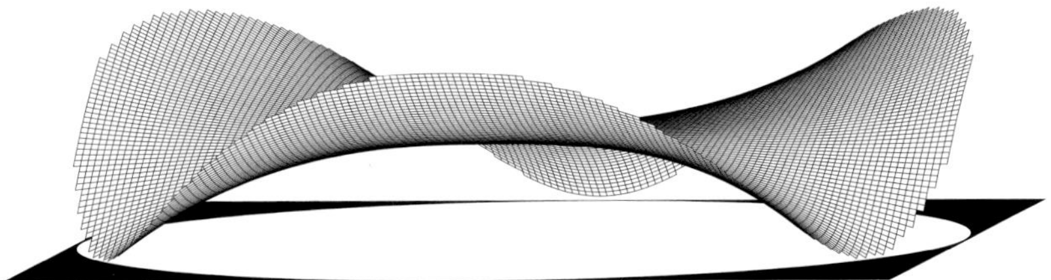

Fig. 8.2 An aberration function of pinched or deformed optics.

have a tripartite symmetry. Optics, of course, can deform in other ways than the three-lobed pattern. However, since mounting cells usually have a three-sided aspect, such deformation is common.

Without physical justification, one can choose a softened[1] cycloidal dependence in angle and a third-order radial dependence to yield the aberration function shown in Figure 8.2 The attractive feature of this pupil function is the asymmetry of the deformation; it ranges from creases in the valley to flat-topped humps. The aberration function could have been made symmetrical in the valleys and peaks, sort of like "three-lobed" astigmatism, but we will see astigmatic behavior later, so it would be redundant to dwell on it here.

8.3 Filtering of Pinched Optics

Figure 8.3 depicts the modulation transfer functions of two unobstructed pinched systems. Because the pupil function is not neatly symmetrical, the transfer varies with the angle at which we place the lines of the MTF target. The figure shows three transfer functions, corresponding to bar patterns at the vertical, horizontal, and 45° angles. The MTF for this particular surface deformation is very similar to that for defocusing.

The least pinched curves of Figure 8.3 yield a Strehl ratio of 0.8 (the same as ¼-wavelength spherical aberration error). Because of the details of this aberration function, however, the total pinching aberration is 0.4 wavelength.

Doubling the total aberration to 0.8 wavelength, the curve is stretched even further downward. At low spatial frequencies, such a telescope preserves contrast about as well as a perfect aperture of 60% its diameter.

8.4 Diffraction Patterns of Pinched Optics

In Figure 8.4, we see the star-test patterns expected from a warped reflect-

---

[1] The form used was a "curtate cycloid". See the Glossary.

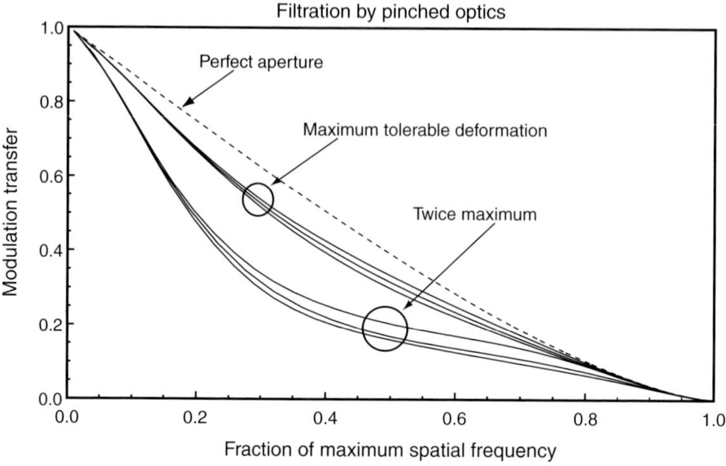

Fig. 8.3 MTF degradation for three-lobed deformed optics.

ing telescope that has a 25% central obstruction. At focus, the diffraction rings are squeezed together into knobby radial spikes.

Such perfectly balanced patterns will probably not be visible in real telescopes. One pressure point seldom induces precisely the same aberration as any other. Also, each deformed surface will display its own unique properties. One tricky pattern not handled by this model is a 6-sided spiking on one side of focus and a muted polygonal shape on the other. I witnessed this problem years ago when a mirror was screwed too tightly into a 9-point cell.

Nevertheless, this problem is easy to diagnose using the characteristics mentioned above:

1. *See if the warping changes with time.* If it is a fixed, unchanging image distortion, it cannot be caused by local heating, tube currents, or atmospheric effects. Look again in 15 minutes. Air optics change dramatically; pinching does not (unless the squeezing changes markedly with temperature).

2. *Count the points around the perimeter.* If you count 3 or 6, the mechanical supports are probably to blame. Three or 6-fold symmetry just isn't ground into the glass that often. Two-fold symmetry indicates pressure from one direction, or it could be an actual astigmatic error in the glass. If two-fold asymmetry is found, try to identify the axis of the astigmatism by putting your hand into the beam the same as you did during alignment. If the axis of the warping is precisely vertical, you can usually assume that either you are incredibly unlucky or the astigmatism results from gravitational forces applying non-uniform pressure.[2]

3. *Shift the telescope to look at a star near zenith.* The easiest confirmation that your difficulty is caused by mechanical supports is simply to change the direction of force on the optics. If the deformation lessens or changes, it could mean that the supports are warping the disk. In any case, it is rare not to change at least the magnitude, if not the flavor, of warping merely by changing the angle of the telescope. An exception is a pinched Schmidt-Cassegrain corrector plate. Because the plate is held so firmly, it gives the same distorted image at any angle.

4. *Rack through focus.* Some warped star-test patterns are similar to astigmatism on opposite sides of focus. If "points" in the image on one side become flat regions on the other, something could be straining the optics. Other deformations do not exhibit this behavior, however. It did not appear in Figure 8.4, so it is not always a reliable indicator of pinching.

8.5 Fixing the Problem

The rest of this discussion will assume you have a reflecting telescope suffering from poorly held optics. Such a problem is rare in refractors, and you can do little about it anyway, except return the lens to the manufacturer for servicing.

If you see a two-lobed image warping in a Newtonian reflector, you could have astigmatism in either the primary mirror or the diagonal. You need an easy way to isolate the problem to the offending mirror. Rotate the tube by 20° to 30°. If the deformation seems to be fixed at the same angle in the eyepiece, the problem is probably contained within the diagonal, because it rotated too. On the other hand, gravity has changed its direction with respect to the tube. If the pattern seems to have rotated by the same angle, or has otherwise changed, it is probably in the main mirror supports.

Nowadays, the telescopes that are often in the most danger of suffering from warped optics are large, thin-mirror Newtonian reflectors. Tube angle cannot be changed in many such instruments, so you can't isolate the origin of the aberration by the simple expedient of rotating the tube. Fortunately, patterns seen in thin-mirror reflectors should change drastically with elevation. Try comparing the appearance of a defocused star just above the horizon with one nearly overhead. If the appearance changes markedly, the most suspicious support is the one carrying the most weight—the main mirror's.

If the telescope is troubled by warping, you are better off to try mirror cell repairs blindly than to do nothing. First, you should verify that the mirror clips are not pressing down on the surface. Optimally, mirrors are not

---

[2] See Chapter 14.

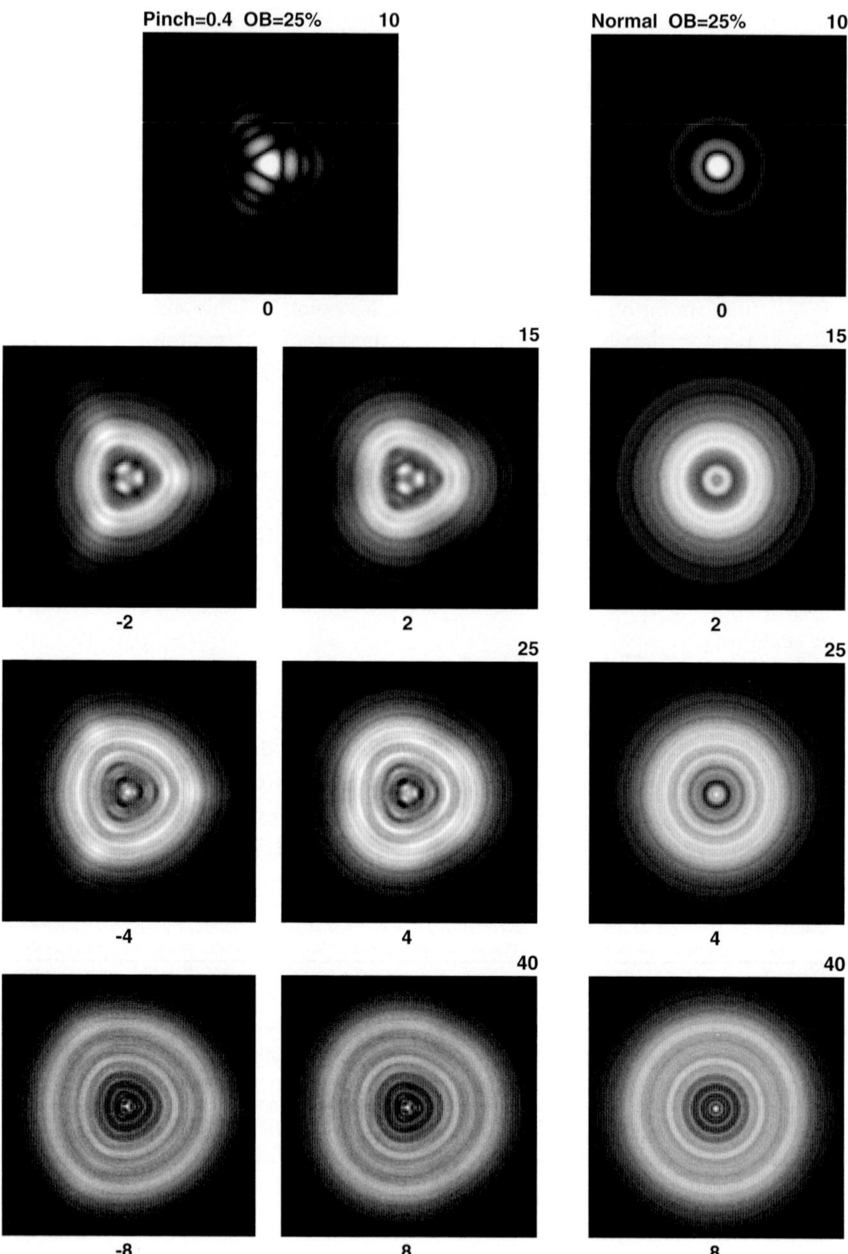

Fig. 8.4 The star-test appearance of a deformed or pinched aperture of 0.4 wavelength total aberration. Normal behavior is in the column to the right. Obstruction is 25%.

even held by such restraints. The purpose of mirror clips is to prevent catastrophic damage during transport and storage, acting as stops so the mirror does not fall face down in the tube. When the telescope is in operating form, these clips should hover above the surface.

You should make sure that the side supports are slightly loose. In many mirror cells, the clips and the side supports are the same units. Since most mirror cells are made to fit a nominal-sized mirror, the side supports are somewhat adjustable, either using a plastic set screw or by adding thin shims to the clip mount.

What you want to avoid is a situation where the mirror is held by excessive force, other than its own weight. One mirror cell I inspected a few years ago had too many shim washers removed from the combined clip and edge supports. This clip was attached with a screw, and the mirror was halting the screw before the metal of the cell did. In effect, the owner was tightening the side of the mirror down with a wrench.

If the mirror rattles a little when you shake the cell, it is mounted properly. Try for less than 1 mm of motion in any direction and even less vertical travel.

Thin, large mirrors in altazimuth Dobsonian mountings should not be held at the edge by two point-like restraints. Their own weight induces unacceptable stresses. Thin mirrors should hang in a strap or a carefully designed edge wiffletree. But even if your mirror is held by a sophisticated edge support, it may still be maladjusted. In one telescope I saw, the strap had slipped and the mirror was sitting on the side of the box. Also, the belt should never support more than 50% of a mirror's circumference or hold it too tightly. Some mirrors are held with a strap that is tightened by a screw, similar to an automobile radiator hose clamp. Such mounting methods should be replaced or modified, because they squeeze the optic excessively.

Of course, few acceptable cell designs exist for large thin mirrors supported in equatorial mountings. As the instrument tracks across the sky it also executes a rolling motion. Thus, the direction of gravity changes orientation with respect to the cell. Besides the mercury bag solution mentioned earlier,[3] makers have tried many other tricks to avoid undue pressure at points on the edge. The most common of these is gluing the mirror to the underneath supports, thus dispensing with edge holders and clips entirely. Even flexible aquarium cement occasionally suffers problems. Gluing the mirror down this way opens the possibility of other pres-

---

[3] This method involves large amounts of toxic mercury, a procedure that can be used safely only by professionals.

sures being applied by deformed mounting plates. Perhaps you may be able to tune such a mirror to work properly in some areas of the sky, but most likely you will have to live with at least a little warping. Compromises are sometimes unavoidable when choosing to mount such a large, floppy mirror equatorially.

Chapter 9
Obstruction and Shading

Modifying the transmission of the aperture pupil causes changes to the diffraction pattern. Obstruction or varying transmission are not deformations of the wavefront in the same sense as aberrations, since the thing affected is not the phase of the pupil but its ability to transmit light. Transmission effects can occur in officially "perfect" apertures. Nevertheless, they can affect the perceived image quality in a similar manner to aberrations.

Five main points are made in this chapter:

1. Central obstructions below 20% of the aperture are indistinguishable in practice from an unobstructed aperture, and for obstruction under 25%, performance can be very good.

2. Reckless efforts to reduce central obstruction can lead to even worse images than those resulting from obstruction.

3. A spider in front of an aperture hurts the image only for dim objects next to bright sources of interference or for low contrast objects imbedded in an extended field. For most dark-field observing, the spider's effect is only cosmetic.

4. Darkening the outside portion of the aperture results in contrast improvements at low spatial frequencies, but only at the expense of high spatial frequencies. Sometimes, but not always, this trade-off improves the image.

5. Dust and scratches are cosmetic errors, similar to spider diffraction except for observing dim objects near bright ones.

9.1 Central Obstruction

The most obvious and potentially the most damaging kind of transmission change is caused by the centrally placed diagonal or secondary mirror. Some observers are almost fanatical about central obstruction. In various astronomical publications, they have made blanket statements such as "obstruction reduces contrast" without giving the spatial frequencies at which that reduction occurs. They imply that obstruction so severely damages the image that no amount can be tolerated.

However, the negative consequences of central obstruction can be readily and precisely calculated. We will see that they worsen considerably

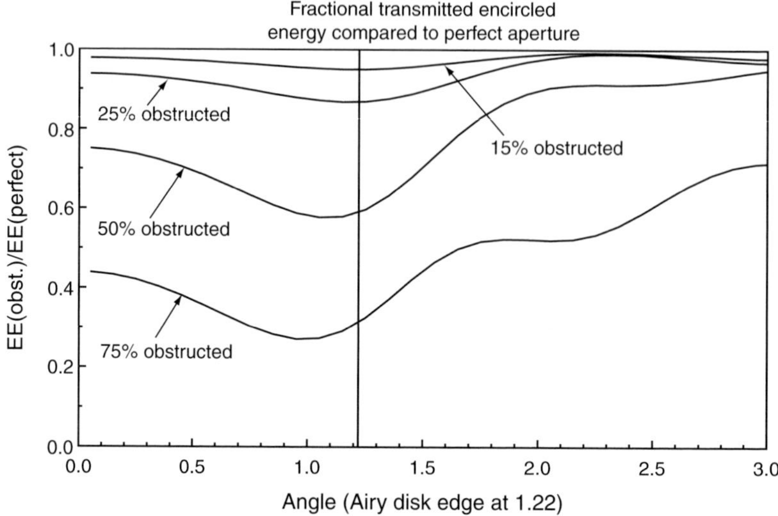

Fig. 9.1 The transmitted encircled energies of centrally obstructed apertures divided by the encircled energy of an unobstructed aperture (later called the encircled energy ratio or EER). The performance drops off sharply at 20–25% obstruction.

beyond a linear obstruction of 20 to 25 percent of the aperture. As long as the obstruction is kept inside that fraction, the image closely approximates that of an unobstructed telescope.

Figure 9.1 does not illustrate contrast but the closely related topic of encircled energy. In Figure 9.1, the normalized encircled energies of the obstructed apertures are divided by the normalized encircled energy of a perfect circular aperture. This means that the encircled energy of all apertures will approach 1.0 as the radius increases toward large values. Each image is thus judged by the shape of the diffraction pattern and not its brightness.

As the circle in which energy is accumulating approaches the radius of the unobstructed Airy disk at 1.22, these ratios fall. The spot size is smaller in obstructed instruments, and the energy robbed from the core diffraction spot is mostly deposited in the first one or two diffraction rings. While the obstructed pattern is crossing the minimum between the central disk and the first ring, it encloses little additional energy and the unobstructed encircled energy gets ahead of it. Not until the circle encloses the first rings of both patterns does the ratio begin to catch up.

In the focused diffraction patterns of Figure 9.2, the intensities of the rings increase as obstruction is increased. By the time a 75% obstruction is reached, all pretense of optical quality is lost. Paradoxically, the radius of the diffraction disk is smaller. All of these images are calculated at the

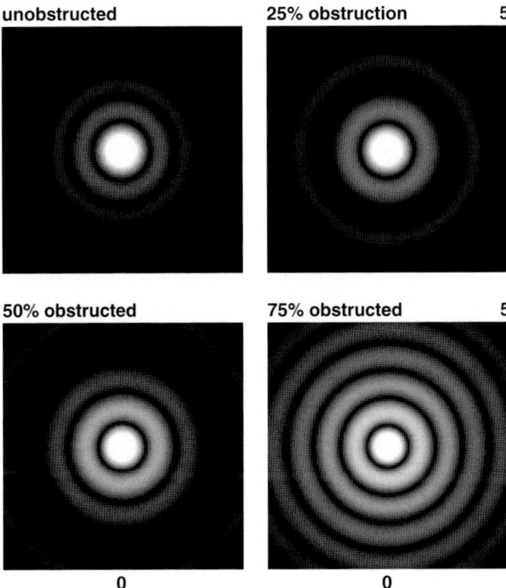

Fig. 9.2 The in-focus diffraction patterns resulting from central obstruction.

same scale and printed with the same central intensity, so this shrinkage cannot be explained as an artifact of the reproduction. This phenomenon is real. In fact, the narrowest central disk is found in an aperture that is almost entirely obstructed, but the powerful rings caused by such an aperture render it useless for fine imaging.

There are a number of ways in Fourier optics of deriving the purely diffractive Strehl ratio of obstruction, or the Strehl ratio that is corrected for the messy double-counting of the extinction paradox (van de Hulst 1981). One of them is to note that it is the value of this normalized encircled-energy diagram as it approaches a radius of zero. At 15% obstruction, it is about 0.977, by 25% obstruction it is 0.96, and by 50% obstruction it is 0.75. The upswing of the encircled energy plot as it approaches zero gives this generalized Strehl ratio an unrealistic tolerance for obstruction. It is not until the obstruction reaches nearly 44% that a Strehl of 0.8 is reached. I have said before, and it bears repeating, that no single-number criterion should be the exclusive judge of optical performance. It is obvious from this diagram that an encircled-energy criterion out near the radius of the classical Airy disk ($r = 1.22$) gives results more commensurate with observation.

Filtering appears in Figure 9.3. Again, the bottom does not drop out at middle frequencies until the obstruction is beyond 25%. An obstruction diameter under 20% of the aperture can be viewed as acceptably small. The

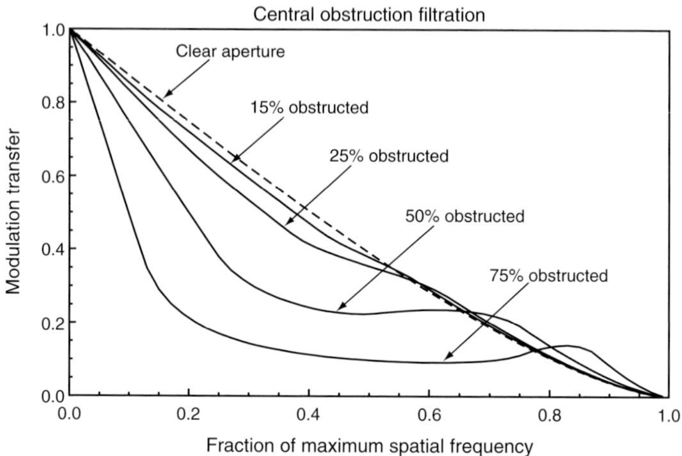

Fig. 9.3 MTF curves of simple central obstructions.

narrower spot size shows itself as an increase in the MTF at high spatial frequencies; contrast here exceeds even the value for a perfect aperture.

These curves demonstrate that a little breathing room is available between 20% and 25% of the aperture. The negative effects of central obstruction have begun to show themselves, but they are saving their full fury for obstructions beyond 30%. Any aperture that is 25% obstructed can be very good, and telescopes that block 20% of the diameter can be excellent examples of their aperture. Instruments that have been modified to achieve less obstruction than 20% are obtaining very little contrast gain and are risking other optical problems.[1]

In the absence of hard knowledge about the tolerable obstruction, telescope makers often cure this difficulty by immoderate measures. Commonly, they build Newtonian telescopes with overly squat focusers and minimum-size diagonals. Unfortunately, reducing diagonal size often forces a compromise with the operation of other useful features of the telescope. Perhaps it even cuts into a safety margin of which the user is unaware. For example, a long focuser tube baffles external light. Some telescopes are designed with eyepieces set so low that light can enter them directly, a problem especially common in open-tube telescopes. Another difficulty arises when a Barlow lens must be used. The focus should be far enough away from the optical path that the Barlow won't jut out in front of the mirror.

The diagonal is also prone to curvature near its edge, so telescopes

---

[1] See the end of Chapter 10 for more information on the degradation caused by obstruction in the presence of spherical aberration.

that require every bit of the diagonal for on-axis imaging often have reduced quality. Certainly, small diagonals are more prone to vignetting, and observers must be careful to make sure that the outer parts of the field of view are adequately illuminated for their favorite objects.

Thus, a well-meaning effort to improve contrast by reducing the diagonal size might have the opposite effect. The instrument may be so malformed by such efforts that contrast is much worse than it would have been with a slightly larger obstruction. This comment is not meant to imply that a small secondary mirror is not a laudable goal. If telescope designers are aware of the risks, they can avoid them. The key is not to reduce the secondary size to the exclusion of every other optical consideration.

9.1.1 The Systems Viewpoint of Central Obstruction

One of the persistent myths of central obstruction is that it irretrievably reduces contrast. For most people, lowered contrast means that a low level of light has been widely scattered throughout the field of view. However, the trend of obstructed MTFs in Figure 9.3 toward low spatial frequency is no different than that for a somewhat smaller aperture. That is, they continue linearly back to an intercept of one at zero spatial frequency. There is no base

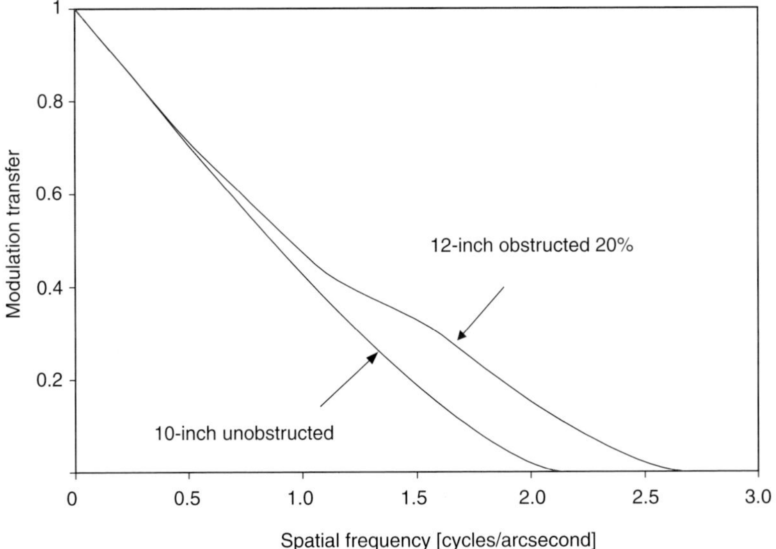

Fig. 9.4 An otherwise perfect 12-inch telescope transfers every spatial frequency of an unobstructed 10-inch aperture with equal or greater contrast. It is therefore capable of more than precisely reconstructing every feature of the smaller aperture's image, including the dark background. Compare with Figure 3.12, where a similar argument is made for spherical aberration.

level of broadly scattered light, just overall reduced resolution. It is as if we were using a somewhat smaller aperture. In fact, this aperture is easy to calculate. If D is the outside diameter and D_{obst} is the diameter of the obstruction, to a very good approximation, the effective diameter D_{eff} has a lower limit of (Zmek 1993)

$$D_{eff} \geq D - D_{obst} \, .$$

9.1

We can plot the performance of a perfect unobstructed 10-inch telescope together with an otherwise perfect 12-inch aperture obstructed with a 20% secondary in Figure 9.4. The low spatial frequency portions of these curves are virtually on top of one another. It is the lower part of the MTF that contains the blackness of the background, so the obstructed aperture is just as dark, far away from the stars, as the unobstructed aperture.

9.1.2 Unobstructed Systems

The previous section seems to infer that unobstructed systems like the the Kutter schiefspiegler, Leonard's Yolo (Mackintosh 1977), and the Stevick-Paul telescopes (Stevick 2006) lack any real benefit proportional to the effort required to make them. After all, a slightly larger ordinary reflector is always much easier to make than an unobstructed reflector. However, the real benefit of such systems has little to do with their being unobstructed.

There are relatively few types of purely reflective, cylindrically-symmetric telescopes. Single curved-mirror systems are pretty much limited to the Newtonian, which has uncorrected off-axis coma. Dual-mirror systems appear in the various flavors of Cassegrain optics. Although one of these (the Ritchey-Chrétien variant) has very good off-axis correction, it has severe field curvature. There are a few other pure mirror astrocameras where the field is almost inaccessible visually. Because of the paucity of degrees of freedom, designers are limited to a few well-trodden paths.

With tilted, unobstructed systems, designers have broken the constraints that previously bound them. The real advantage of tilted-component telescopes is that—in principle—all of the aberrations and optical conditions are once again correctable. The unobstructed feature is an advantage, to be sure, but it is not the only feature or even the chief advantage.

9.2 Spider Diffraction

The support hardware that holds a secondary mirror causes some light diffraction. For linear spider vanes, the pattern takes the form of two or more

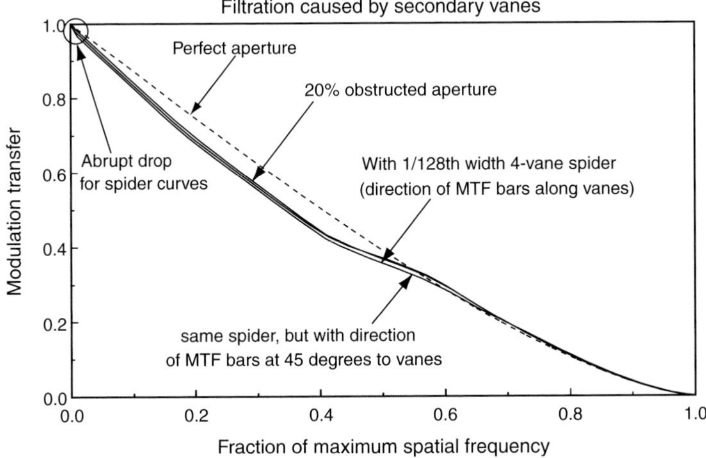

Fig. 9.5 Modulation transfer function for spider diffraction. The degradation is slight compared to a spiderless 20% obstructed aperture. The lowest curve is for an MTF pattern at 45 degrees to the vanes.

radial spikes away from a point image. Spider diffraction takes a bright, extended image and smears it to either side.

Contrast is reduced, but by how much? Clearly, if the area of the vanes is not large when projected against the mirror, the spider cannot be scattering much light. That light *looks* brighter because it is blurred in only a few directions.

The MTF of a spider with vanes of thickness $\frac{1}{128}$ the diameter of the mirror appears in Figure 9.5 (this MTF is compounded with the degradation of a 20% obstruction). The thickness is $\frac{1}{16}$ inch (nearly 2 mm) for an 8-inch (200 mm) mirror. Usually, small telescopes have vanes less than half that thick. Even so, the contrast is degraded only about 1.6% at first. The area of the vanes is also 1.6% that of the aperture. Thus, a simple relationship exists between the area of small-scale obscuration and the amount of abrupt degradation in the MTF.[2]

Spider diffraction causes the quick drop of the MTF, and thereafter the degradation is about a constant fraction of the normal behavior.[3] One can see from the very slight decrease that spider diffraction is a cosmetic defect for most dark-field observation. Only in exceptional situations does it significantly affect image contrast. If, for example, a dim object resides

---

[2] A similar abrupt drop in the MTF would have been visible if the central obstruction had been capable of scattering much light throughout the field.

[3] For modulation patterns at 45°, the MTF will recover at 71% the maximum spatial frequency and even slightly exceed the value expected in its absence. MTFs for horizontal or vertical bar targets display less violently-changing structure. See Zmek 1993.

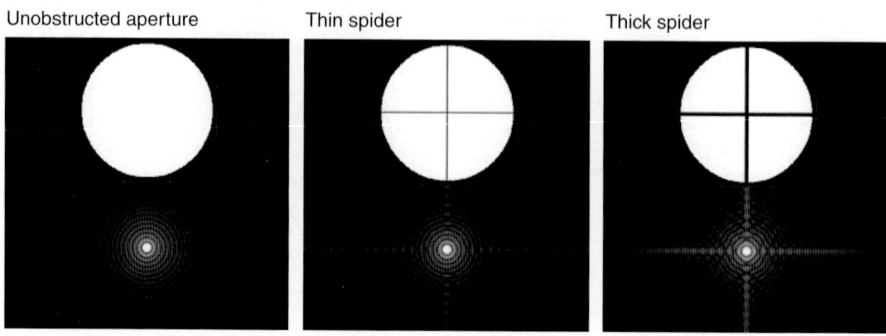

Fig. 9.6 "Overexposed" monochromatic spider diffraction pattern compared with the same pattern calculated with no spider.

in the spider diffraction spike of a bright nearby star, this light could become a problem. The spider diffraction does not cause difficulty if the dim object is alone or at a different angle to the spike, but because the nearby star is so bright, the spike may illuminate the whole area around the dim object, thus reducing contrast. Fortunately, many telescopes allow rotation of the spider or tube.

On extended bright objects such as planets, spider diffraction causes a much subtler degradation. Each point on the image has its own spikes, which then are mixed together. Low contrast details can be washed out by weak light diffracted from the spider. A realistic thin-vaned spider has a surface area of about 0.5% the area of the aperture. Thus, the total scattered light is as much as a factor of 200 below the signal. Luckily, much of the stray light is beyond the edge of planets. If the vanes are as narrow as 0.5 mm, the width of the diffraction spikes is about 200 arcseconds. Spider-vane diffraction would greatly exceed the size of even a large planetary image, such as the 50 arcsecond angle subtended by Jupiter during oppositions. Signal-to-background might exceed 1000 to 10000 for planetary observation. Light diffracted from spider vanes becomes troublesome, if at all, chiefly for lunar and solar observations where the extent of the target image is larger than the angle of diffraction.

Despite the conclusions from Figure 9.5, spider diffraction can be an important design consideration. If the diagonal support vanes are much too thick, the diffraction spikes become brighter (see Figure 9.6). More area of the mirror is intercepted, and more light is diverted. Thicker vanes also have shorter diffraction spikes. Since the light diffracts into a smaller area, it is relatively brighter.

You can see the effect of a wide spider vane by stretching a strip of black electrical tape across the front of your instrument (this experiment

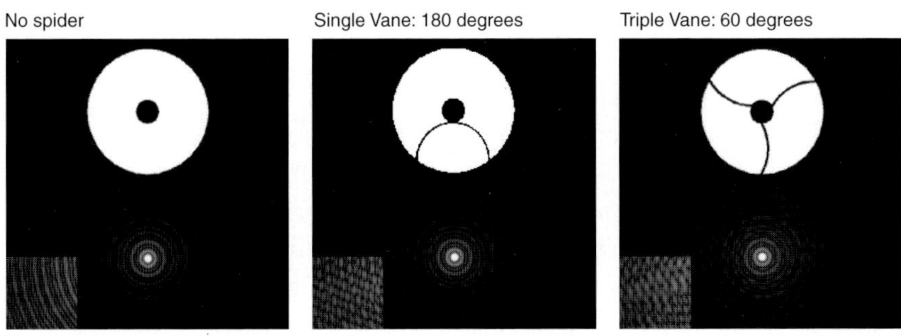

Fig. 9.7 Various curved-vane options. Insets are extreme contrast stretchings showing the fine structure of the image.

even works for refractors and Cassegrain style instruments—just keep the tape away from the lens). Direct the telescope towards a bright star and use a medium-to-high power eyepiece. You should see a bright spike of light at right angles to the tape. This spike glistens with sparkles of color. If the star is bright enough, the spike fades out with increasing distance from the star and then brightens again, perhaps doing so many times. You're observing the side peaks of spider diffraction, similar to the rings of circular aperture diffraction.

Another design mistake is the use of extremely thick, curved-vane spiders or a diffraction-spike suppression mask. Two such masks were suggested by A. Couder using curved-edge vane covers (Ingalls, Book 2 1996, pp. 358–360). These throw a great deal of light around the image. The most important factor is the *fraction* of the mirror such vanes cover, because contrast is lowered by this fraction. The only acceptable change to spider vanes is *thinning* them, using curved-vanes, or eliminating them entirely by supporting the secondary on an optical window.

Various curved-spider images, together with the pupils that produced them, are shown in Figure 9.7. All of these renderings (as did those of Figure 9.6) have an intense contrast shift to make dim details visible. Although the image of the 180-degree curved single vane seems smoothest, it is the 60-degree three-vane spider that is likeliest to hold the secondary satisfactorily. For more detail on spiders and curved-vane spiders, see Suiter and Zmek 2003 and the excellent paper Harvey and Ftaclas 1995.

While spider diffraction is only a minor annoyance for small telescopes, it can become a major object of concern in the design of large instruments. Because mechanical structure does not scale linearly, the support requirements of large reflector secondaries can be very great indeed. Heavy Cassegrain secondaries must be supported rigidly enough that alignment does not suffer. Thick vanes are necessary, and diffraction from the vanes becomes considerable (Beyer and Clune 1988). For example, if vanes expand to 2 cm wide (not unusual in huge professional instruments) then the radius of the first and brightest side lobe is compressed to only 5.8 arcseconds. This value is well inside the disk of major planets.

One should make certain at the end of coarse alignment that the secondary support vanes are turned to present the minimum interception area to incoming light. This maintenance costs nothing, and can improve the image substantially.

9.3 Shading or Apodization

Diffraction rings are among the most objectionable features of the perfect circular diffraction image. This problem is perhaps most observable at the boundary between a bright area and a dark sky background. If seeing is sufficiently good, this boundary does not appear sharp, but shows "echo" images, one or more thin lines at the edge. Observers must be aware of this phenomenon, or else they will report spurious detail in the image. It is particularly prominent in images that are strongly colored or viewed through color filters. Planetary disks display limb-darkening, so this effect is difficult to notice at the edge of a planet. However, the problem is observable elsewhere and is always a source of interference with any clustering of spotty or linear detail. Several diffraction rings may add to create another spot or line where one didn't exist before.

Changing the transmission characteristics of an aperture is called *apodization*. "Apod-" literally means "without feet" and refers to a shading of the entrance pupil that results in lowered diffraction rings. Apodization existed informally before it was named. Jacquinot and Couder made a one-dimensional apodizer in the 1930s to suppress the dim side lines that appeared next to bright lines on spectrographic plates (Jacquinot 1958). A simple square or diamond shape can be used to brush away the rings if it is desired to observe a dim star next to a bright one. See Figure 9.8.

R.K. Luneburg widened the definition of apodization by suggesting a set of generalized problems that don't necessarily lead to smaller rings but induce modifications of the diffraction pattern to optimize any given characteristic.[4] Based on Luneburg's investigations, it can be stated with some confidence that the best pupil shape *in most cases* is an unobstructed one (Luneburg 1964, pp. 344–359). One can set any sort of condition on the diffraction image, however, as long as some other parameter is allowed to

---

[4] For this reason, I prefer the term *shading,* which comes from antenna and acoustic array theory.

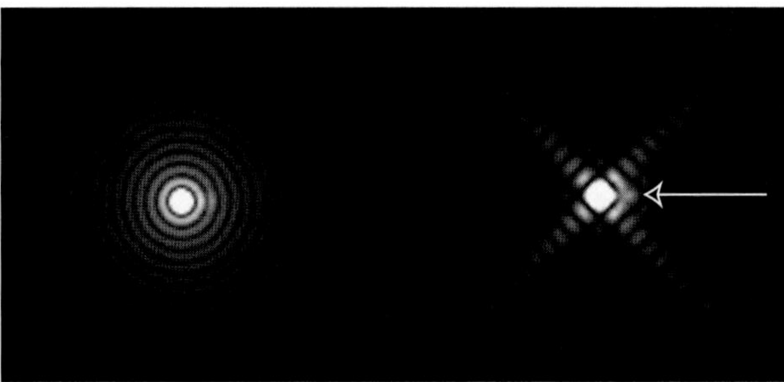

Fig. 9.8 One-dimensional apodization can be demonstrated by placing a simple square opening over an aperture and rotating it until a dim companion star is better shown. The square is oriented parallel to the spikes.

swing freely.

For example, the diameter of the central spot can be minimized if the height of the diffraction rings does not matter. In fact, G. Toraldo di Francia designed a complex pupil shading that results in arbitrarily fine resolution and suppression of diffraction rings out to a specified radius (di Francia 1952). Unfortunately, such a shaded pupil is fantastically inefficient for all realistic apertures, diverting most of the energy of the beam to a bright ring beyond the specified radius. A field stop must be positioned inside of this radius to prevent the bright ring from dazzling the super-resolved star.

Similarly, C.L. Dolph (1946) derived a shading technique for linear-array radar antennae that balances resolution and diffraction. It also works for slit apertures. Dolph's technique features arbitrary suppression of side lobes to a specified level surrounding the central peak. This method is also inefficient but helps most for strong sources where interference from multiple directions is troublesome.

An early systematic investigation into partially transparent coatings of lenses to achieve enhanced resolution was made by Osterberg and Wilkins in 1949. They were able to theoretically achieve a central spot diameter only 77% that of the usual Airy disk. The Strehl ratio of such an aperture is 0.21, so this resolution comes at considerable optical costs. The first diffraction ring is about $\frac{1}{10}$ as high as the central peak. This behavior is similar to that of the obstructed apertures mentioned earlier, but it is carefully optimized to do the least damage and at the same time achieve the highest resolution.

All of these advanced solutions have the same general features: high-resolution pupils seem to have soft-edged obstructions, while low diffraction-ring pupils become darker toward the outside in a slow taper (Barakat

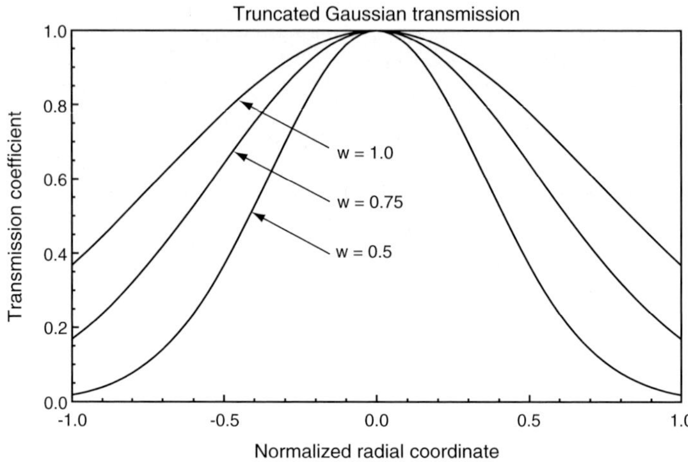

Fig. 9.9 The transmission coefficient of the Gaussian pupil as it varies with radius. The outside of the aperture is at a radius of 1.0.

1962; Jacquinot and Roizen-Dossier 1964).

Also, these advanced techniques generally don't employ simple shading, such as would be provided by a variable strength neutral-density filter (particularly for resolution enhancement). They shift the transmission coefficient back and forth between positive and negative. Admittedly, the notion of negative transmission sounds odd. One imagines light springing from the eye and going back through the telescope in a reversed direction. The truth makes more sense—a negative transmission coefficient refers to places on the aperture where the phase is reversed, or areas that have a uniform aberration of ½ wavelength. Of course, jerking the aperture to both sides of the phase severely reduces central spot intensity. Such filters are also difficult to make.

The ideal transmission shape for ring suppression is a truncated Gaussian function. The Gaussian is the familiar bell-shaped curve representing statistical deviations in measurements. It is similar to the curve teachers sometimes consult when they assign students' grades. A Gaussian aperture starts with full transmission at the center and gradually tapers off until the edge is reached. The transmission coefficient is modeled by the following formula:

$$T(\rho) = e^{-\rho^2/w^2}, \qquad\qquad \textbf{9.2}$$

where ρ is the radial coordinate and w is related to the width of the Gaussian. Figure 9.9 shows this transmission pattern. The word "truncated" refers to the sharp drop at the outside edge of the aperture. As the width is

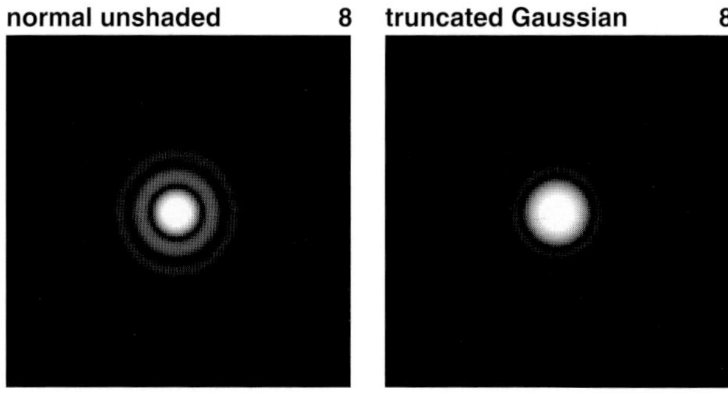

Fig. 9.10 In-focus pattern of perfect unshaded and truncated Gaussian pupils.

decreased, this drop is less important, but the clear area in the center of the aperture also decreases with smaller widths. If the transmission at the edge of the aperture is small, the usable window is also smaller, and the effective aperture is lower. This Gaussian shading must not be overdone lest the resolution be destroyed too.

The Gaussian function has a unique mathematical property. When one calculates the diffraction pattern of a perfect circular aperture, the result is a complicated expression that goes through many oscillations—the cause of diffraction rings. When the same calculation is made for an untruncated Gaussian-shaded pupil, the result is another Gaussian. Once a Gaussian function dies away, it does not rise again. Therefore, the diffraction pattern has no rings around it.

If the Gaussian function has a small drop-off at the edge (as it does for the truncated examples in Figure 9.9), the rings reappear, but they are strongly suppressed. Figure 9.10 shows the focused appearance of a truncated Gaussian with $w = 0.75\rho$. First, note that the pattern is definitely a

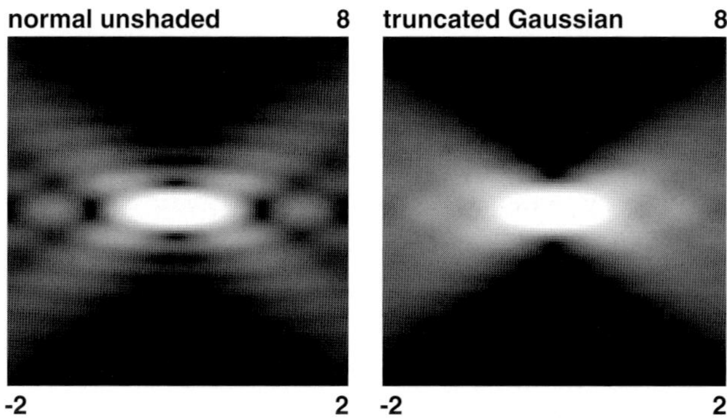

Fig. 9.11 Longitudinally sliced image pattern of a normal and Gaussian pupil.

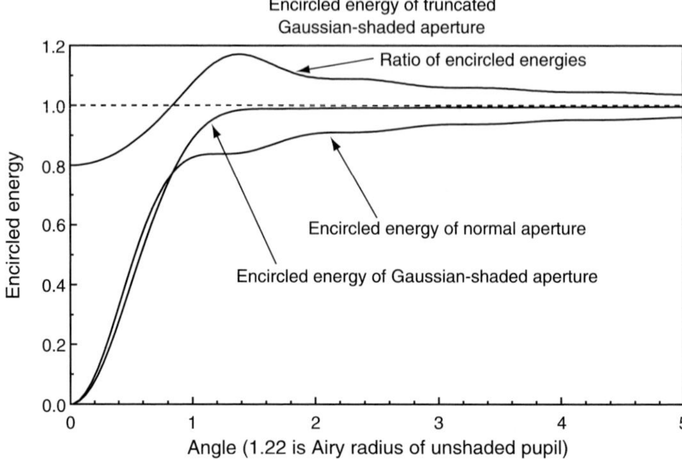

Fig. 9.12 Encircled energy transmitted by a truncated Gaussian pupil compared to a normal unobscured pupil.

bit larger. Second, muted rings remain around the central spot. If we turn the pattern sideways, the longitudinal slice of the diffraction pattern from this truncated Gaussian pupil is seen in Figure 9.11. We still see nodes, but this behavior is severely diminished. Another interesting characteristic of this diagram is the boxy appearance of the central lozenge. Compared to the regular aperture, the image seems to display a cohesiveness more tolerant of defocusing.

What sort of changes to the image take place? The encircled energy of a $w = 0.75\rho$ truncated Gaussian appears in Figure 9.12. Remember, this diagram is corrected for the simple darkening of the aperture. Thus, the ratio of the encircled energies goes to 1 as the circle becomes very large. Amazingly, the ratio for the Gaussian-apodized aperture sweeps upward to enclose more of the transmitted energy than the normal aperture over most of the range.

This strange behavior is related to the suppression of diffraction rings. For a Gaussian pupil, all of the stray energy that normally is in distant portions of the image has already been gathered. A normal pupil encloses only 83.8% of the energy in the Airy disk, 91% inside the edge of the first ring, 93.8% inside the edge of the second ring, 95.2%, 96.1%, and so forth. Gaussian pupil transmission is the same as taking a broom and traveling around the image, sweeping all of the remaining intensity toward the center. This sweeping is not as tidy as one would like, so the energy raked inward is piled up at the edge of an enlarged diffraction disk.

Figure 9.13 shows the effect on filtering. Gaussian transmission

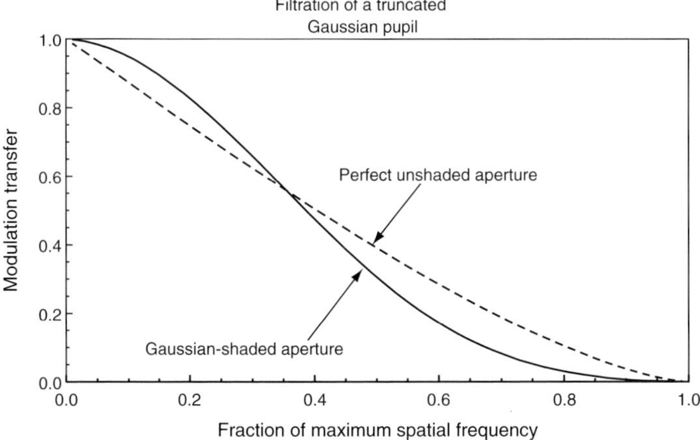

Fig. 9.13 The MTF of a $w = 0.75\rho$ truncated Gaussian pupil. The normal MTF is plotted also for comparison. The Gaussian shifts the frequency response from high spatial frequencies to low ones.

enhances low spatial frequencies at the expense of high ones. The lowered response at high spatial frequency is understandable if one remembers that the central spot is larger. The enhancement at low frequencies is related to the gathering of energy from distant portions of the image. Low frequency modulation targets have wide bars. If the energy is piled close to the center of the diffraction disk rather than spread out, less light leaks from the light areas into the dark ones.

Thus, Gaussian apodizers will not show detail as well near the resolution limit of the telescope, even though the diffraction rings are suppressed. On the other hand, much of an image's content is at low spatial frequency. The Gaussian filter will help to show features that don't require every scrap of the resolving power. Another useful benefit of lowered diffraction rings is resolution of unequal double stars just outside the resolution limit, where the bright ring is inconveniently coincident with the dim star (similar to Figure 9.8 but this time occurring for a Gaussian filter). For general extended images, however, making the rings less apparent does not help high resolution.

Apodization came to amateur astronomy with a series of letters and articles in the "Amateur Astronomer" column of *Scientific American* in the early 1950s. These articles culminated with a suggestion for pupil shading using layers of periodic screening (Leonard 1954). This suggestion was a clever and practical way of achieving a peaked pupil shape. The encircled energy ratio of the zeroth-order image in Figure 9.14 is very nearly as good as the encircled energy ratio of the Gaussian of Figure 9.12, even though

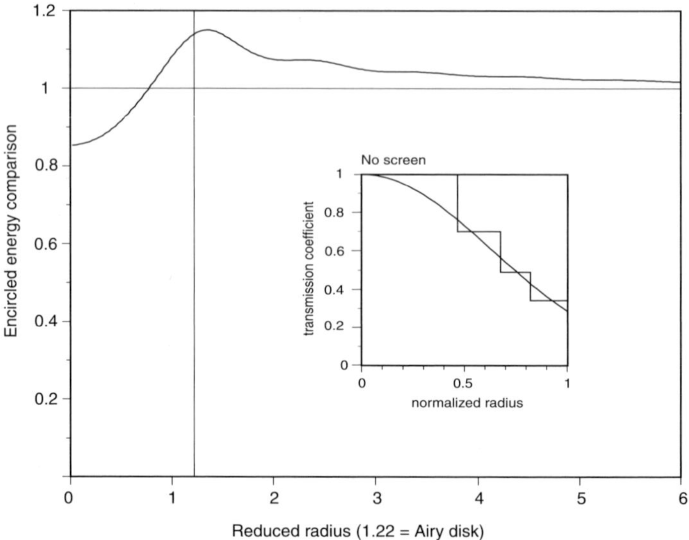

Fig. 9.14 The encircled-energy ratio plot of the three-layer screen depicted in the inset. The smooth curve through the screen depicts a truncated Gaussian as a guide to the eye.

it originates with the stepped three-layer transmission coefficient depicted in the inset. At the edge of the Airy disk, the encircled energy ratio is 1.14 times that of the perfect aperture.

In spite of their mixed performance on the MTF chart apodizers have been used in planetary viewing for many years. They have been vigorously promoted by several amateurs, who claim that they are useful as "seeing" filters, although their explanations of the cause differ (e.g.,Van Nuland 1983; Gordon 1984). This popularity, even though it has been confined to a small number of observers, is difficult to dismiss.

Is there a mechanism by which these filters can act to steady an image? Edberg (1984) suggested some indirect benefits, including stopping down the aperture, lessening the dazzling brightness of the planets, and covering poor optical fabrication. This view is supported by examination of the patterns of Figure 9.15, which show the manner in which apodization could obscure a turned-down edge. One other possibility is the way eyepieces work better with narrower light cones.

Another article suggested that a Gaussian filter could shift contrast rendition from high spatial frequencies—where turbulence was destroying the image anyway—to low spatial frequencies. Thus, coarser details that were still visible through the turbulent atmosphere were being rendered with higher contrast (Suiter 1986b). The MTF of a circular aperture trou-

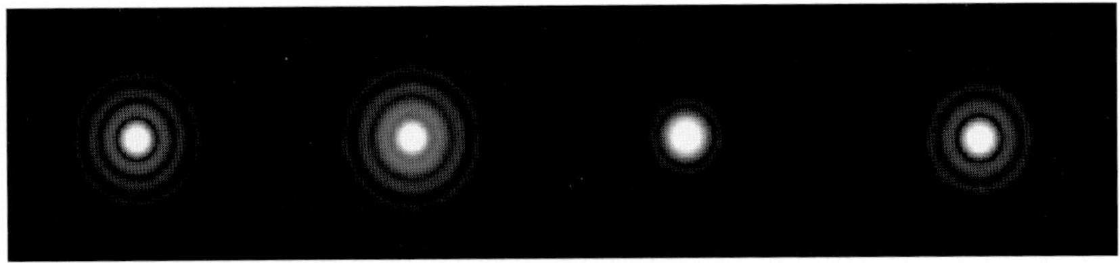

Fig. 9.15 Apodization used to cover poor figuring. At left is perfect image, overexposed somewhat to emphasize ring structure. Next right is the image of a 1 wavelength rolled edge that starts at 90% of the aperture. Next right is a strong Gaussian apodizing filter which diminishes rings as well as covers the aberration. At right is the behavior simply if we had stopped down the aperture beyond the rolled edge.

bled by turbulence provides evidence for that claim.

Figure 9.16 illustrates the Gaussian pupil transmission MTF for a single moment of turbulence. This figure shows how contrast is being shifted. Below a spatial frequency of about 0.4 maximum, the Gaussian-filtered system performs better than the open telescope. Below 0.1, the Gaussian-filtered instrument transfers contrast better than a perfect unaberrated system.

For example, a perfect 10-inch (250-mm) telescope can resolve MTF targets with bars separated by about ½ arcsecond. However, when troubled by 0.15 wavelengths RMS turbulence, the MTF curve fluctuates wildly for bars separated by about 0.8 arcseconds and below. We may then choose to

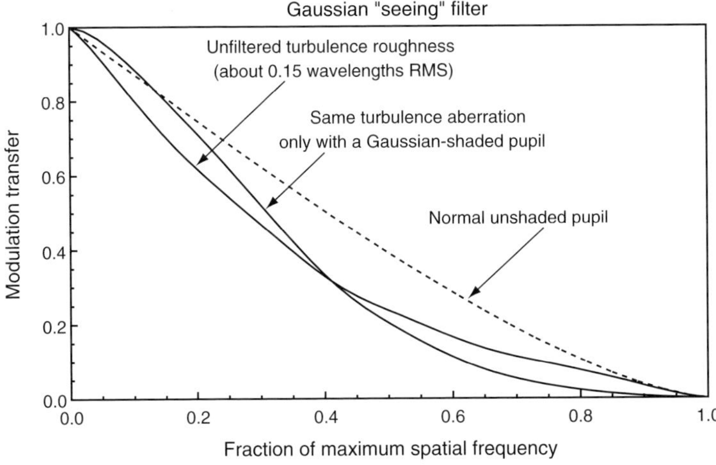

Fig. 9.16 Improvement of the contrast transfer of low spatial frequencies for a $w = 0.75\rho$ Gaussian-filter pupil suffering from 0.15 wavelengths RMS of turbulence aberration. The same aperture when unshaded is also shown. A single MTF orientation angle during one snapshot is depicted. At higher spatial frequencies, the curve is unstable.

throw away contrast at high resolutions—where the behavior is unreliable anyway—and shift it down to spatial frequencies where it can be profitably used. With Gaussian filtration, bar separations of about 1.5 arcseconds and higher are imaged with higher contrast than the unapodized aperture. Above 4 arcseconds, the apodized aperture is behaving better than a perfect circular aperture on a steady night.

Thus, a Gaussian or a step-Gaussian apodizing filter does seem to help with details already within reach during bouts of poor seeing, but it does so only in a backhanded manner. The apodization filter is perhaps useful in order to transform bad seeing to passable, but for critical resolution nothing less than steady skies and full aperture are necessary.

9.4 Dust and Scratches on the Optics

We can formally use the admittedly esoteric topic of superposition of apertures to explain the effect of central obstruction. The central obstruction can be mathematically replaced by a negative oscillating aperture that just cancels the positively oscillating area of the full aperture on top of it— Babinet's principle. This procedure works the same for lesser obstructions such as dust particles, only in this case, the dust grains or scratches act like a myriad of pinhole or short slit apertures.

Tiny pinholes and slits do not have good resolution. They emit a diffuse glow that is spread throughout the field of view. This behavior is observable in certain benchtop experiments. If a bright image is barely hidden from view by a straightedge (as in a Foucault test), dust flaws can actually be seen shining on the darkened aperture.

We can predict the curve of the modulation transfer function without calculating it. It is very similar to the MTF calculated for spider vanes. If an outrageous fraction of the mirror or lens were covered with dust or scratches (say, 1%) the contrast would lurch down suddenly by 1%. After that, the degradation of the MTF would remain fairly constant. The reason for the sudden drop is simple. The MTF must always start at 1, but the diffuse glow affects narrow bars and wide ones equally. Not until the bars are very wide does the leakage from a bright bar not extend over the dark region.

Apertures with 1% of their area covered with dust are very grimy optics indeed. Nearly everyone keeps lenses and mirrors cleaner. Imagine how spotted the optics would look if, for each square centimeter, 1 square millimeter (about the area of the head of a pin) was blocked. A 200-mm aperture has an area of over 314 cm^2. Sprinkling salt on the mirror or lens could scarcely produce 1% obstruction. Frequently, however, enthusiastic mirror-makers will aluminize mirrors before they are truly polished. They are either covered with a haze of pits or have a fuzzy ring out toward the edge. Unfortunately, such mirrors cannot be "cleaned" at all and must tube returned to fine grinding or polishing.

Images from dusty mirrors don't suffer the diffraction spikes caused by conventional spiders. An increased level of scattered light is still there—it's just spread around to all angles. In imaging of extended objects (such as planets and nebulae), it doesn't matter whether the light is confined to a diffraction spike or has been averaged over various angles.

Dust and scratches, like spider diffraction, are mostly cosmetic errors except for unusual situations that feature bright, non-interesting objects close to the object to be observed. If a dim deep-sky object were observed quite near a dazzling star (NGC 404 behind Beta Andromedae or NGC 2024 near Zeta Orionis, for example), then the diffuse glow surrounding the bright star might invade the image of the dim object. Observing tricks, such as obscuring the star behind a field stop, can diminish scattering in the eyepiece and the eye, but if the main lens or mirror is dusty, the damage is already done. The stray light will peek out beyond the edge of the stop anyway. Excessive dust on the optics can also damage the detection of low contrast details on bright extended objects, such as planets, because the scattered light exists in a general haze.

The maximum amount of dirt the observer should tolerate on the optics is about $\frac{1}{1,000}$ of the surface area. We have already seen in Chapter 3 how this can lead to signal-to-background ratios as low as 1000. Luckily, most of the halo extends beyond the edge of a planet. Hence, much of the scattering is ugly but not harmful. As with spider diffraction, the worst signal-to-background ratios are reserved for very large bright-field objects, such as the Sun and Moon. It is difficult to estimate the fraction of the optics covered by dirt, but $\frac{1}{1,000}$ of the area is the size of a single obstruction about $\frac{1}{30}$ of the diameter. On a 200 mm mirror, the accumulated grime would cover a spot 7 mm across—slightly smaller than the size of your little fingernail. Even a telescope that is that dirty will not contribute much additional scattering to a reflector with spider vanes, and the tolerance for dirt could perhaps be relaxed somewhat for such instruments.

The observer should heed one additional warning. Some telescope owners, after reading the above comments, might be tempted to clean their mirrors or lenses too frequently. Optics possess delicate coatings for which the safest prescription is to *leave them alone.* Overuse of even the most gentle of cleaning materials leaves a myriad of tiny scratches in coatings. Don't decide to clean mirrors on the basis of shining a light down the tube at night. All mirrors fail such a harsh inspection.

The best procedure for clean optics is not washing but prevention. Keep them covered and dry. Clean them only when you believe they have been chemically attacked or when the dust begins to visibly affect the images. Of course, specialized observers might need cleaner surfaces at all times, but no doubt these people have already learned of the considerable risks and expenses involved (more frequent aluminization, etc.). By following good maintenance procedures, you should not have to clean optics very often.

9.5 Conclusions

I hope you have learned some general principles in this chapter, the most important of which are:

1. Central obstruction can be a problem, true, but many things associated with reducing obstruction can be a larger problem. Try to keep obstruction below 20 to 25% linear diameter, and recognize that anything smaller has benefits unlikely to be noticeable.

2. Viewing the optical system as a black box, diffraction from central obstruction can be entirely cured by moderate increases in aperture.

3. Unless one is familiar with the concept of spatial frequencies and which spatial frequencies are attacked by the unique forms of obstruction, one cannot reliably predict the effects on the image. Certainly, blanket statements like "obstruction reduces contrast" have insufficient information or give false impressions, and may cause observers to perform useless modifications to their instruments.

4. Spider diffraction and diffraction from dust specks or very small obstructions like mirror clips have effects in the image at such different spatial frequencies that they can be considered separate problems.

5. Spider diffraction may be obscured by a curved vane, but it is still there. Only decreasing the fractional area covered by the vane lessens spider diffraction. On planetary imaging with a small, thin, spider this form of diffraction is more cosmetic than harmful, because most of the spike light is outside the radius of the planet.

6. Apodization has subtle but real effects on the image. However, anyone claiming too spectacular a result is probably influenced by other effects, like dimming the image or using the eyepiece at a higher effective aperture ratio.

Chapter 10
Spherical Aberration

The most dominant error on the glass is called spherical aberration. It resides to some degree on all surfaces and need not become debilitating, though that depends on its severity.

The star test for spherical aberration is sensitive and easy to interpret. This chapter will make four points:

1. Wavefronts deformed by simple spherical aberration (fourth-order curves) are recognizable by noticing that light approximately follows the behavior of the ray-optics caustic. On one side of focus, light is taken from the outer parts of the out-of-focus disk and deposited at its center. On the other side of focus, light taken from the center brightens the outermost rings.

2. The strength of lower-order spherical aberration can be roughly estimated by using the tables appearing in Chapter 5. Set the defocus precisely and the resulting patterns may be compared with patterns in this chapter. Correction error must be the only contributing aberration for this estimation technique to be valid.

3. By calculating the energy deposited within an angular radius near the edge of the diffraction disk, the telescope user can usefully add together the combined effects of spherical aberration and central obstruction. This encircled energy ratio yields a number similar to the Strehl ratio, but it includes the degradation of obstruction in a single standard.

4. A quarter wavelength of correction error delivers images of extended objects that are barely acceptable, but noticeably soft.

10.1 What Is Spherical Aberration?

Because they seem to have the form of shallow bowls or protrusions, it is natural to assume that perfect mirrors and lenses are sections of spheres. However, spheres are not necessarily the best geometrical form with which to produce an image. The following thought experiment makes this point obvious. Imagine the inner surface of a hemisphere as a mirror and light as incident from infinitely far away on the right, as in Figure 10.2. Light striking the edges is hardly deflected; it kisses the interior and bounces around the half dome. Indeed, such rays are more familiar acoustically in "whispering galleries," where a whisper is audible right next to a curved wall for

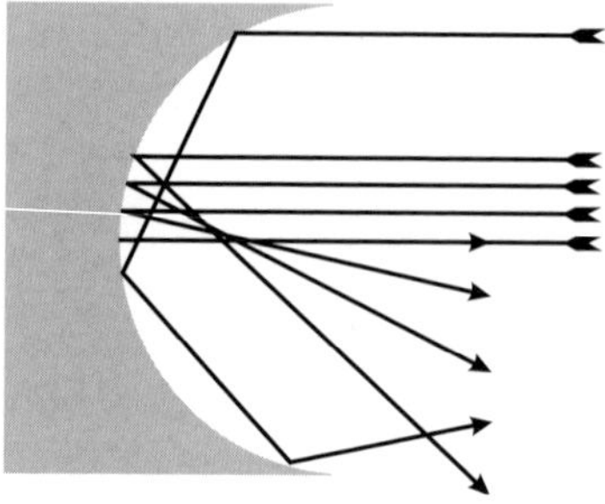

Fig. 10.1 Half-dome spherical mirror. The focus becomes more poorly defined as the incoming ray approaches the edge. The ray bouncing around the circle shows a nearly limiting case.

long distances, but they are seldom seen in imaging optics. Light incident more toward the center is deflected towards a sort of focus, but the light striking near the axis is directed the farthest away. The focus region is expanded in Figure 10.1 to demonstrate that the focus is bent from a geometric point to a horn-shaped caustic envelope.

The original meaning of *caustic* was "burning," so an optical caustic is at or near focus. In ray tracing, the term acquired a slightly different meaning. A caustic is a curve or surface along which rays seem to pile up. It refers to locations where the geometric spreading or convergence of a ray bundle does not give the correct value of the intensity. Caustics are places where one must resort to diffraction theory to model actual intensities.

The type of spherical aberration shown in Figure 10.2 is called spherical undercorrection. With overcorrection, the order is reversed. Centrally incident light crosses the axis too near the objective, and light incident at the edge crosses the axis at the most distant point. In overcorrection, the horn points the other way.

The best surface configuration to image light varies from telescope to

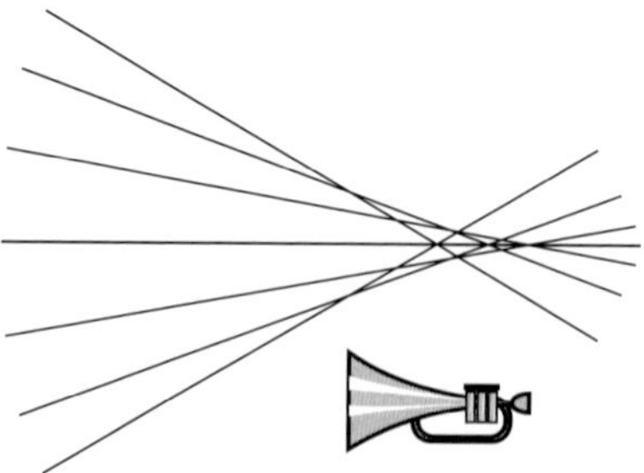

Fig. 10.2 The caustic "horn."

telescope. Some objectives, such as compound refractor lenses, correct for spherical aberration by bending or separating spherical surfaces. Here the designer uses a trick to retain the easily made spherical shape. A similar bending occurs in the Maksutov design. The meniscus shell doesn't really possess much focusing power since the rear surface has about the same curvature as the front. The real purpose is aberration modification. Other telescopes share the burden of correcting the main focusing element's spherical aberration with an oddly deformed corrector plate, as in the common Schmidt-Cassegrain telescope.

For telescopes with only one focusing mirror (i.e., Newtonian reflectors), the proper geometric form can only be a paraboloid. The flat secondary diverts the beam so the observer's head doesn't get in the way, but it doesn't actively participate in forming the image. The paraboloid is one special case of a family of surfaces called *conic sections of revolution.*

Classical Cassegrains combine a paraboloidal primary with a hyperboloidal secondary to achieve the requisite spherical correction. One can take out the secondary mirror of a classical Cassegrain and install a diagonal to use it as a Newtonian. The secondary mirror must correct its inherent spherical aberration independently. One might reasonably guess that the secondary would have to be a convex paraboloid, but the secondary doesn't do the same job as the primary, so it must be curved differently.[1]

Other Cassegrain-style instruments, such as the Dall-Kirkham type, only correct part of the spherical aberration of the whole system in each mirror. The convex secondary remains spherical. This small mirror adds a component of spherical aberration of opposite sign to a concave spherical primary mirror, but the amount is not sufficient to correct the system completely. As a consequence, the properly made primary mirror of a Dall-Kirkham is deformed to a prolate spheroid (between a sphere and a paraboloid). Dall-Kirkhams are popular with telescope makers because the spherical secondary is easy to make. However, they suffer severe off-axis coma.

The Ritchey-Chrétien-Cassegrain design, which has penetrated into amateur ranks in recent years, goes the other way. By putting stronger hyperboloidal curves on both the primary and secondary mirrors, the designer can achieve a degree of coma correction superior to the classical Cassegrain and vastly superior to the Dall-Kirkham. However, these telescopes are somewhat more difficult to make, and are usually of interest only to professional observatories. For the relatively mild sizes and sec-

---

[1] One peculiar Cassegrain-style design, the afocal Mersenne, has a paraboloidal secondary mirror. This rarely seen telescope doesn't require an eyepiece, though another telescope on the back end would be required in any useful implementation. (King 1955, pp. 49–50.)

ondary magnifications made by amateurs (say an *f/4* primary times a 4 power secondary) yielding an *f/16* system, the Ritchey-Chrétien is a pyrrhic victory. Any advantages are reduced by the strong curvature of field of any visual Cassegrain system. It is not until the designer is attempting really fast and large systems for photography or astrometry, and is willing to add field-flatteners, that the advantages are great enough to justify the extra effort.

10.2 Generalized Spherical Aberration

From the point of view of the star test, you don't have to think about the form of the optical surfaces, or even remember their long names. Consider only the shape of the final wavefront as subtracted from a perfect sphere. This difference can be expanded in the form of a simple polynomial function:

$$W(\rho) = A_0 + A_2\rho^2 + A_4\rho^4 + A_6\rho^6 \cdots,$$

10.1

where ρ is the radial coordinate with range 0 to 1. The symbol $W(\rho)$ stands for the deviation of that wavefront away from a sphere with a center at the focus. If $W(\rho)$ is zero, then the curves are the same. Let's look at each of these terms and the coefficients in front of them (the "A's") and discuss what they mean.

The first is a constant, A_0, that only advances or delays the wavefront. We can think of this number as the time or phase constant, and it should be chosen so that the comparison sphere is not too far removed from the wavefront as it passes through the pupil. In actual use this constant, called "piston," is used to reduce the RMS deviation to the minimum value possible in the context of the higher terms.

The term $A_2\rho^2$ is a smooth bending, either pushing the wavefront in a small amount or pulling it out somewhat. If our reference sphere has its center placed at the wrong focal point, this term bears the brunt of the increase. Thus, $A_2\rho^2$ is called the "defocusing aberration" here and is the same one appearing in Figure 4.15.

The A_2 defocusing term can be conceptually included in the spherical aberration expansion, and from the perspective of star-testing defocusing should be considered as just another aberration. However, defocusing is not a feature of the glass, so it isn't customary to refer to it as a spherical aberration term. Defocusing is so important that it is set aside and considered separately.

The fourth-order term, $A_4\rho^4$, is what is usually thought of as *spherical aberration*. Another name for this term is *primary spherical aberra-*

tion. Errors here are said to be "correction" errors, such as undercorrection or overcorrection.

The terms $A_6\rho^6$ and beyond are usually small but can become important with zonal errors. You can see by the ellipsis that this expansion goes on forever, but for most surfaces each factor A_n is usually much smaller than the previous coefficient. We can regard the A_6 coefficient as the last important term in this chapter. Even with advanced aspheric optics, the deliberately-applied coefficients rarely advance beyond A_6.

Spherical aberration can be expressed by a number of equally valid conventions. Another way of referring to the fourth- and sixth-order wavefront terms is to call them third- and fifth-order spherical aberrations. These names are derived from the slope of the wavefront, not the wavefront itself, and how far the light is shifted sideways from the center of the diffraction spot. Some authors find it more convenient to consider the residual changes in focal distance with ρ, or "longitudinal aberration" (Kingslake 1978, p. 114). In this way of looking at errors, primary aberration is the coefficient of the ρ^2 term (this is the way Foucault testing views spherical aberration). Thus, we can find the same primary spherical aberration expressed as fourth-, third-, or second-order coefficients, depending on whether the polynomial expansion refers to the wavefront, to the residual slope, or to the longitudinal aberration respectively. In this book, we will usually refer to the wavefront.

10.3 The Aberration Functions

The aberrations are usually measured at the position of best focus — also called "diffraction focus" — because the diffraction disk is smallest there. It is what we think of when we say "the telescope is focused." When A_2 is zero, Equation 10.1 has the position of focus arbitrarily set at what is called "paraxial" focus, or the focus of the center of the mirror or lens (the narrow end of the caustic horn above). Although a convenient place for the mathematics of wavefront shape, it has little to do with the location where we visually perceive the tightest focus.

If we make the constant A_4 non-zero, for example, we find that some defocus must be added to chase best focus as it scoots away. If we change A_4 a second time, we must move the focus again. The term A_6 creates even more complications. If we wish to consider pure higher-order spherical aberrations at the best focus position, we must subtract out just the right amounts of lower orders. Removing these terms can be tiresome, although the process is computationally straightforward.

It is much more convenient to encapsulate just enough of the lower-order aberrations to automatically cancel them out of each term as it is increased. This step was taken, in a considerably more complicated form, by Frits Zernike, and the resultant terms are called the orthogonal Zernike polynomials on the unit circle (Born and Wolf 1980). The lower terms, the constant and defocusing, are much the same as in the simple expansion. Only with higher orders do the polynomials become more complicated. The interesting terms are limited here to 4th- and 6th-order, but they exist for higher orders as well:

$$W_4(\rho) = 4A_4{'}\left(\rho^2 - \rho^4 - \frac{1}{6}\right)$$

$$W_6(\rho) = -\frac{A_6{'}}{2}(20\rho^6 - 30\rho^4 + 12\rho^2 - 1).$$

10.2

The primes are placed on the coefficients to indicate that they are not the same size as in Equation 10.1. These complicated-looking equations simplify somewhat when displayed as aberration functions (see Figures 10.3 and 10.4). All focused patterns appearing below are referenced near the focus positions implied in these functions. The reader is reminded that the factor ρ is a normalized radius; it reaches 1 at the edge of the optic.

The functional form of the fourth-order Zernike polynomial has just one doughnut-shaped ring, whereas the higher order has an elevated center. The 8th-, 10th-, and higher-order polynomials add one extra curl each. In fact, as we go higher, the pattern appears increasingly corrugated. But remember, the sum of these polynomials represents smooth surfaces. The signs and amplitudes turn out so that the washtub appearance normally goes away. Only for the zonal defects of Chapter 11 do anomalous contributions of high-order polynomials add up to give a significant result.

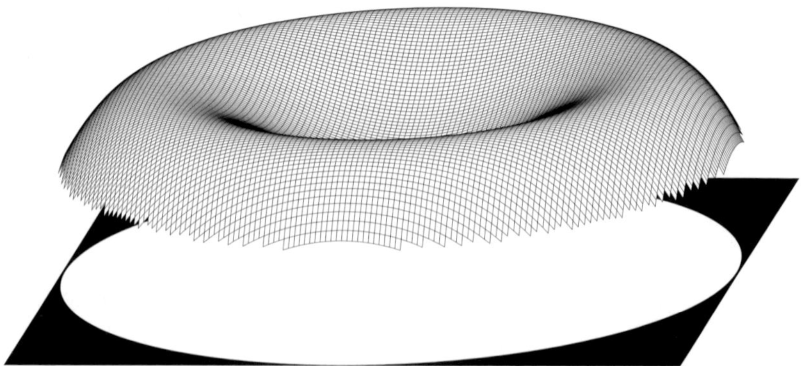

Fig. 10.3 Aberration of Zernike function $W_4(\rho)$, illustrating overcorrection at best focus.

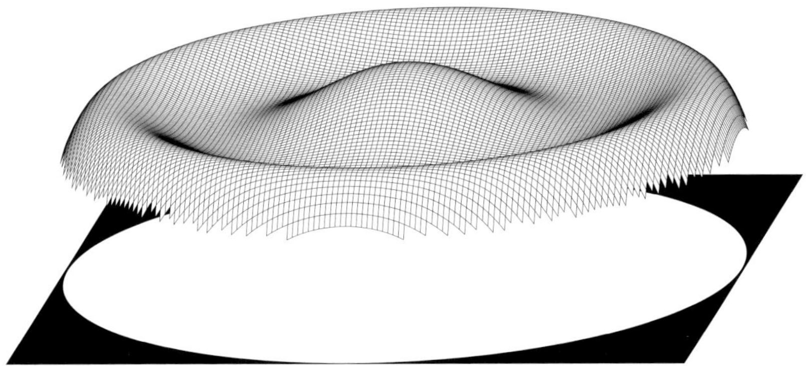

Fig. 10.4 Aberration of Zernike function $W_6(\rho)$, depicting higher-order spherical aberration.

Although it is not shown, the 6th-order function has a more involuted caustic shape. Imagine if a very powerful individual reached into the bell of the musical horn of Figure 10.2 and pulled the mouthpiece halfway back through it. The resulting damage would look much like that caustic (Cagnet *et al.* 1962). (We will discuss higher-order spherical aberration more thoroughly in Chapter 11, grouping it with zones. In fact, another name for this error is *zonal spherical aberration*.)

Another minor point is that the position of diffraction focus moves slightly if the telescope is obstructed. Most of the patterns appearing in this chapter take into account this change. The longitudinal slice patterns are an exception.

10.4 Correction Error (Lower-Order Spherical Aberration)

10.4.1 Filtering of Spherical Aberration

The modulation transfer function (MTF) is depicted in Figure 10.5. The drooping lines show the way contrast is reduced as spherical aberration is steadily worsened. One interesting feature is the mild increase that occurs for small aberrations near a spatial frequency 0.5 to 0.6 of maximum. This recovery corresponds to a frequency where the separation of the target bars is about the distance to the first diffraction ring. It sags on either side.

The MTF worsens significantly when the aperture pupil goes from ⅛ to ¼ wavelength of total aberration, emphasizing that optical quality begins to fail at around ¼ wavelength of correction error.

Spherical aberration shifts light from the central disk to the outer parts of the diffraction pattern. A peculiar feature of spherical aberration is that it leaves the central core of the image alone (until the aberration is quite

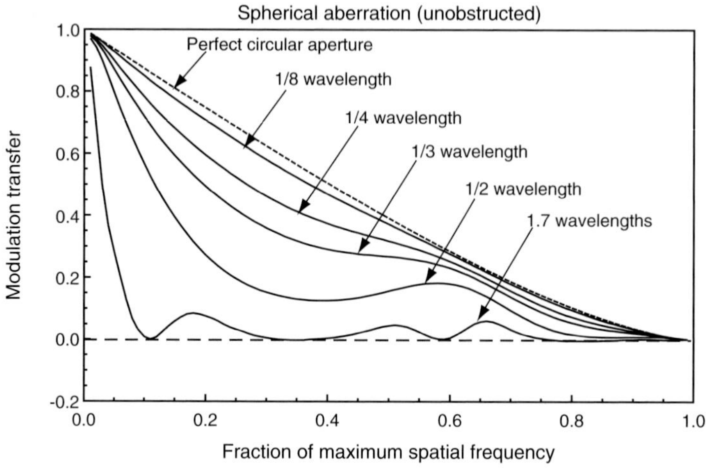

Fig. 10.5 MTF characteristic of correction error.

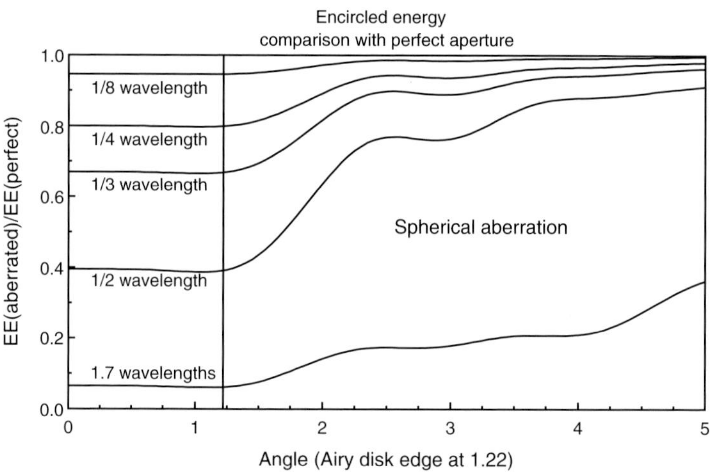

Fig. 10.6 Encircled energy of unobstructed apertures suffering from spherical aberration divided by the encircled energy of a perfect circular aperture.

strong) other than to sap it of its intensity. Spherical aberration drains energy from the central Airy disk and feeds it to the rings. In Figure 10.6, the encircled-energy ratio of increasing low-order spherical aberration is shown plotted against reduced angle. Note the flat or near-flat appearance of the encircled energy ratio out to the radius of the Airy disk. This signature identifies spherical aberration of any order (even mild defocusing displays this peculiar behavior). Energy has been removed from the central disk of the aberrated pupil, but it holds the unaberrated shape fairly well until out beyond the first ring. Also in Figure 10.6, the encircled energy ratio for ¼ wavelength of spherical aberration intercepts zero radius at 0.8, just where the Strehl ratio says it should.

The effect of draining the energy in the core image is to maintain a more or less equivalent (but dimmer) Airy disk surrounded by a blurring that lessens with distance. Planetary detail suffers considerably, because low contrast dark markings are often very close to bright areas that bleed or wash over. The MTF chart indicates that the worst effects of correction error begin to occur for surface markings having separations of about ⅓ to ½ the resolution limit of the instrument. Thus, an aberrated telescope capable of resolving stars separated by 1 arcsecond shows planetary detail (such as banding separated by less than about 2 or 3 arcseconds) with reduced contrast.

10.4.2 Monochromatic Star-Test Patterns of Correction Error

When star testing, you may wish to view the image through a color filter to avoid color mixing effects. Because the maximum sensitivity of the human eye is around the 550 nm wavelength of green, you probably can derive the most information from a filter centered on this color. If you use an artificial source (such as a flashlight) some serendipitous filtering is caused by the lower color temperature of the filament. You may discover for yourself the helpful effects of filtration when star testing a telescope on Arcturus during a hazy night. The star is naturally yellow and so heavily colored by passage through the haze that it becomes orange. Few color filters cut out all of the lights of other bands, however.

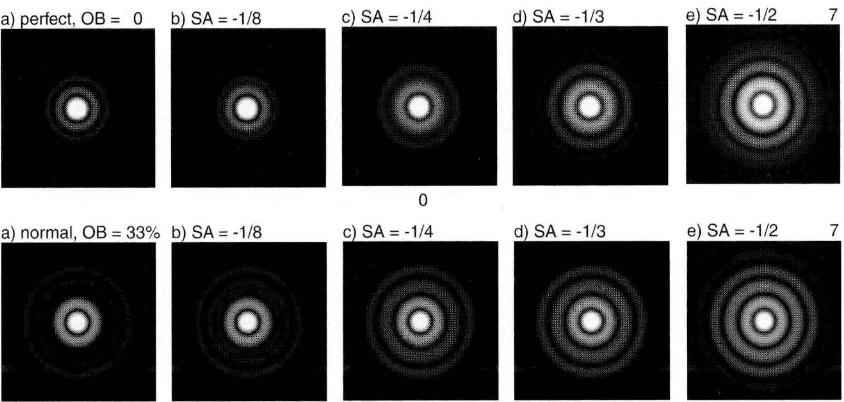

Fig. 10.7 In-focus patterns for a) 0, b) ⅛, c) ¼, d) ⅓, and e) ½ wavelength of lower-order spherical aberration. The sequence in the top row is unobstructed; the sequence in the bottom row has a 33% circular obstruction centered in the pupil.

Of course, you are welcome to try other filters, but be aware that as you go from blue to red the wavelength error lessens. Red filters may help subtract small amplitude trash from a diffraction-pattern image when you are trying to see broad and coarse deformations of the wavefront, but it lessens the overall error too.

The behavior of the monochromatic in-focus image is depicted in Figure 10.7, where maximum refocused spherical aberration runs from 0 to 0.5 wavelength undercorrected. All correction errors are inferior to the perfect aperture, but the rings brighten badly for aberrations greater than ¼ wave. Since the use of a secondary is so common, a similar comparison is made in the bottom row of Figure 10.7 for a 33% obstructed aperture. In focus, the images are identical whether they are undercorrected or overcorrected.

Out-of-focus behavior appears in Figure 10.8 with each successive pair having slightly worse undercorrection. Each of these patterns is calculated for ±10 wavelengths defocusing aberration. For overcorrection, just imagine these figures reversed, so that the fuzzy disk appears inside of focus. We are viewing slices of the horn-shaped caustic in these figures. On one side of focus, the horn is sliced near the flaring bell. As a consequence, most of the energy is confined to a thin outer ring. On the other side of focus, the horn is sliced near the mouthpiece, so much of the energy is concentrated near the center. Still, a lot of energy spills out into the surrounding area to make the disk appear blurred.

10.4.3 Polychromatic Star-Test Pattern of Low-Order Error

The monochromatic patterns of the previous section do a remarkable job of illustrating most of the appearance changes inherent in correction error, but they display a detailed structure not seen in the ordinary white light image. Polychromatic images show a softening of these patterns, particularly in the center and especially for defocused images. Figure 10.9 depicts the focused patterns, and they differ little from the monochromatic cases. However, in the 17-color photopic polychromatic grayscale images of Figures 10.10 through 10.12, the softening effects become apparent. Also note another characteristic. As the aberration droops below ¼ wavelength, the star-test differences become easier to read at ±5 wavelengths defocus instead of ±10. As the aberration diminishes, the defocus values at which it can be detected get smaller.

10.5 A Compact, Uniform Standard for Optical Quality

Consumer telescope makers and observers alike tend to divide aberration

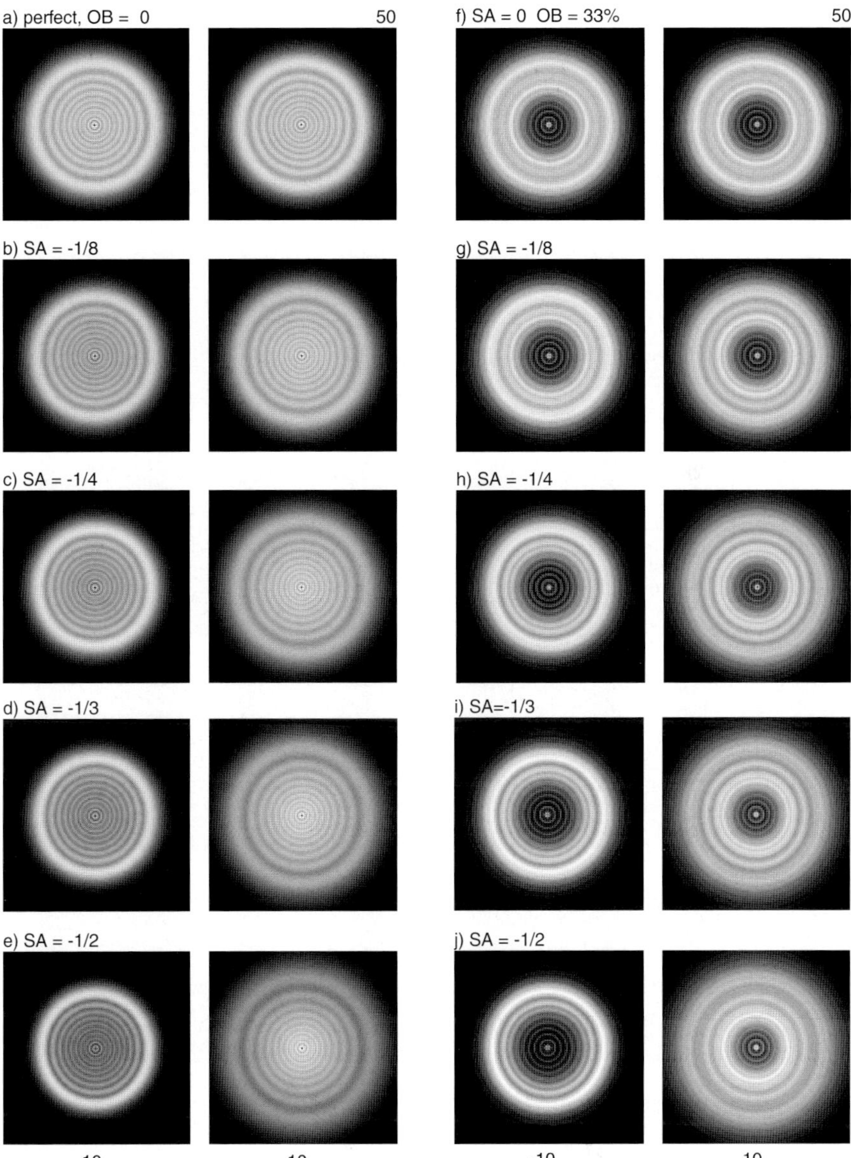

Fig. 10.8 Monochromatic defocusing aberration ±10 wavelengths (negative inside focus and positive outside focus): a) 0, b) $-\frac{1}{8}$, c) $-\frac{1}{4}$, d) $-\frac{1}{3}$, and e) $-\frac{1}{2}$ wavelength lower-order spherical aberration. In a) through e) aperture is unobstructed; in f) through j), we see the same cases if they are 33% obstructed. All of these apertures are undercorrected; for the overcorrected case, merely reverse the directions of defocus.

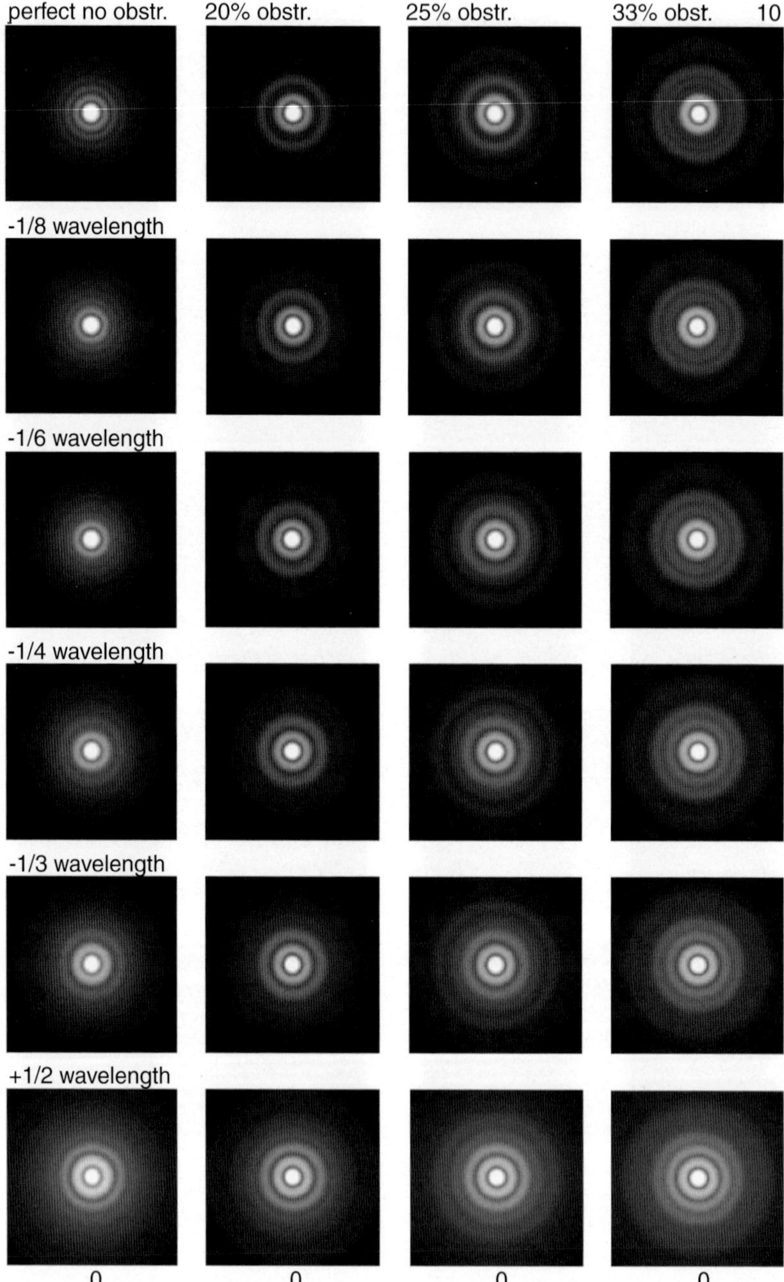

Fig. 10.9 Focused patterns for 0, $-\frac{1}{8}$, $-\frac{1}{6}$, $-\frac{1}{4}$, $-\frac{1}{3}$, and $+\frac{1}{2}$ wavelengths of lower-order spherical aberration. The aperture is unobstructed at left and has progressively 20, 25, and 33% obstruction moving to the right. These are fully polychromatic calculations.

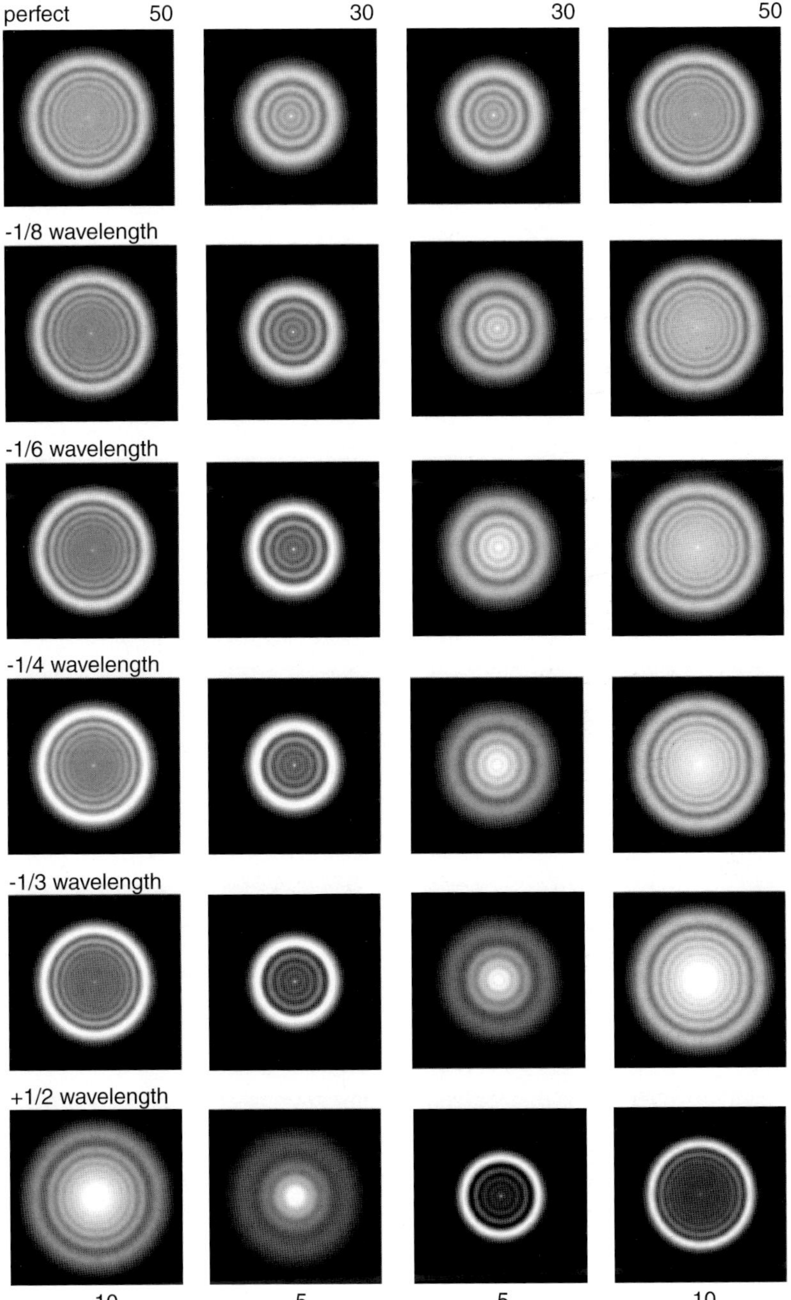

Fig. 10.10 An unobstructed aperture with low-order spherical aberration of 0, $-\frac{1}{8}$, $-\frac{1}{6}$, $-\frac{1}{4}$, $-\frac{1}{3}$, and $+\frac{1}{2}$ wavelength. From left to right, defocus ranges from -10 wavelengths (inside focus) to +10 wavelengths (outside focus). Polychromatic calculation.

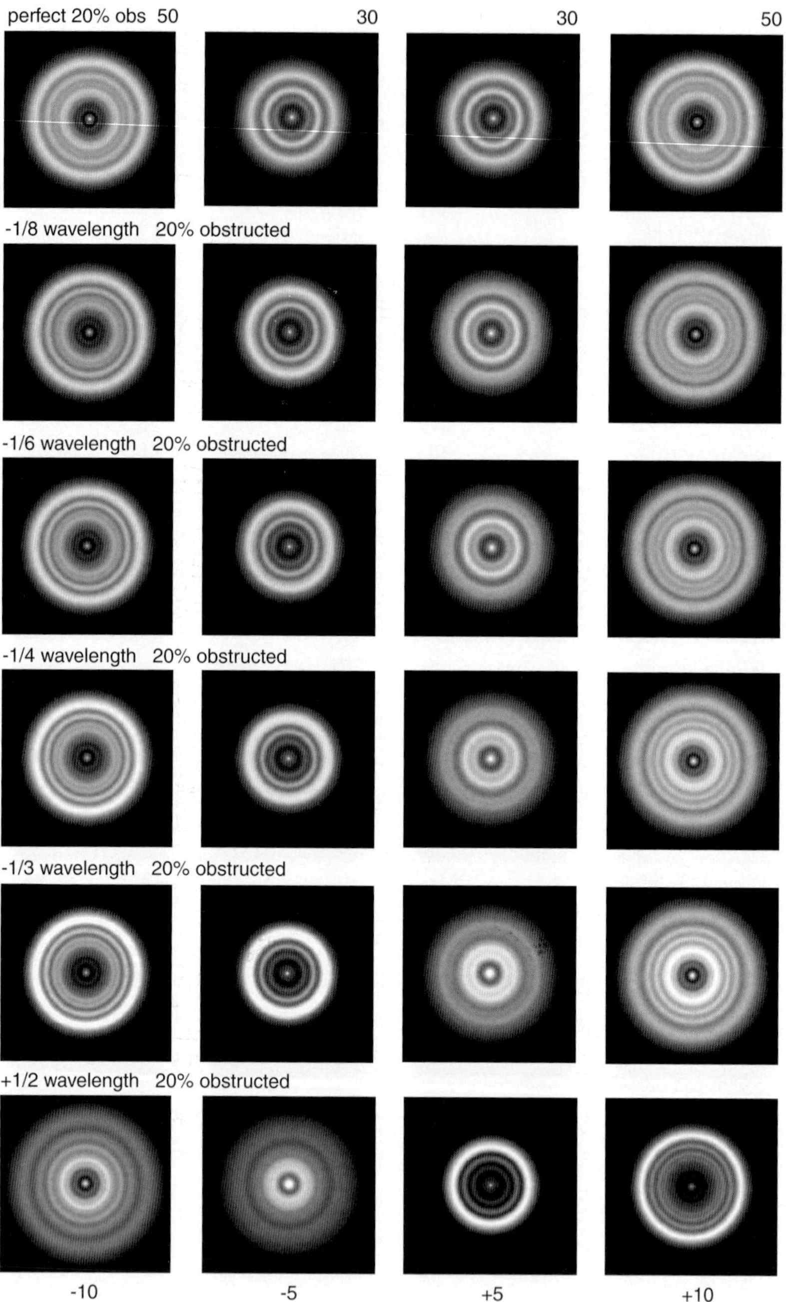

Fig. 10.11 A 20% obstructed aperture in the same situations as in the previous figure. The amount of defocus is measured from the minimum amount of RMS deviation in the lit annulus. The calculation is polychromatic.

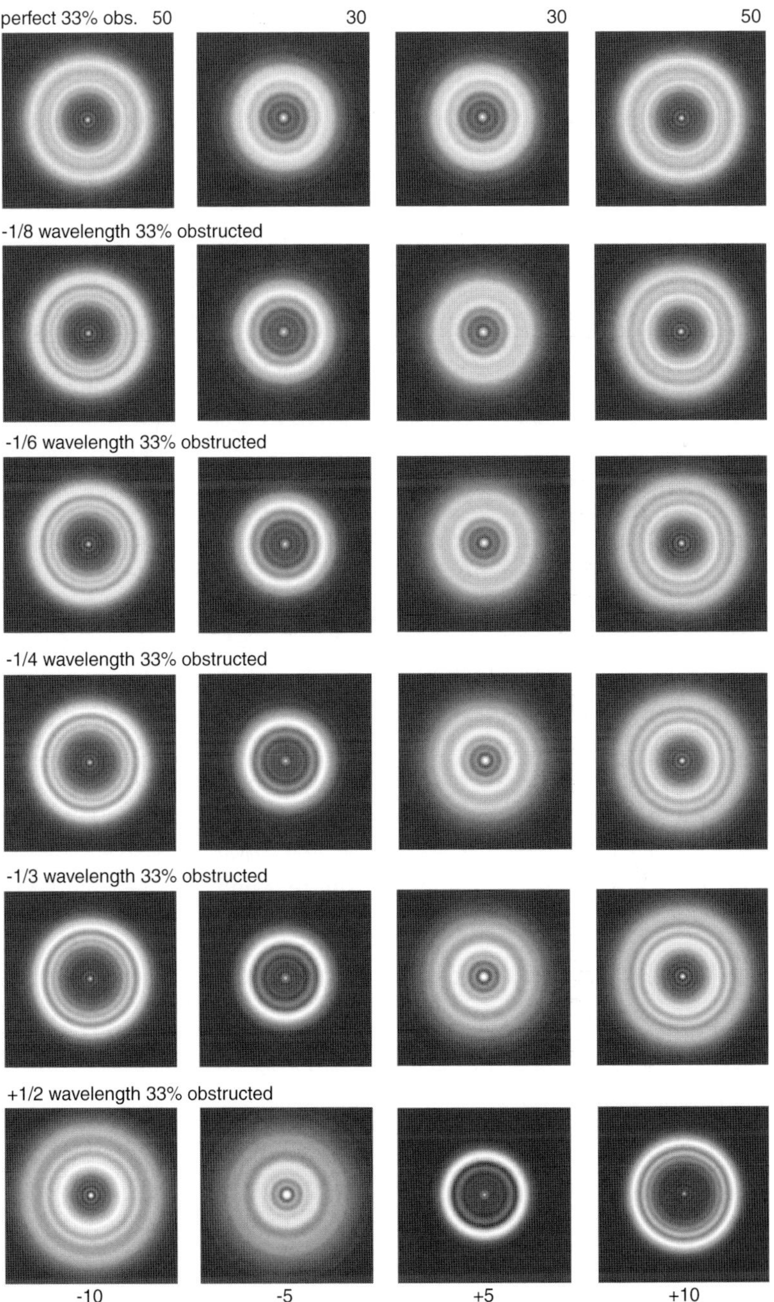

Fig. 10.12 The same as Figure 10.11, except now the obstruction is 33% of the diameter.

and obstruction into separate compartments, treating the two as incomparable phenomena. However, a single standard can easily be defined to cover them both. It is based on the *encircled energy ratio* (or EER(θ)). The encircled energy ratio provides a way of comparing these two degradations on an equal footing.

Please note that all these ratios go to unity as radius goes to infinity. They are corrected for any simple dimming of the aperture by a secondary mirror, apodization, or reduced transmission. The degradation of the image caused by diffraction is more pertinent than a simple transmission loss. Remember the extinction paradox of van de Hulst mentioned in Section 3.2.4. About half of the lessening of the central intensity is caused by shadowing (light that hits the rear of the secondary), and half is caused by the damaging effects of diffraction. Judging a telescope for an effect that does not increase the point spreading is unfair, so it is normalized out. Besides, if we are going to talk of the absolute transmission, we also must know details of the coatings and internal reflections (a real problem in the case of solar telescopes). These are unknown in the generic case, so leaving in the dimming caused by the secondary is an incomplete treatment.

Here's the way such ratios are calculated: First, we find what fraction of the total energy from a point source is focused by the imperfect telescope on a tiny circle of specified angular radius at the focal plane. The fraction always normalized to unity at large radius. This number is then divided by the same normalized fraction for a perfect, unblocked aperture of the same diameter. For example, a moderately obstructed telescope that also has a trifling amount of spherical aberration encircles 72% of its energy at a certain angle and a perfect aperture encloses 84% at the same angular radius. The encircled energy ratio would then be 72/84 = 0.86 at that angle.

The encircled energy ratios appearing in Figures 9.1 and 10.6 are complete curves. For a single number that represents a quality criterion, one needs to take the EER(θ) value at a specific value of θ. The question arises: what angle is best?

Unfortunately, no one angle is the last word on optical quality. We could choose an angle (or circle) very near the center of the image, or EER $(\theta \rightarrow 0)$. This number is the same as the normalized brightness ratio at the center of the diffraction disk. In fact, it is identical with the Strehl ratio in unobstructed apertures, and might be thought of as a generalized extension of the Strehl ratio that includes the diffraction effects of obstruction. EER taken near the center of the image, however, seems excessively tolerant of obstruction, as Figure 9.1 demonstrates. EER $(\theta \rightarrow 0)$ does not dip below 0.8—the cutoff point of good optical

Table 10.1

EER(1.22) for apertures with lower-order spherical aberration. The wavefront is refocused for best RMS on the annulus. Obstruction is the fraction of the diameter covered.

Peak-to-valley correction error on unobstructed aperture

Obstruction	0	$\frac{1}{8}\lambda$	$\frac{1}{6}\lambda$	$\frac{1}{5}\lambda$	$\frac{1}{4}\lambda$	$\frac{1}{3}\lambda$	$\frac{1}{2}\lambda$
0.00	1.00	0.95	0.91	0.87	0.80	0.67	0.39
0.15	0.95	0.90	0.87	0.83	0.78	0.66	0.41
0.20	0.91	0.87	0.84	0.81	0.76	0.65	0.42
0.25	0.87	0.83	0.80	0.78	0.73	0.64	0.43
0.30	0.82	0.79	0.76	0.74	0.70	0.62	0.44
0.33	0.78	0.75	0.73	0.71	0.68	0.61	0.44
0.40	0.70	0.69	0.67	0.66	0.63	0.58	0.45
0.50	0.59	0.58	0.57	0.56	0.55	0.52	0.45

quality in the Strehl ratio—until obstruction is above 45%. One could also define the quality factor as EER within a circle of radius $\theta = 1.22\lambda/D$ (Equation 1.1), or the edge of the Airy diffraction disk. In the general case, we could define two or more criteria at multiple radii, but here the radius is taken at the edge of the Airy disk.

EER(1.22) of apertures mixing the two optical problems of obstruction and refocus lower-order (Zernike) spherical aberration have been collected together in Table 10.1. We see a very similar behavior to the Strehl ratio in the unobstructed top row. A quarter wavelength of spherical aberration still results in a degradation of EER(1.22) to 0.8. It is the second axis that is most interesting, however. It is possible to compare the loss of encircled energy ratio of obstructed, but otherwise perfect apertures. EER(1.22) = 0.8 for obstructions slightly less than 33% of the full diameter.

Notice that obstruction does not always diminish quality. The case of ½-wavelength correction error shows a curious flattening, with increasing obstruction serving (together with refocusing) to *cover up* the poor figuring.

My experience with a large number of telescopes having various amounts of correction error suggests the following empirical ratings. These cutoffs are necessarily hazy, and the "good" point is deliberately chosen to match the Strehl ratio (i.e., 0.8), above which optics are conventionally called "diffraction-limited."

1. 0.90–1.00 excellent to perfect

2. 0.80–0.90 good to excellent

3. 0.70–0.80 poor to good

The only acceptable instruments with EER(1.22) below 0.70 are special-purpose ones, such as astrocameras or richest-field telescopes. No instrument having ⅓ wavelength of correction error, even if unobstructed, reaches up to this minimum standard. No aperture with an obstruction slightly larger than 40%, even if figured perfectly, ever meets it. Keep in mind, though, that these degradations apply only within a single aperture. Considering the system concepts discussed in section 3.4, we see that introducing differing apertures complicates the comparison, particularly for small obstructions and minor aberrations.

10.5.1 Tolerable Errors

All telescopes are made with some spherical aberration. The perfect Newtonian paraboloid, for example, is an unattainable goal between an infinite number of prolate spheroids and an infinite number of hyperboloids. The question is whether the telescope suffers under the load. Once EER(1.22) is over 0.90 or so, spherical aberration is gratifyingly small and the optics could justifiably be called excellent.

We saw in Chapter 3 how modulation transfer functions stacked individually. Most obstructed telescopes are teetering on the brink already. It takes very little to push them over. By such logic, we should be intolerant of any correction error, but that attitude is unrealistic.

Commercial telescope optics have always been corrected to a tolerance of about ¼ wavelength. The way that accuracy is stated has changed, but commercial telescope makers still fabricate the same ¼-wavelength (or 0.8 Strehl ratio) optics they always did.

Let's recognize a simple fact. Making objectives to higher accuracy than ¼ wavelength is expensive. The scaling of price with quality is similar to the scaling of price with diameter. Incremental improvements in surface accuracy cost much more because we are paying not for glass but for an optician's valuable time. For better or worse (usually worse), buyers use price as a strong deciding factor.

Is more accuracy really needed? In informal tests, a telescope with a ¼-wavelength correction error has been found difficult to distinguish from a very good telescope unless seeing is excellent and the observer is skillful (Ceravolo *et al.* 1992; also see Chapter 15). For most people who observe under average skies, a ¼-wavelength correction error represents an acceptable compromise between quality and the price of optics.

In the previous section, we defined those apertures with EER(1.22) greater than 0.9 as excellent. We see this designation only applies to the upper left corner of Table 10.1; i.e., to obstructions less than about 21% or correction errors less than ⅙ wavelength. Notice that a 25% obstructed aperture with only ⅙ wavelength of correction error is still "good" at 0.8, but that a 15% obstructed aperture with a ¼-wavelength error is below the cutoff at 0.78. The lesson is clear. Accurate figuring allows the telescope to get away with other difficulties.

Personally, I find the images of optics that are nudged against the Rayleigh limit a bit too soft. However, of the telescopes I've tested, most of them that obviously didn't perform well on the sky have been much worse than Rayleigh's limit. A quarter wavelength of correction error is barely acceptable if it is the only significant problem. With a reasonable 25% obstruction, such an aperture has EER(1.22) = 0.73, and has a transfer function better than a perfect, unobstructed aperture half to two-thirds of its size. Even with optical problems of this magnitude, a 6-inch *f*/8 reflector is at least as good as a perfect 3- to 4-inch unobstructed refractor. At some spatial frequencies, it is better, and it certainly is brighter.

10.6 Estimation of Low-Order Spherical Aberration

Low-order spherical correction error of sufficient amplitude creates a marked contrast between the inside-focus and outside-focus star-test patterns. An experienced observer under excellent conditions can certainly detect errors smaller than ¹⁄₁₀ wavelength and possibly ¹⁄₂₀ wavelength (Welford 1960). Ironically, the star test conducted in high-sensitivity mode for spherical aberration is almost too sensitive. It is so revealing that nearly any telescope fails casual inspection. It is not until the tester carefully measures the amount of defocus that a proper sensitivity balance is achieved.

A high-resolution light detection system would allow measurements over the focused stellar disk and determination of exactly how the aberration affects the telescope. Because the eye is a terrible radiometer, it cannot be trusted to measure absolute brightness of a focused image.

People who use the eye to determine the magnitude of variable stars are only successful if they follow a careful procedure using similar comparison stars. Estimating brightness of extended objects (like defocused star disks or even diffraction rings) is hopeless. Meticulously calibrated light sensors have been used to perform this job, but such a solution requires precise knowledge of the geometry. It is not practical for those wishing to do a fast test. (For an example of these difficult measurements, see Burch 1985, and for analysis methods see Fienup *et al.* 1993.) One would also think it would be easy to measure the brightness of the first diffraction ring. If it measured between 0.0316 and 0.0175 as bright as the center of the disk, the telescope would be excellent. But the disk and ring system is always jumping around and the ring is much smaller than most electronic arrays. For the lone observer, a method should be chosen that uses the strengths of vision instead of its weaknesses, some sort of tool that does not rely on the eye's absolute ability to determine brightness.

10.6.1 Estimates of Large Amounts of Low-Order Spherical Aberration

The shadow spot gives testers a way to estimate the aberration. Figure 10.13a shows a longitudinal slice through a perfect aperture's focus point. The objective lens or mirror is to the left, the outside-focus direction is to the right (for an explanation of labeling, see Appendix D). Except for the spot activity along the axis, looking like beads on a string, the out-of-focus profile is almost smooth and uninteresting. In Figure 10.13b, we are looking at the same situation with a 33% obstruction. The aperture is otherwise perfect, and this situation is symmetric. We shall make 33% our standard obstruction, because it is larger than the obstruction in almost all astronomical instruments and can be achieved in cases of lower obstruction with a mask, whereas the reverse is impossible.

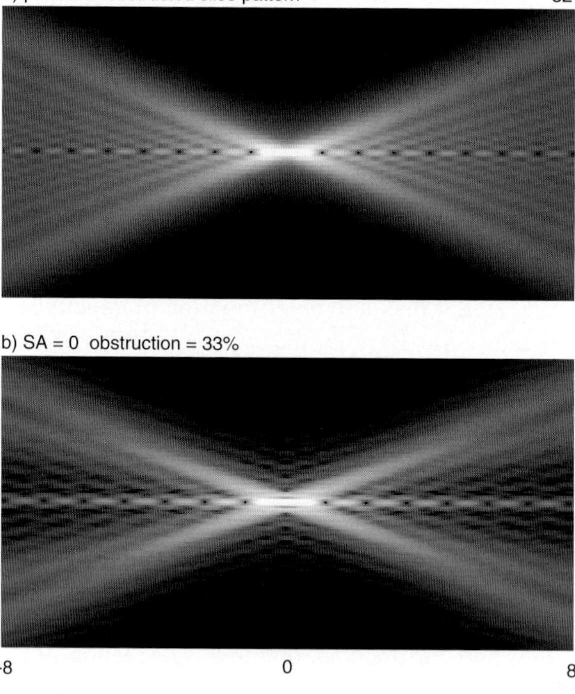

Fig. 10.13 A longitudinal slice through the focus of a) a circular unobstructed aperture, b) a 33% obstructed aperture. Neither pattern has any aberration associated with it. The slice is taken from defocusing aberration of –8 wavelengths to +8 wavelengths. The corner angle of $32\lambda(f/D)$ corresponds to the ray-tracing edge of geometric shadow ±8 wavelengths defocused. Thus, the picture has been squeezed until it resembles the cone of an f/1 system.

In dark cones emerging from the center, the shadow of the diagonal seems to break out of the defocused image at a finite distance close to either side of focus. Since the image is quite small, the bright spot at the center delays the steady appearance of the central obstruction until the defocusing aberration is close to 2 wavelengths on either side. (See Chapter 5 for the conversion of defocusing aberration to focuser motion.) The eyepiece must be moved a little more until the spot is clearly defined. Nevertheless, notice that the breakout points for a perfect mirror are balanced; they are the same distance on either side of focus.

What happens when we add some undercorrection to the obstructed aperture? The answer is shown by Figure 10.14.

The first point of interest is that the best focus point slides a little with progressively worse undercorrection. The aberration was entered as a Zernike polynomial, but even those functions have a slight focus shift for obstructed apertures.

The next peculiarity is the small size of the disk inside focus compared with that of the outside. This condition is caused partly by the obstruction focus shift, but it is noticeable in Figures 10.8 to 10.12 above, which have been corrected for this shift. No unique focal point exists for the aberrated wavefront. Approaching focus, the wavefront must buckle and change shape, manifesting itself in these different sizes.

Energy conservation also plays a role. In the longitudinal slice figures, a vertical line cannot be drawn that doesn't intercept an illuminated region. The intensity is never allowed to turn itself off everywhere in a sliced plane. In fact, if we very carefully keep track of the total energy at any value of defocus, we discover it to be the same as the total energy that passed through the aperture. The gnarls and knots are just an arrangement. An appearance of a bright ring is counterbalanced by a dark ring showing up elsewhere in the sliced plane.

The dark cones of the secondary shadow are no longer at equal offsets in the presence of correction error. In the ¼-wavelength diagram, the 33% central obstruction does not show itself until it is about twice as far from best focus.

Two effects are conspiring to offset the secondary-shadow breakout point. One is the clustering of energy around the rim of the horn-shaped caustic on one side of focus. This fierce excavation of energy from the center allows the secondary shadow to burst forth more quickly. The other effect is the pile-up of energy toward the mouthpiece of the horn on the other side of focus. This intensity fills in the secondary shadow and retards its reappearance. Not until the eyepiece has been moved well past the focus

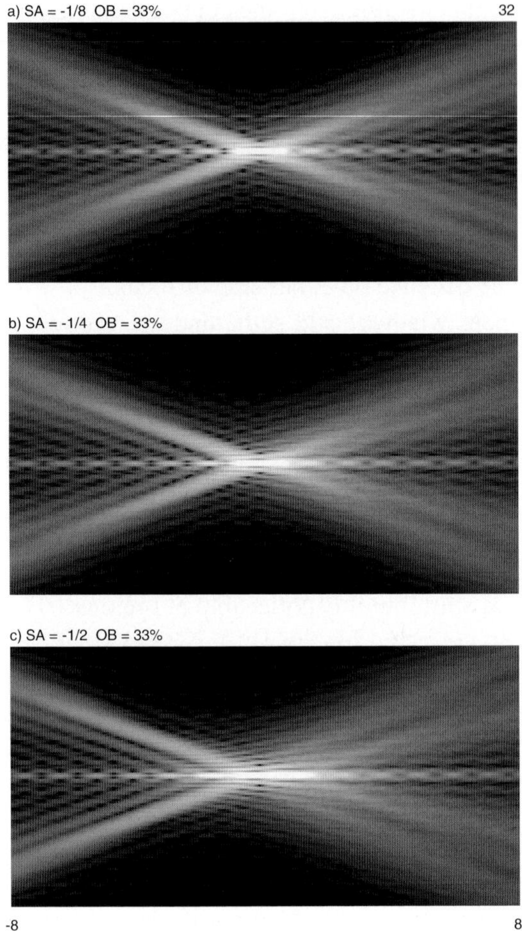

Fig. 10.14 33% obstructed apertures show the differing distances of the emergence of the secondary shadow from the center of the diffraction disk. Correction error is a) $-\frac{1}{8}$ wavelength, b) $-\frac{1}{4}$ wavelength, and c) $-\frac{1}{2}$ wavelength (all undercorrected).

region is the secondary shadow allowed to emerge.

The reappearance of the shadow can depend on the brightness of the star, the seeing, and the admixture of other aberrations. The tester must take into account the general performance behavior of the telescope before rejecting it for failure of the "2:1 test" alone. It must be performed only on apertures for which the low-order spherical aberration has been fairly well diagnosed (in other words, it fits the behavior previously described in this chapter – a doughnut on one side of focus and strong center brightness on the other), and only a yes-no decision is needed about whether the aberration is too severe. The 2:1 tolerance is satisfied pretty well for errors less than ¼ wavelength, but beyond ⅓ wavelength or so it always seems to be much worse, with the shadow appearing almost instantaneously on one side and remaining invisible for a long distance on the other.

10.6.2 Method of Estimating Small Correction Errors (Based on Technique of Ellison)

The purpose of estimating small amounts of low-order spherical aberration is to confirm that the aberration is indeed minimal and to get a fresh perspective on an inconclusive 2:1 test. Again you must have previously diagnosed that the aberration fairly well follows the star test images earlier in the chapter. Although diagnosis is easy, estimation is difficult. You should not expect high accuracy from this method. You certainly should not look to this method to provide bragging rights. Rather, you should employ it to confirm other information, like a manufacturer's specification or bench-test data. The goal is to have more independent information than you possessed when you started the star test. If a large inconsistency exists, then your bench test may be in error, for example.

A hint of the method appears in *The Amateur's Telescope*, by Rev. William F.A. Ellison, which was reprinted in *Amateur Telescope Making Book One:* (Ingalls 1976)

> It is easy enough to see, by the out-of-focus images of a star, what is the state of correction of the mirror. A truly corrected mirror, out-of-focus, will give an expanded disk, uniformly illuminated except for faint traces of diffraction rings, having a clean, sharply defined edge, and a round black spot in the center. This black spot is the shadow of the flat, and it should be *the same size at equal distances inside and outside focus.*
>
> If it is larger inside focus, the mirror is under-corrected. If it is larger *outside,* it is over-corrected. And many a time on a night when temperature was variable, the writer has watched a mirror *change through all these phases* within not very many minutes, the changes of the black spot answering faithfully to those of the thermometer... [italics in original].[2]

Ellison didn't possess a method of calculating the theoretical diffraction patterns, so he didn't know how much asymmetry to expect, but we do. We can use the information in the polychromatic star test in Figure 10.12 to make estimates of the amount of aberration.

Again, use the standard 33% obstruction. Make a mask if your telescope is obstructed lower than this amount. Even refractors may be

---

[2] As it turned out, Ellison's comments were imbedded in an argument that seemed critical of star testing. Perhaps many of Ellison's readers were confused by this discussion into thinking that the star test was ineffective. Ellison's point, however, was valid. The plate glass mirrors common at that time were untestable in an environment that was rapidly varying in temperature. Any test would have failed in this situation.

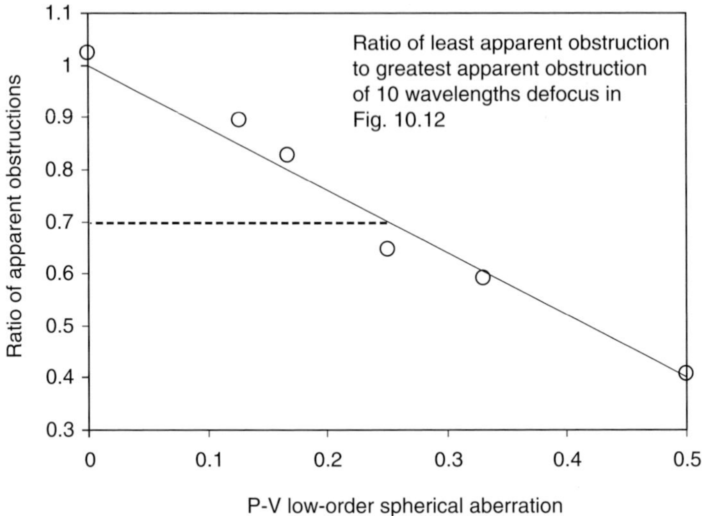

Fig. 10.15 Measurement of the ratio of the apparent obstruction of 33% obstructed case.

obstructed temporarily by attaching a mask on a string crossed over the dewcap. Defocus the instrument plus or minus 10 wavelengths by measurement (see Chapter 5).

Measurements were taken of the apparent obstructions of Figure 10.12, ratios of them were taken and result appears in Figure 10.15. For example, I measured the apparent obstruction fraction of the +10-wavelength defocus case of ¼-wavelength correction error to be close to the actual obstruction at 34% and the fraction of the −10-wavelength defocus case to be 53%. The ratio of these is 34/53 = 0.63. I do not expect all such curves to approximately follow a line, but note for this value of defocus and obstruction that a line does approximately fit.[3]

By estimating the fraction of the disk covered at each setting of defocus, and taking the ratio of the least over the greatest, we arrive at a number less than unity.

If the estimation error is 5 percent on each fractional obstruction, the propagation of error on the ratio (Pugh & Winslow 1966) simplifies to 7% of the ratio. Thus, your error in a fractional estimate of, say, 0.8 at the eyepiece might be within a window of 0.74 to 0.86. In this case, you would be in the range of safety from the point of view of the aberrations alone,

---

[3] For small values of correction error, the ratios are more likely to expand quadratically. For larger values, energy conservation would demand that the light is packed into the outer ring, so it would converge to about the inverse of the defocus.

although from the perspective of Table 10.1, it would depend on the true value of obstruction.

If you truly cannot tell the difference between the apparent obstructions on either side of focus at 10 wavelengths defocusing aberration, move to 5 wavelengths and make the estimate from there. As can be seen in Figure 10.12, the imbalance in the central obstruction at 5 wavelengths defocusing aberration goes bad somewhere between 0 and ⅛ wavelength of low-order spherical aberration (Strehl ratio between 1 and 0.95). Indeed, merely noting the amount of defocus where the error is still easily detectable in the 33% obstructed case is a first-order method of detecting it: if you have difficulty seeing the imbalance beyond 5 wavelengths defocus it means excellent correction, 10 wavelengths defocus means possibly mediocre correction, and clearly visible low-order spherical aberration at 15 wavelengths defocus means poor correction. Of course, if you do not put on a 33% mask, these numbers shift, as Figure 10.11 demonstrates.

10.7 Conclusion

At the end of the introductory survey of the star test (Chapter 2) I made the comment that you would repeatedly encounter telescopes having correction errors as you gain experience with the star test. This is because of the way that errors in the glass are applied. Correction errors are the most common low-order errors that are left there by deliberate application of a tolerance. Azimuthal errors of large amplitude — such as astigmatism, coma, and trefoil — are more common with external factors such as misalignment or poorly supported optics. They can actually be polished into the glass, true, but their presence is something that is usually diminished to well beyond the limits by good fabrication technique, which most manufacturers would hopefully develop. Only astigmatism in large, thin mirrors is seen as often as correction error, and it is easy to distinguish from the latter.

Perfection must be paid for, and no doubt there are some willing to do so. On the other hand, people who consider their telescopes a system must be prepared to accept a certain amount of correction error as a trade-off for finite cost. Techniques presented in this chapter enable the end user to sort their instruments into three categories: excellent, transitional, and poor. What has not been presented is a way of accurately quantifying the degree of error. For that you need to pay someone to do a bench test of the optical train, a move which may entail unacceptable costs, or you must learn the techniques of Appendix A to perform the test yourself. I encourage people who wish to learn shop testing to investigate these various techniques, some of which are fascinating, and some of which possess some equally fascinating blind spots. Those merely wishing to sort the optics into the above broad categories will find that the star test is perfectly adequate.

Chapter 11
Circular Zones and
Turned Edges

This chapter discusses higher-order spherical aberration, zonal defects, and a common type of zonal error, a turned edge. A zone is a circularly-symmetric ring or trench in the wavefront that is not characteristic of the surface as a whole, but is confined between a small radius difference. Higher-order spherical aberration is a whole-surface defect, but shares characteristics with the other zonal errors. Six chief points will be made here:

1. Higher-order spherical aberration, also called *zonal spherical aberration*, is not obvious in the star test as long as the coefficient is kept below 0.2 waves (⅛ wavelength peak-to-valley). The key factor in the star test is to recognize this aberration to verify that it is acceptably small.

2. Higher-order spherical aberration, although noticeable in star tests of certain designs in which it is the remaining slight residual aberration, practically never appears in strong, unadulterated form.

3. On mirrors of amateur size, interior zones are seldom large enough to be troublesome.

4. Zonal defects can be identified by defocusing a larger amount than is usual.

5. Turned edge is a persistent problem that yields contrasts worse than the smaller aperture inside the turned annulus.

6. Narrow turned edges can be treated by masking or painting the edge.

11.1 Higher-Order Spherical Aberration

Recall that in Equation 10.2, we gave the form of the first two terms of the Zernike expansion that describe spherical aberration. The higher-order equation bears repeating here.

$$W_6(\rho) = -\frac{A_6{}'}{2}(20\rho^6 - 30\rho^4 + 12\rho^2 - 1)\cdot$$

Sometimes the $A_6{}'$ coefficient is neglected or uncorrected. In most telescopes, this aberration makes no difference because the coefficient is very small, but it can be significant for some unusual instruments.

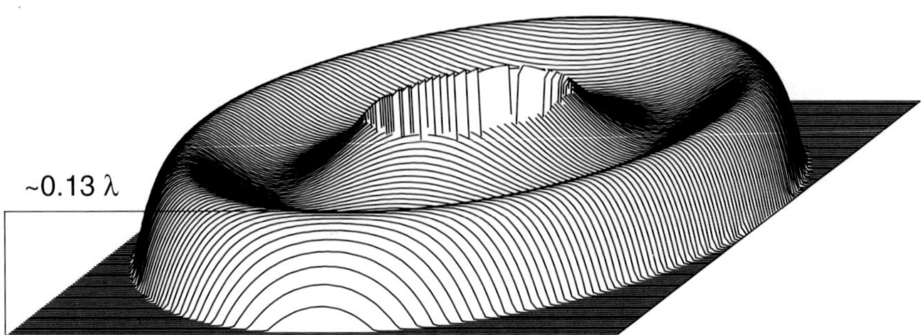

Fig. 11.1 The wavefront of a 152-mm f/12 all-spherical Maksutov-Cassegrain described in Section G.3. In spite of appearances, this aberration's effect is mostly that of a rolled-edge zone.

For example, the designed shape of Schmidt corrector plates, being limited to vacuum-pan deformations or hardware constraints, is often only fourth order. Fast mirrors contain many orders of spherical aberration, but because the corrector has only fourth-order terms, it cannot correct the wavefront of 6th-order and higher polynomials (see central wavelength of Figure G.14). This residual amount of aberration is negligible in an f/10 instrument, but it becomes more important for very fast Schmidts.

Various refractor designs can also add trifling amounts of "secondary" spherical aberration (i.e., sixth order on the wavefront [Kingslake 1978, p. 114]). This residual spherical aberration is enhanced by fast lenses, which makes fast apochromats the likeliest site of this aberration in all-lens telescopes (see the central wavelength of Figure G.2). Still, the spherochromatism—that is, variation in low-order spherical aberration with wavelength—in most apochromats dominates the slight amount of high-order spherical aberration. It is visible in pure form only at the central wavelength, as was the case with the Schmidt.

The most common source of noticeable high-order spherical aberration is in consumer aluminized-spot Maksutov-Cassegrains, in which every color has about the same amount. First, a word of explanation about the "consumer" name is necessary. From the beginning days of Maksutovs, it has been a well-known fact that they are troubled by higher-order spherical aberration (Maksutov 1944). The question of how this aberration is handled is important. It can be

    a. hand-retouched (as in the f/15 instrument of Gregory 1957), with a great deal of meticulous work,

    b. redesigned into a more elaborate telescope (which costs money), or

    c. ignored (thus being consistent with mass production and low prices).

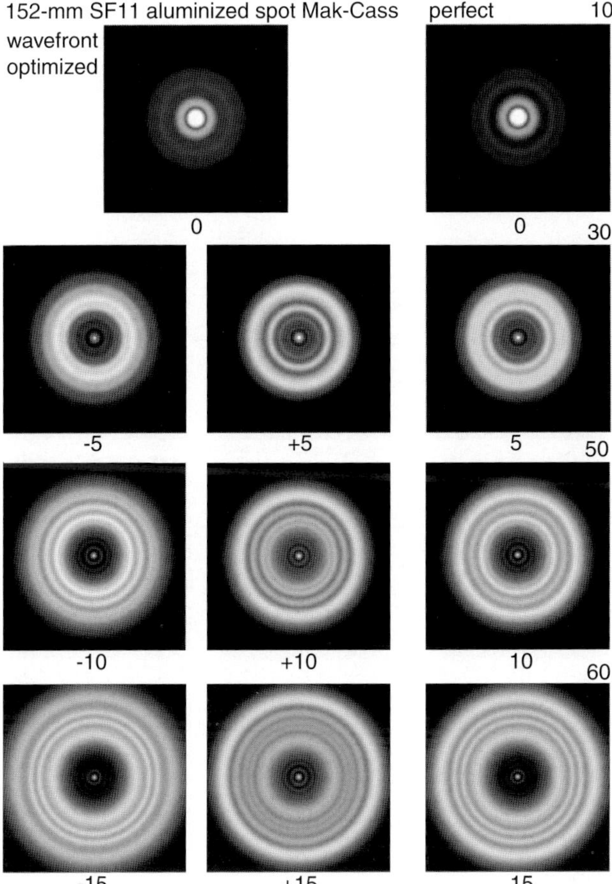

Fig. 11.2 152-mm $f/12$ aluminized spot Maksutov-Cassegrain designed by optimizing the wavefront. Considerable differences can be detected in this aberrated aperture. Image is grayscale polychromatic.

Strategy "c" works well when the telescope is small or slow (90 mm and of aperture ratios greater than $f/12$), but is less successful when the attempt is made to scale the instrument to large size or low aperture ratio. The testing, reworking, or more elaborate design alternatives ("a" and "b" above) represent the difference between "consumer" Maksutovs and the premium kind.

11.1.1 Star-Test Pattern of Consumer Maksutov-Cassegrains

To start off the examples, let's look at the first 152-mm $f/12$ Maksutov-Cassegrain, with the design discussed in Section G.3. The top of Figure 11.1 is the best-focus wavefront of this system. This instrument is

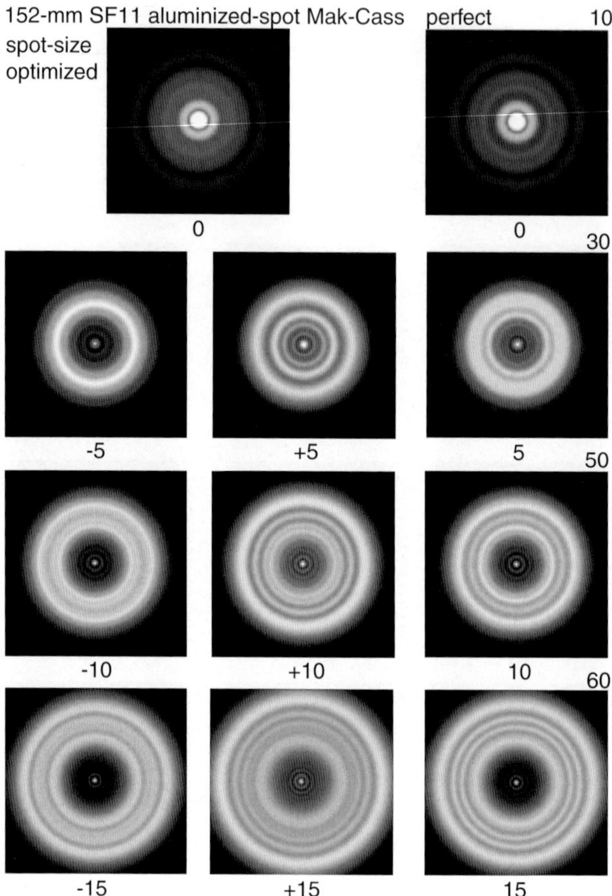

Fig. 11.3 Similar 152-mm f/12 all-spherical aluminized spot design, this time optimizing on RMS spot size instead of wavefront. This design choice lessens the apparent turned edge and increases the low-order correction error. Grayscale polychromatic calculation.

somewhat beyond the limits of what is realistic or advisable. To get the aberration as low as it is, I had to design it with a notional and unlikely high-index corrector (i.e., Schott SF 11). With this unusual element, at 560 nm it still has a peak-to-valley amplitude of a little over ⅛ wavelength. Most of the wavefront deviation has the same effect as a rolled-down outer zone. It should be emphasized that this design is not very good when pushed as large 152-mm. It is very much a "c"-type telescope and is much more likely to be found with an aperture of 90 mm, where the residual aberrations diminish next to the enlarged Airy disk. It appears here to make it directly comparable to the other 150-mm Maksutovs, although at least one inexpensive 6-in f/12 Maksutov-Cassegrain has been seen on the market. The way this commercial instrument reduces the considerable geometric aberrations inherent in a simple crown-glass $f/12$ design is unknown to me. The star-test pattern of a consumer all-spherical Maksutov-Cassegrain appears in Figure 11.2.

The design of Figure 11.2 was optimized on the RMS deviation of the wavefront. Here we see considerable inside-to-outside differences in the star-test pattern, and some may cry foul that an aperture that has a Strehl ratio of around 0.95 should show differences so readily in the star test. However, there are many good features of this star test, too. Although there are differences in the numbers of rings, the ±5 wavelength cases are surprisingly balanced. At ±10 wavelengths, it may look as if a trifle of low-order spherical aberration is showing, but that may well be the case since the wavefront RMS optimization did not take account of the obstruction. At +15-wavelength defocusing aberration, the disk displays a nicely flat distribution except for the characteristic outer-ring brightening of an edge zone. Any tester who possessed such a Maksutov could look for this behavior and ignore it, and in viewing the patterns a few seconds later rather than side-by-side, many observers would ignore it anyway. The −15 wavelength image is so flat that the diminished intensity of the outside ring would probably go unnoticed.

By optimizing on other criteria, different mixes of the aberration can be achieved. One such design appears in section Section G.6.1. The net affect of these different mixes diminishes the rolled outer-zone effect in a trade-off toward introducing more low-order spherical aberration. It is equally detectable in the star test (shown in Figure 11.3)

11.1.2 Premium Commercial Maksutov-Cassegrains

The instrument in the previous section does not represent the fine optical capabilities that the name Maksutov has come to mean. It is a compromise with manufacturability, and must be judged on the basis of cost. Its combined EER(1.22), including the diffraction of the 30% obstruction, is 0.86. It is a good telescope, not a perfect one, and it star tests imperfectly.

If we want better, we are going to have to design better. There are a number of things that can be done, but two of the most popular are to separate the secondary from the back of the corrector (giving it a different curvature) or to put a conic section of revolution on one of the optical surfaces. This is different from retouching one surface until the wavefront is flat, because the target conic constant is part of the design.

Separating the secondary from the back of the corrector and viewing its curvature as a new degree of freedom leads to a great improvement of

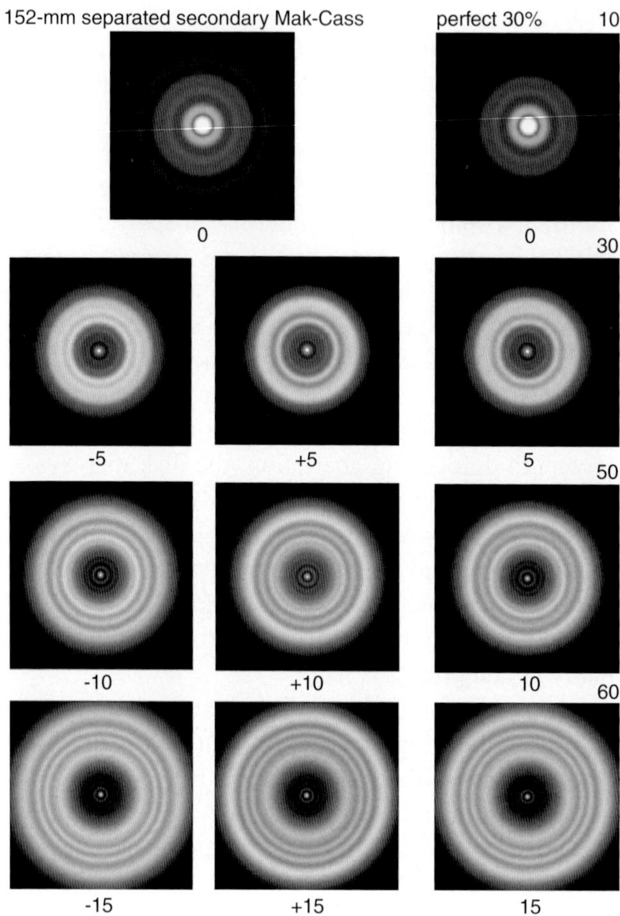

Fig. 11.4 Star-test pattern of 152-mm $f/12$ separated secondary Maksutov-Cassegrain. Differences between inner and outer defocused images are very slight. Grayscale polychromatic images.

the optics. The trend is to call such telescopes "Rumaks," after their popularization in the book *Telescope Optics* by Rutten and van Venrooij, but they first appeared in the drawings accompanying Maksutov's English-language 1944 paper. The peak-to-valley wavefront error inherent in the representative design of Section G.4 diminishes to about $1/20$ wavelength and the polychromatic Strehl ratio is 0.99. The encircled energy ratio at the edge of the Airy disk (including the diffraction effects of the obstruction) is 0.90, reaching the cutoff between good and excellent optics. The diminishment in EER(1.22), in fact, is almost entirely due to the obstruction. On-axis optics are almost perfect in this all-spherical instrument. For the star-test pattern, see Figure 11.4. The inside and outside focus regions are

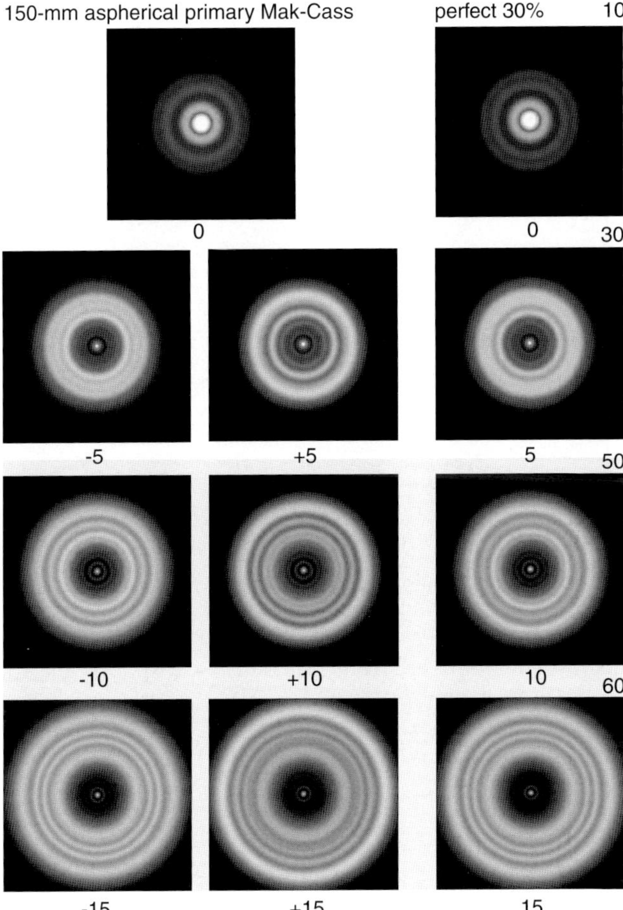

Fig. 11.5 150 mm f/12 Maksutov-Cassegrain with aspheric primary mirror. There are slight differences from the perfect pattern, but they are very difficult to see in practice. Grayscale polychromatic image.

nearly identical, particularly beyond 10 wavelengths.

Another strategy is to deliberately make one of the optical elements aspherical, but retain the aluminized spot on the rear of the corrector. In Section G.6.2, an example of such a design is described, and the star-test pattern is given in Figure 11.5. This star-test pattern may show differences between the inner and outer images, but these are delicate differences that when viewed are more evident in these side-to-side pattern comparisons than in serial order.

This instrument has a Strehl ratio of about 0.98 to 0.99. It is only slightly inferior to the separated secondary Maksutov-Cassegrain on-axis,

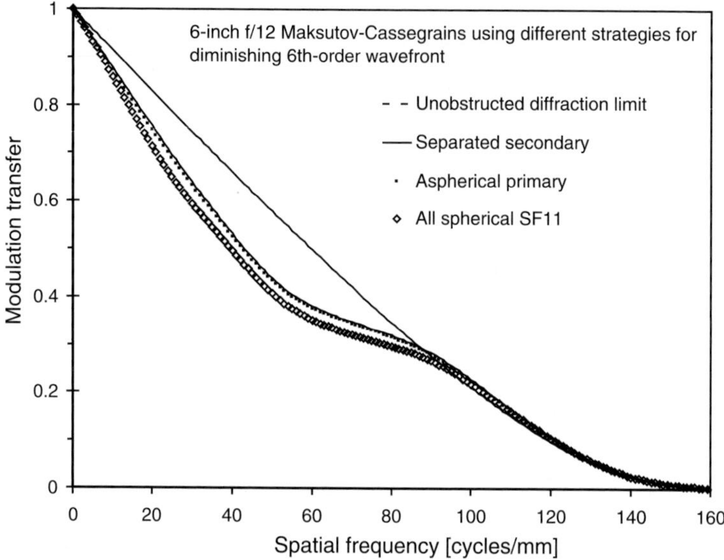

Fig. 11.6 Polychromatic MTF patterns of three 152-mm f/12 Maksutov-Cassegrains. All the instruments are fairly good.

but more resembles an all-spherical Maksutov off-axis; it exhibits similar coma, but this is hardly surprising as the addition degree of freedom (the conic constant of just the primary) has little power to affect it.

11.1.3 Filtering Pattern of the Three Maksutovs

The on-axis filtering of these three variants of the Maksutov-Cassegrain design appears in Figure 11.6. The patterns of the separated-secondary and aspherical Maksutovs are close together and closely hug the obstruction-only MTF (not shown). Only the MTF of the consumer instrument droops somewhat below it.

These patterns differ in one respect from the MTFs characteristic of low-order spherical aberration. The filter graph for higher-order spherical aberration, when it is stripped from the interference of the obstruction, appears in Figure 11.7. Not until the aberration has been increased to 0.4 wavelength does the damage reach a Strehl ratio of 0.80. The worst part of the decrease occurs at a lower spatial frequency than it did in Figure 10.5 for low-order spherical aberration. In this case the sharpest part of the drop occurs at less than about 20% of the maximum spatial frequency instead of 35%. Recalling that the maximum resolution of a 200-mm aperture is about 0.6 cycles/arcsecond, this aberrated system transfers surface details separated by less than 3 arcseconds with reduced contrast.

When the "elbow" of the MTF curve appears farther to the left, it is a

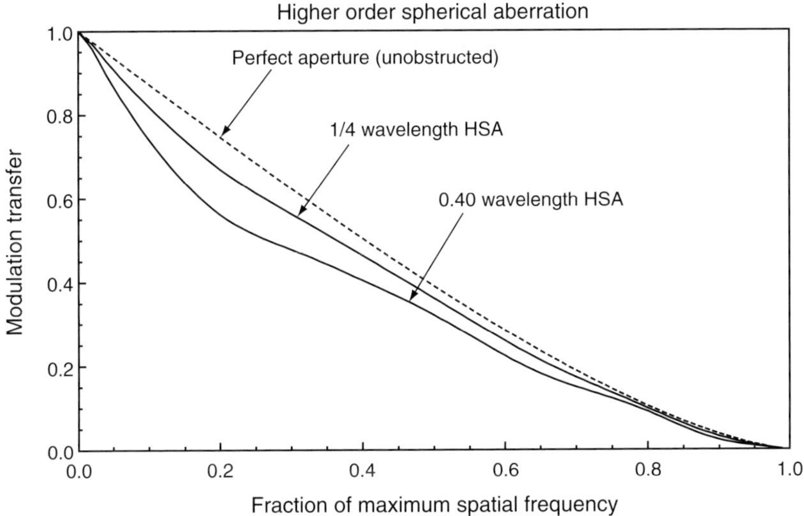

Fig. 11.7 Pure form higher-order spherical aberration without interference from obstruction.

sign of a more corrugated appearance of the aberration function, and it is also a sign that the removed energy is distributed into a wider pattern. The surface error becomes more localized and the diffraction pattern is wider. As the optical error becomes smaller and goes through more oscillations, the corresponding MTF exhibits a sharper decline at lower spatial frequencies. As the optical errors become more localized, the MTF at higher spatial frequencies is also reduced but does not oscillate much. The damage is already done at lower spatial frequencies. However, it should be emphasized that pure higher-order spherical aberration of such a magnitude as to reduce the Strehl ratio to 0.8 is unlikely to trouble ordinary instruments. If the optics exhibit fabrication error, the bulk of the aberration is usually expressed in simple fourth-order wavefront error.

11.1.4 Encircled Energy Ratio of High-Order Spherical Aberration

The encircled energy ratio of higher-order spherical aberration (Figure 11.8) is very similar to low-order (Figure 10.6), but the amounts are superficially different. Even though the coefficient in Equation 10.2 that reduces EER(0), or the Strehl ratio to 0.8, is about equal to 0.4 wavelength, this wavefront can be refocused to a minimum peak-to-valley wavefront of ¼ wavelength. Thus, the minimum peak-to-valley wavefront errors of similarly destructive low-order and high-order spherical aberration (reducing Strehl to 0.8) are each about ¼ wavelength peak-to-valley.

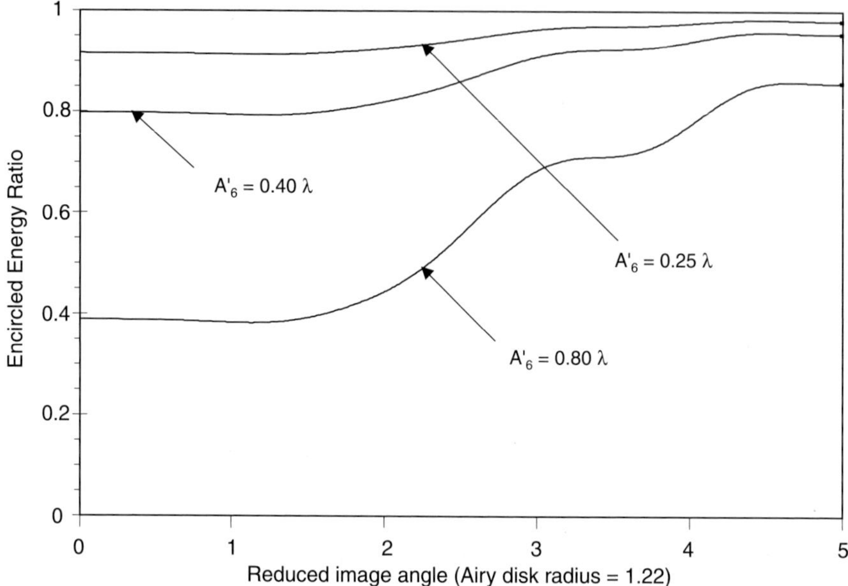

Fig. 11.8 Encircled Energy Ratio of pure high-order spherical aberration (unobstructed) is similar to low order in that the inside of the Airy disk has largely the same shape as the zero-aberration case.

11.1.5 Judging the Amount of Higher-order Error

As with low-order spherical aberration, methods to evaluate higher-order spherical aberration should concentrate not on detection but on an appraisal of whether or not the aberration is excessive. We have already seen how any well-designed telescope should have a Zernike coefficient of no more than 0.2 wavelength (recall Equation 10.2), or a refocused minimum peak-to-valley wavefront error of no more than ⅛ wavelength. Higher-order error is identified by its likelihood (you should look for it in any Maksutov-Cassegrain) and the shape of differences at the edge.

Figure 11.9 displays the behavior of a full 0.4 wavelength of high-order spherical aberration (that is, Strehl ratio dips to 0.8) in another 30%-obstructed telescope. Notice that the aberration is obvious, even at 15 wavelengths. Also, the aberration more resembles low-order overcorrection, and may be mistaken for it.

The primary distinction between the more or less tolerable aberrations seen in the Maksutovs earlier in the chapter, and in this pattern, is in the behavior for the region of 10 to 15 wavelengths of defocus. In the case of the acceptable Maksutov-Cassegrains, the residual aberration has subsided for such defocus values. In the barely-tolerable case, the residual aberra-

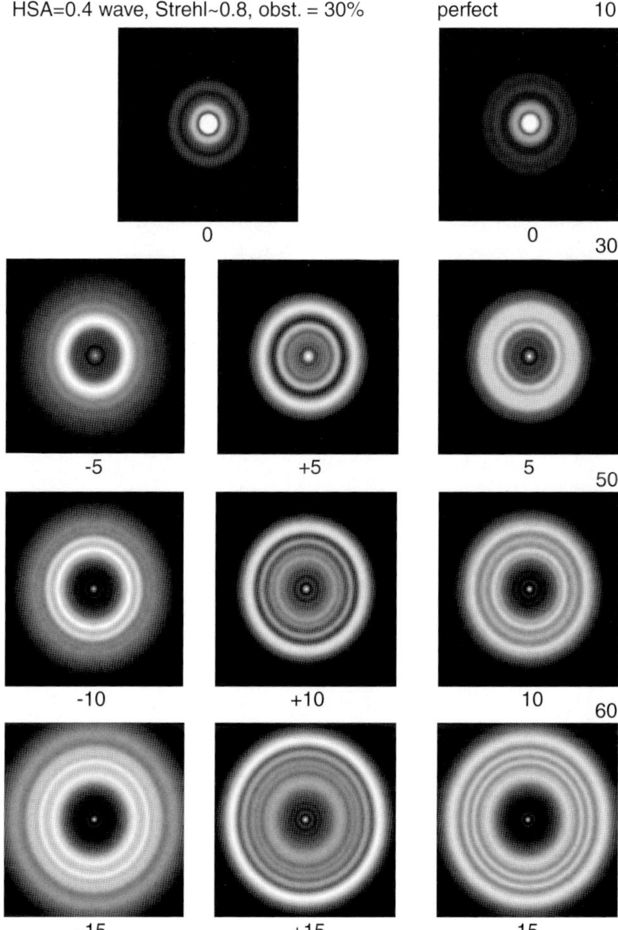

Fig. 11.9 The star-test pattern shows noticeable asymmetries as far out of focus as 15 wavelengths for a coefficient of 0.4 (Strehl = 0.8). Polychromatic calculation.

tion is still in evidence. The lesson is clear. When judging higher-order spherical aberration, using too little defocus is an impediment. Judge it out at 10 or more wavelengths.

Another comment can be made. I have never seen a pure high-order aberration as large as that appearing in Figure 11.9. When the amplitude of higher-order spherical aberration is as large as Figure 11.9, it is generally mixed with other aberrations.

11.2 Causes of Other Zonal Defects

Zones can originate from causes besides uncorrected higher-order terms in the design. The improper use of fast polishing materials or small polishers may lead to zones in optical elements. For example, the lap is typically pressed against the optical piece to achieve uniformity of polishing action. If too little pressing is done, or if part of the lap overhangs during pressing, sections of the polisher can ride the optical piece with more pressure than the rest. Because the lap and the stroke direction rotate with respect to the mirror, this uneven pressure digs a trench around a certain radius of the mirror. Many other mechanisms can also result in zonal defects. When too short a stroke is used, good statistical averaging of the two surfaces doesn't take place. Channeling a lap with a centered pattern will often result in a profusion of thin rings. If a piece of the brittle pitch breaks off and is trapped beneath the rest of the lap, it may plow a furrow in the mirror during the next few minutes until the piece is forced back into the lap.

However, these causes are minor compared to the chief reason for zones. Fast aspherical mirrors demand the use of smaller polishers. The old style of optical work (common in the days of long-focus mirrors) involved the use of two identical disks. Telescope makers modified their stroke slightly for a few minutes, but still used an equal-diameter lap to achieve the aspherical figure. Unfortunately, that method won't work on mirrors of low focal ratio. The sphere is so different from the correct shape that the maker struggles for proper conformance. A lap of equal size won't ever reach the correct form. The natural tendency toward statistical averaging will keep dragging its curve back toward a sphere. Also, full-diameter laps are cumbersome when the optical piece is large, regardless of its focal ratio.

The optician chooses a sub-diameter lap, usually one about half the size of the disk being worked (unless the optical piece is really huge—then it's even smaller). The aspherical shape can be approached merely by rubbing the center of the disk more than the outside. Using a smaller lap is dangerous though. The optician must enforce an artificial randomness on the polishing machine to prevent the unnatural precision of the mechanism from digging trenches at fixed radii and, by implication, leaving ridges at other radii. A partially worked surface must be blended or smoothed.

Happily, interior zones on most commercial mirrors are detectable in sensitive bench tests, but they are usually so slight that they are unseen in the eyepiece. Most common is a small depression or nipple at the center, an error that is largely obscured by the secondary mirror. In leaving this error untreated, the optician is saving time and money by ignoring an error that will not be illuminated. (A central zone is shown in Figure A.3.)

Another common type of interior zone consists of one or more ghostly thin rings appearing about halfway out. They can usually be seen in the

Foucault test when the surface figure is very near a sphere, but such an error is only cosmetic. The slight ringing is evidence that the blending is going well.

On the other hand, if the maker is hurried and testing is inadequate, a severe zone may be left in the mirror. One form of interior zonal defect has two competing radii of curvature—one inside the zone, the other outside the zone. This condition may be more damaging to the image than light scattered from the vicinity of the zone itself. Light deflected from the immediate area of the zone will appear as diffuse glow if the zone is sharp enough, but these large areas of the mirror on either side of the zone are directing a great deal of light onto interfering focal points. They cover sufficient area that light is attempting to come to two tight disks at different focal distances.

The most debilitating form of a zone is turned edge. It can result from even a full-diameter lap. It is caused by excessive wear at the edge of the disk during polishing. If too much pressure is applied while the tool is teetering on the edge of the optical surface, or if the lap is not maintained in good conformance to the shape of the disk being worked, a turned edge can result (Texereau 1984). Turned edge also seems to be a problem associated with a rocking motion of the mirror disk during polishing.

Turned-down edge, because it happens at the very periphery of the optical surface, is not limited in amplitude. Interior zones are temporary intruders. If good contact between the mirror and tool is maintained and the stroke is not too short, the averaging effects of many strokes at many different angles will eventually average the zone out. It will be automatically blended away.

Turned edge, on the other hand, is derived from bad figuring habits or improper use of materials. Once it starts, the cause generally doesn't go away. It just keeps on occurring or even becomes worse. Turned edges are usually deep, and since the edge is on the perimeter of the optics, it covers a surprisingly large fraction of the aperture's surface area. A 5% turned-edge zone is struck by about 10% of the light incident on the aperture.

Zonal aberration is related to the expression for general spherical aberration. Remember, Equation 10.1 for the wavefront involved terms like

$$W(\rho) = \text{constant} + \text{focus term} + A_4\rho^4 + A_6\rho^6 + A_8\rho^8 + A_{10}\rho^{10} + \cdots.$$

Normally, for global figuring errors, the coefficient of ρ^4 is largest, with a smaller coefficient of ρ^6. All of the remaining terms are much smaller. But a zone is a special case where one or more of the higher order

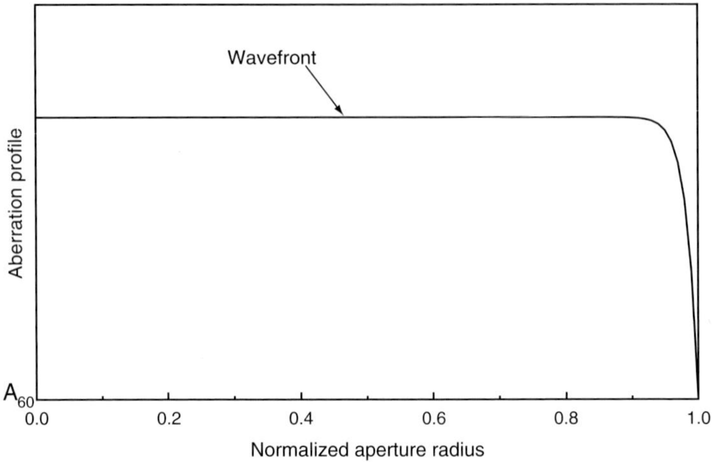

Fig. 11.10 Turned edge, as modeled by a 60th-order term of Equation 10.1.

terms contributes. A zone is like a switch that turns on very high-order spherical aberration, suddenly waking up errors that were best left sleeping.

A simple example is graphed in Figure 11.10. Turned edge is modeled as a term $-A_{60}\rho^{60}$. There is nothing special about the 60th order. Similar results would have been derived from 58th order or 62nd order.

Interior zones are more complicated. They are combinations of many high-order terms. Broad zones are described by lower orders than sharp zones.

11.3 Interior Zones

Let me emphasize at the outset to avoid frightening the reader; interior zonal defects on amateur-sized optics are rarely more than cosmetic defects. Although they are common during brief periods of fabrication, reasonably careful work is sufficient to lessen them. Few opticians would release a small optical piece with a significant interior zone on it. Large mirrors, however, are typically figured face-up with polishers much smaller than their diameters. Such optics often show persistent zones. These mirrors require more careful evaluation to determine if their zonal defects are negligible.

Two types of interior zones are considered here. The first is a narrow trench where the deformation is isolated or does not persist over the rest of the surface. Huge observatory optics often suffer from this type of zone, because the surface is worked by very small polishers. Usually, opticians

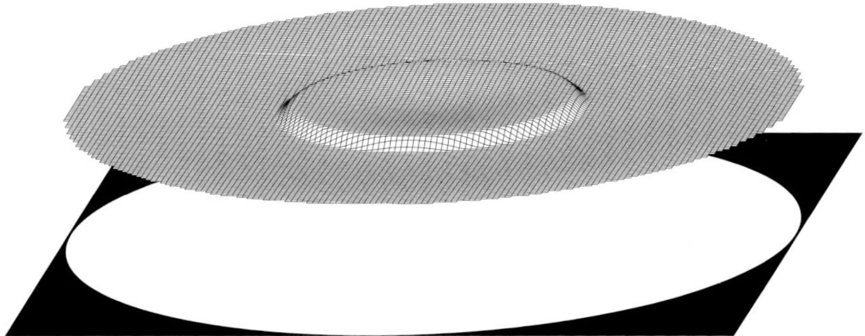

Fig. 11.11 A sharp S-zone at a radius of 40% of the aperture. It divides up the surface into two broad areas with different radius of curvature, and produces a sharply inclined scattering ridge.

working on large optics will not let conditions that generate these zones endure for long. Still, the skeletal remains of zonal grooves will sometimes appear on test photographs from big mirrors.

The star-test pattern in Figure 2.9 was calculated from such a ¼-wavelength trench zone. It was modeled as a narrow Gaussian-function dip in an otherwise flat mirror. The particular extrafocal patterns shown in Figure 2.9 had defocusing aberrations of 20 wavelengths. A general feature of the star test for zones is that one must rack farther out of focus to see their effects well.

A zone commonly produced in the fabrication of small mirrors has a profile that looks like an "S" or "Z," and is called an "S-zone" here. The zone remains after global figuring errors below sixth order are subtracted from the wavefront. This zone can have different radii on either side of its sharpest slope.

11.3.1 Aberration Function of S-Zones

Zones can be described by linear combinations of many spherical aberration terms, but that method is somewhat cumbersome. Instead, the zone is specified here with three parameters: amplitude, width, and radius. The optical surface is then divided into three regions, and the surface is constructed by fitting cubic polynomials through the points defined by those three parameters. At the border of each region, the slope of the wavefront is set to zero. Figure 11.11 shows a narrow zone at a 40% radius. A cupped plateau is inside the zone and a shallow dish surrounds it. Less evident is the slightly different curvature of those areas.

11.3.2 Filtering of S-Zones

Interior zones have two characteristic scales. Light reflected from the vicinity of the zone itself diverges into a broad fuzzy halo, so one anticipates that the modulation transfer function should dip rapidly at low spatial frequencies. When such zones have a large diameter, more or less like a big central obstruction, the MTF oscillates similarly at higher spatial frequencies.

The behavior of the MTF matches both predictions in Figure 11.12 for two radii and two amplitudes. One radius is just the 40% zone described in the last section. The other is the same zonal defect moved out until it is centered on 70% of the radius. Both zones are plotted with total aberrations of ⅛ and ¼ wavelength, and both zones have width equal to 0.1 aperture radii. The ¼-wavelength zones yield a slightly better Strehl ratio than the 0.8 value that marks the Rayleigh ¼-wavelength limit for spherical aberration. The ⅛-wavelength amplitude zones are well inside the 0.8 Strehl ratio tolerance.

The slight oscillations of the zonal defect MTF at high spatial frequency don't concern the tester much. More troublesome is the brisk drop at low spatial frequency. It is this degradation that Danjon and Couder were talking about when they made the distinctions between the slope of the defect and its amplitude (mentioned in Section 1.2.2). A smooth surface is the difference between merely adequate optics and those rare instruments that take the observer's breath away.

Notice another characteristic of these plots. A zone appearing at 70% of the aperture is worse than an equal-depth zone appearing closer to the center. At low spatial frequencies, the contrast drops more precipitously for the larger-radius zone. Lowered performance in this case is purely a function of how much area the defect covers.

Of course, interior zonal defects are not well described by peak-to-valley wavefront error. The RMS deviation is a much better way of characterizing zones. Zones are unacceptable when their RMS deviation is ¹⁄₁₄ wavelength, which corresponds roughly to the ¼ wavelength curve of Figure 11.12 at a radius of 70%. This deviation is much worse than any self-respecting optician would tolerate. For deviations half that size, the MTF only shows a slight dip. This amount of aberration generally is acceptable.

11.3.3 Detecting Interior Zones in the Star Test

In his 1891 work on star testing, Taylor gave a brief description of how he detected a zone in a refractor:

> "In order to detect such zonal aberration, which is caused by imperfect figuring of one or more of the surfaces, it is best to direct the telescope to a very bright star, using a moderately high power, and rack in and out of focus as before, only

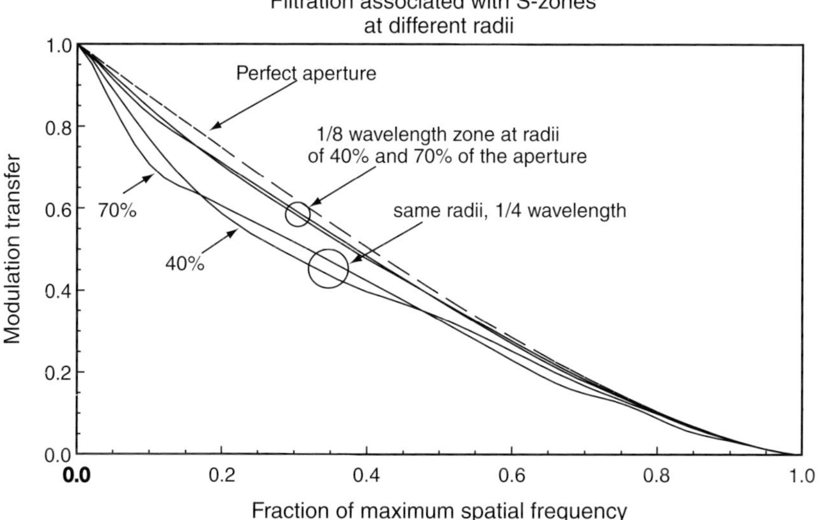

Fig. 11.12 Modulation transfer functions for a single zone. The MTF curves are for two values of zonal aberration (¼ wavelength and ⅛ wavelength peak-to-valley) and two zonal radii (0.7 and 0.4 of the full aperture). The aperture is unobstructed.

> it is best to rack out until 8 to 20 interference rings can be counted, for the irregular zonal effect is most easily detected under such conditions . . . Counting from the edge inwards, it may be noticed, for instance, that the outer ring is poor and weak, while the next one or two appear disproportionately strong, the next two or three weak, while those about the center are strong again. . ." (Taylor 1983).

Even though Taylor was writing about refractors and referring to smoother deformations that those modeled here, he described the two essential techniques to detect zones. First, bright stars are used to test for zonal defects. Second, the telescope is defocused farther than is recommended to detect other types of aberration. Defocusing long distances is especially handy because it lessens the effect of low-order aberrations. Also, because light near focus is still strongly mixed, the zone doesn't tend to isolate itself from the rest of the diffraction structure.

Star-test patterns for unobstructed optics with an S-zone at 40% the radius of the aperture and ⅛-wavelength total aberration on the wavefront are shown in Figure 11.13 (corresponding to one curve of the MTF chart in Figure 11.12). The defocusing aberrations are 10 and 20 wavelengths, whereas most of other aberrations appearing in this book are all depicted inside of 8 or 10 wavelengths. The telescope is a long way out of focus, but still this zone is easily visible.

The zone is most distinct at ±20 wavelengths, where it appears near

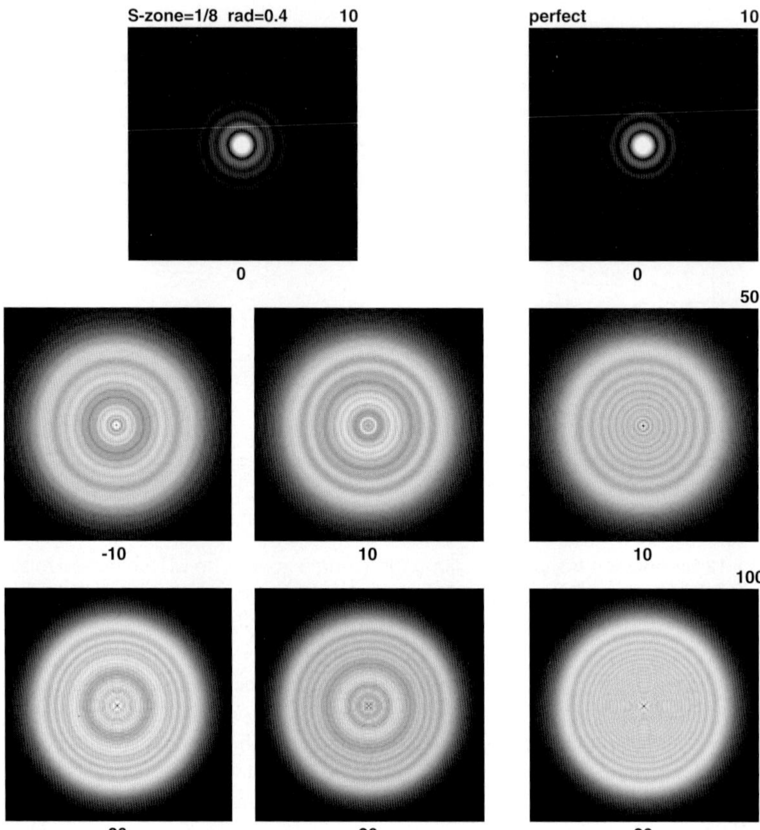

Fig. 11.13 Star-test patterns for an S-zone at 40% radius with amplitude of ⅛ wavelength. The perfect patterns are in the column on the right. The aperture is unobstructed.

its proper radius, but it has some anomalous features even here. It dips before it crests. In fact, the zone seems to be located at the crossover point between the extra bright ring and the depressed ring, presumably because it is an S-zone. Trench or hill zones are located at the correct radius but are bracketed by opposite-brightness rings. A dark area is always associated with a bright one because energy must be conserved. Energy that makes one ring look brighter always must be scooped from a nearby ring.

The patterns at ±10 wavelengths defocusing aberration show another interesting zonal property noticed by Taylor—the pattern at 10 wavelengths seems to be the complement of the −10 wavelengths pattern. At this value of defocus, the reversal is not perfect, only approximate. In portions of the disk far from the edge and close to the radius of the zone, the pattern seems to be a negative of the pattern on the other side of focus. Far

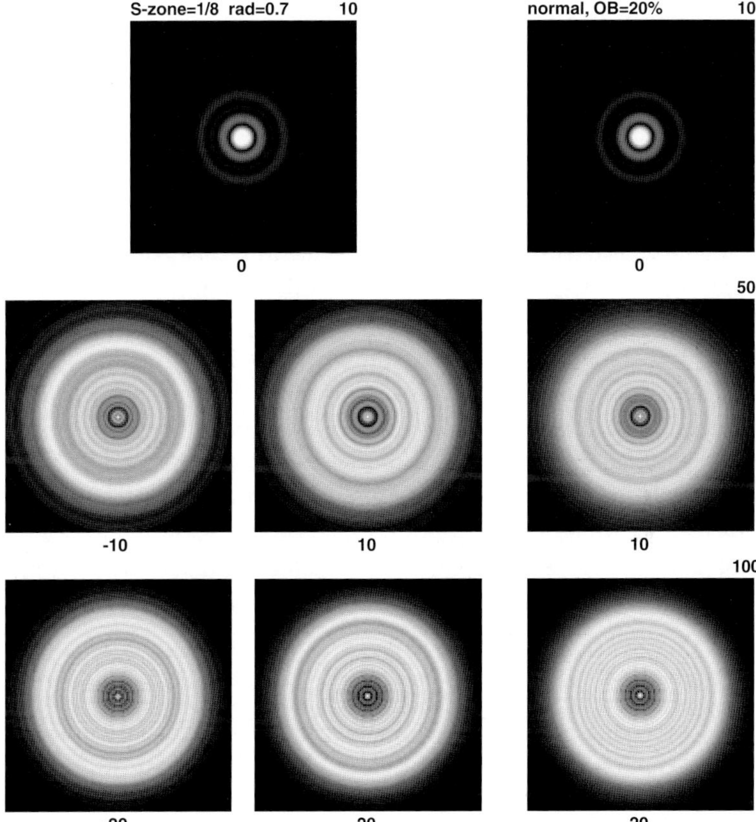

Fig. 11.14 Star-test patterns of an S-zone at 70% of the disk radius. Total aberration is ⅛ wavelength. Normal patterns are in the column on the right.

from focus, where diffraction effects are small, the reversal is more per-fect.

The in-focus pattern indicates that such a zone will frighten a tester more than it will disturb an image. The Strehl ratio is reduced only to about 0.95, well within the tolerance for excellent optics.

Another pattern appears in Figure 11.14. This time the zone has been moved to a radius of 0.7 of the full aperture and the aperture is obstructed by 20%. Here we see really complicated behavior. The localized effects of the aberration have not yet unmixed from the image at 20 wavelengths defocus. This fact teaches something important about identifying zones. Zones are easy to detect, but their radius and number are hard to pin down. The most we can say, unless very clean separation is seen, is that the sur-face exhibits zones.

In casual conversations, one hears star-testers speaking confidently about locating zones, saying things like "I detected two zones, one at 50% and the other at 75% of the radius." These claims assume that the structure of the out-of-focus pattern is completely unmixed, which is probably false. Figure 11.14 is a calculated pattern from one zone at a known radius, yet this aperture seems afflicted with many zones. The multiple ringing will separate out at 30 or 40 wavelengths defocusing aberration (not shown) but here they appear in an obscure manner. Since this zone is so far out on the aperture, it interferes with the ever-present dark ring on the inside of the bright outer ring.

One final conclusion can be drawn from both Figures 11.13 and 11.14. The star test is almost too sensitive to interior zonal aberrations. The filtration of ⅛-wavelength total aberration of either situation is mild. The in-focus patterns are nearly the same as the unaberrated optics. Yet the zones look severe when defocused. If you can only barely detect the existence of zones with defocused optics, then you have nothing to fear. They will not damage the focused image in most dark-field observing situations. In fact, I have never seen a zone as bad as the one depicted in Figure 11.14.

11.4 Turned Edges

Turned-down edge is common enough in both amateur and commercial mirrors although it usually takes different forms. In amateur-made mirrors, it is often wide and shallow, starting at a radius somewhere around 80% or 90% of full aperture and rolling gradually toward the edge. In commercial mirrors, turned-down edge is usually right at the perimeter, but it is steeper. The reasons for this dichotomy are varied and obscure. Perhaps amateurs, who usually work by hand, are capable of putting less pressure on the mirror, or maybe they use a pitch with different working characteristics. Commercial mirrors are likely polished with stiff pitch and hence are not prone to rolled edge, but the machines are capable of putting greater force on the tools. It all depends on each optical worker's peculiar methods.

For the purpose of this chapter, which is less concerned with mirror making than mirror testing by observers, the type of turned edge described is the narrow one. Wide turned-down edges will be classified as zones that happen to be at the boundary of the mirror. Their behavior at best focus is similar to simple spherical overcorrection. Since lower orders of spherical aberration (4th and 6th) have already been discussed earlier in this chapter and in Chapter 10, this section will concentrate on the other limiting case.

Turned edge seems to be more prevalent in fast, large, or thin mirrors, but this rule is not rigid. It appears often enough in slow, small, or thick

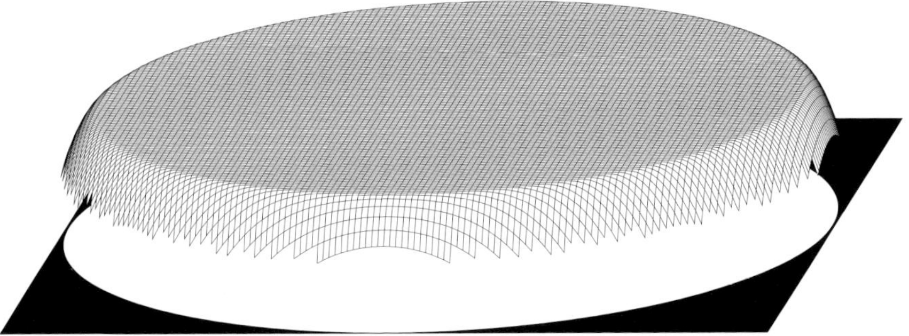

Fig. 11.15 An example aberration function of a turned edge.

optics. One mirror-making author advises that during design of large thin-mirror telescopes, observers plan a larger surface than will be expected in the final instrument. Then they can cheerfully (and somewhat fatalistically) mask the far edge (Kestner 1981).

11.4.1 Aberration Function

Again, a turned edge can be handled by a very high-order term in the spherical aberration equation. Here we choose a much easier path. It is much more convenient simply to allow the mirror to be flat out to a certain radius and then start a quadratic fall toward the edge. The quadratic nature of the downward trend is not based on any physical evidence or theory of the way these edges are put on optical surfaces. It is chosen arbitrarily. Other ways to describe the descent were tried, and the results didn't change much.

Figure 11.15 shows such a turned edge as a skirted table. This figure is not quite accurate. Best focus for turned edges demands that the inner flat area become a very shallow bowl. The aberration functions actually used to generate the patterns were modified so that they had a minimum variance.

11.4.2 MTF of Turned Edge

The filtering of a turned edge is shown in Figure 11.16. This aberration is very compact and highly sloped, so it is no surprise that the light reflected from the edge is diverted far from the center. The MTF drops quickly to reflect the damage to the widely spaced bar patterns of a low spatial frequency target. At the other end of the scale, the contrast preservation at high spatial frequency is much like that of a smaller aperture with a radius the size of the unturned area. Light is driven so far from the central core that it doesn't interfere with the focused spot, but the outer portion of the aperture is not contributing appreciably to the image. A telescope with a

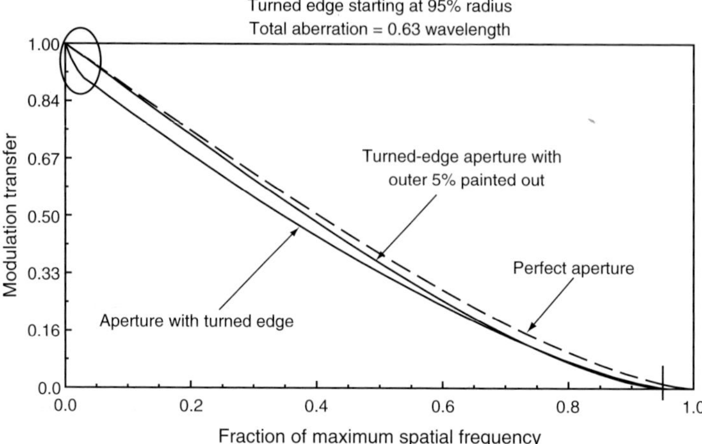

Fig. 11.16 Filtration of an unobstructed aperture caused by an example turned edge that reaches 0.63 wavelength at the far edge. The flat area inside covers 95% of the diameter. The Strehl ratio is slightly higher than 0.8, so this error is about the maximum tolerable. A comparison is made to an aperture with the turned portion masked or painted out. The MTF goes to zero at about 0.95 of the maximum spatial frequency just as the smaller aperture would.

turned edge behaves no better than a smaller telescope. Worse, at low spatial frequencies, the spurious light from the edge region actually harms the image.

11.4.3 Image Pattern of Turned-Down Edge

Figure 11.17 shows a focus run of the previous section's turned edge. Because refractor lenses are necessarily oversized and stopped down by their retaining rings, the effects of sharp turned edges are covered and not visible in the out-of-focus stellar images. Hence this figure has been generated assuming the usual case of a reflector with an obstruction. Gentler rolled edges could still extend beyond the retaining rings of a refractor, so be vigilant in watching for a variation of this pattern. In a refractor, the soft edges show outside of focus, the reverse of this pattern.

A diffuse glow spreads over the field of view inside focus. Contrast between the rings is appreciably lessened. The reverse is true as well—the contrast between the rings visible outside of focus is increased. Of course, this effect is easier to see with a filter that passes only one color (perhaps a deep green or a crimson red), but using such a filter requires a very bright source of light. Also, the edge of the diffraction disk softens inside focus. Turned edge, like other zones, is easier to detect at long defocus distances.

At greater distances inside focus, the hazy glow of turned-down edge

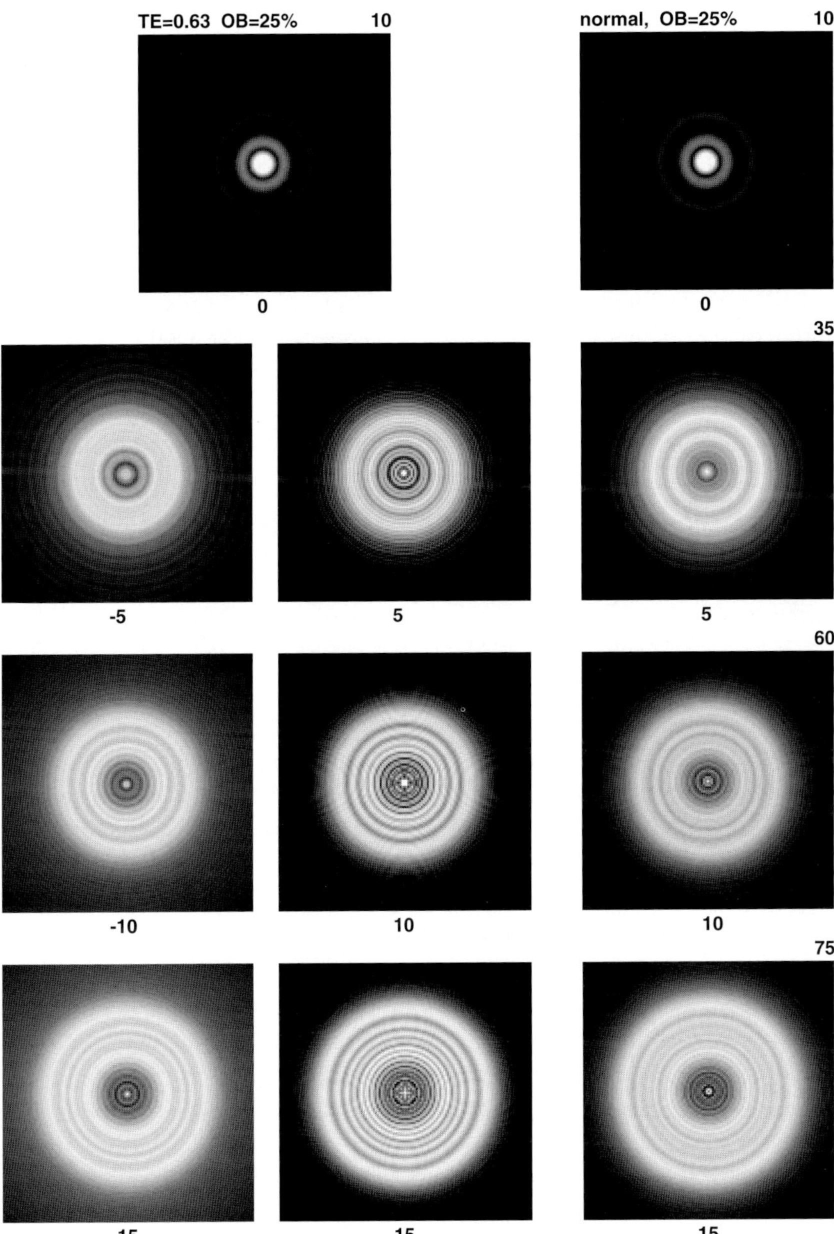

Fig. 11.17 Image patterns of turned edge starting at 95% the radius and having the value of −0.63 wavelength right at the edge. Obstruction is 25%, and the normal unaberrated patterns appear in the column to the right. Monochromatic image.

condenses into a smaller bundle, and the edge of the disk seems to bleed light into the outside. This behavior appears in Figure 11.18, where the defocus is shown at 30 wavelengths inside and outside of focus.

Other aberrations also render the diffraction rings more distinct on one side of focus than the other—an example is lower-order spherical aberration—but none of them show a uniformity of disk illumination. One might well wonder if a turned edge is detectable in the presence of spherical aberration or if it is too difficult to tease out of the confused image.

If, say, a turned edge as severe as Figure 11.17 is added to ¼ wavelength of spherical aberration, the turned edge would be visible beyond 20 wavelengths defocus (image not shown). The spherical aberration becomes hard to detect with 20 wavelengths defocusing aberration, but the turned edge is still there to see. Thus, turned edge can indeed be detected in the presence of mild correction error.

When the image is focused inward a long way, the low-brightness soft edge of the disk becomes hard to see. Low contrast in the rings is a surer indicator of turned edge, and the delicate appearance of the boundary is further evidence.

Ellison and subsequent authors pointed out that the edge of the disk inside of focus looked "hairy" (Ingalls 1976). It would not be surprising if the uncontrolled way a turned edge is applied to a mirror would lead to some structuring in the scattered light, thus causing a "hairy" edge. However, I have never been sure such an effect is generated by the turned edge or is caused by the common turbulence-induced aberration acting on slight low-order spherical aberration.

With the wider and less deeply turned edges characteristic of amateur-made mirrors, some modifications of these patterns should be expected. The turned area diverts light to lower angles and looks more compact. Also, the optimum distance for detecting a soft edge is somewhat nearer the focus than the 20-wavelength number recommended above.

Refractors can have turned-down edges, too, but the appearance is reversed. Contrasty rings are found inside focus; the soft edge and low contrast rings are outside. However, refractors don't tend to exhibit turned edge unless it is wide. Normally, the lens cell hides the far edge of the objective behind retaining rings. In a way, this advantage more than compensates for not being able to hide a central zone behind a secondary. Few refractors are made so badly that they show noticeable turned edges. If a turned edge is seen, do further diagnosis with a thin annulus of cardboard taped to the end of the dew cap.

The in-focus images of Figure 11.17 display the almost negligible effect of turned edge on high-resolution applications. The diffuse glow is still there, but is unseen compared to the dazzling stellar image (particu-

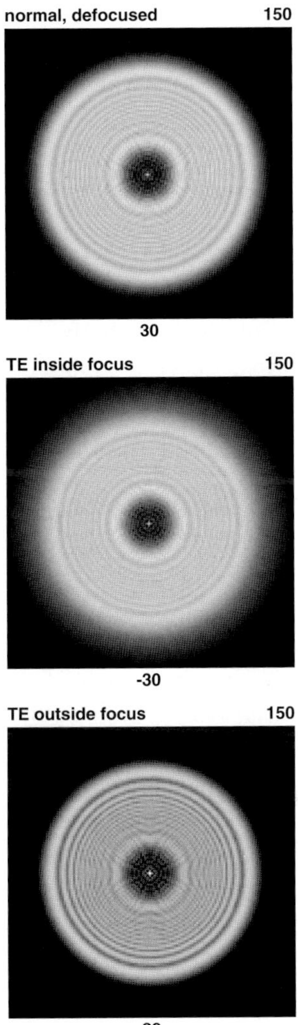

Fig. 11.18 Turned edge ±30 wavelengths out of focus. Normal appearance is also shown. The Maltese crosses are a moiré effect caused by the sampling rate of the calculated image. They would not appear in a real view.

larly in this paper and ink rendering). Other than the slightly reduced effective aperture, turned edge disturbs low spatial frequencies (or wide details) preferentially.

11.4.4 Signal-to-Background Ratio of a Turned Edge

The effect of turned edge is very similar to dirty optics in that it "scatters" light throughout the field of view. Therefore, it is instructive and revealing to make an equivalent signal-to-background calculation. In most dark-field observing situations, turned edge is like dusty optics or spider diffraction in that small amounts matter little. A tiny fraction of an already dim object's light is negligible. However, in the case of lunar-planetary astronomy or attempted observation of a dim object next to a bright one, turned edge can become a serious source of trouble. How much can be tolerated?

For planetary astronomy, two things are important: 1) the diameter of the induced halo of the turned edge and 2) the amount of light removed from the image. The diameter is easy to estimate. Consulting the MTF chart in Figure 11.16, we see that the halfway-down point in the initial sharp drop (marked with oval) occurs at about 2% the maximum spatial frequency. This number inverts to a radius of about $50\lambda/D$, or about 40 Airy disk radii for a 5% turned edge. In a 200 mm aperture using yellow-green light, the radius of the glow is about 30 arcseconds. For a turned-edge amplitude of only ⅛ wavelength, the amount of energy removed from the image (and reappearing as noise) is slightly less than 1% of the total energy. Thus, the signal-to-background ratio can be as bad as 100.

The really disturbing thing about turned edge is the relative compactness of the halo. It doesn't throw the light as far from the image as dust does. Thirty arcseconds means that much of the offending light is *still inside planetary disks*. Turned edges are stealthy errors that remove contrast far in excess of their nominal magnitudes. The amplitude of a 5% turned edge zone must be decreased to ¹⁄₂₅ wavelength before it reaches the signal-to-background ratio of 1000 that was given as the very conservative tolerance for dust. That amount is scarcely measurable. Of course, a turned edge of such small magnitude hardly behaves in the manner of a turned edge. It has begun to compete with interior zones and roughness errors.

The best way of reducing turned-edge diffraction is to make sure that it is extremely narrow. The turned-edge halo will then appear much larger and correspondingly dimmer. It will divert light outside planetary disks and diffract less light to begin with. For a 2.5% turned edge of the same ⅛ wavelength amplitude as the case above, a signal-to-background ratio of 250 is diffracted to a halo with twice the angle. Now, because much of the stray light misses the planet and the area of the turned edge is smaller, the true signal-to-background ratio increases to 500 to 1000.

11.4.5 The Width of the Turned Edge

Usually, when turned edge is less detectable than in Figures 11.17 and 11.18, you have very little to worry about. But if you have a serious problem with an edge zone, the radius at which the zone begins to roll is a useful bit of information. You can perhaps do further tests with edge masks to try to determine the turning radius and severity of the zone, but such checks are difficult to interpret.

Turning radius is easier to determine using a variation of the Foucault test. Try an occluding knife outside the focus of an artificial-source star test done at night. The source should be very bright, so move the flashlight close to the sphere or use a larger sphere. Mount the knife over one half of an empty tube of the same diameter as an eyepiece. (A fragment of a playing card or opaque slip of paper works almost as well as a true knife and presents less opportunity for accident.) You can do fine adjustments by rotating this "eyepiece." Defocus should be 5 mm or more. Arrange the telescope so the knife's shadow covers either ¼ or ¾ of the aperture, but not half. Do not replace this knife edge by a Ronchi ruling, unless it is a very coarse one. The side-order images of the grating disturb interpretations of the edge.

If the shadow is perfectly straight on the aperture and seems rock-solid, try to put the knife nearer to the focus. If you see a blurred, quavery shadow that is disturbed by the slightest touch, you are probably too close to focus. Set the knife farther away. You are looking for a slight curl of the shadow very near the edge of the aperture. The curl becomes severe beyond the turning radius.

If you can easily detect the width of a turned edge in this manner, then the rolled region is probably too wide to paint as suggested below. What you want to see is little or no evidence of a turned edge from this crude test. You only want to paint a narrow turned edge. If you detect a wide turned edge, you will find it easier to mask the offending region. Use the turning radius determined here to calculate the size of the aperture through which you wish to allow light transmission.

11.4.6 Remedies for Turned Edge

As was shown in the MTF diagram in Figure 11.16, the mirror with a turned edge performs no better than a smaller perfect aperture at high spatial frequencies and worse than the smaller aperture at low spatial frequencies. Clearly, the last bit of aperture on the periphery is doing less than nothing for the imaging performance. Turned edges gather worthless, imperfectly focused light and corrupt otherwise perfect images with it.

Masking is not an irreversible step, but it requires a little mechanical skill. Generally, it involves constructing a narrow annulus and finding a way of holding it over the mirror. A mask holder mounted above the mirror works best for open-tubed, Dobsonian mountings. Access to the region just above the mirror is easy with such a telescope, and installing and adjusting an edge mask is convenient.

Those mirror owners who have a definite diagnosis of turned edge and who are willing to accept the considerable risks involved, may significantly improve the performance of a telescope mirror by painting it. Painting should not be attempted with wide turned edges. The definition of "wide" depends on aperture diameter. Ten millimeters sounds wide for a 150-mm mirror and narrow for a 500-mm one.

Painting the mirror, however, works with any telescope and requires only a steady hand. Remove and place it on a lazy-susan or other rotating platform. Make sure the mirror is larger than the rotating table under it. Spend some time carefully centering and leveling it on the pivot, so that it doesn't wobble or rotate eccentrically.

Slowly spinning the disk, brace your hand in one spot and introduce the brush point gingerly onto the mirror's surface, working in from the edge. You are in no hurry, so try to lay down a clean line over many revolutions. I don't advise using a permanent marker but if you do, be advised that the solvent tends to wipe up existing marks, so you may have to make repeated passes before the ink stays on the mirror. Brushes are less controllable, but painting in this manner results in a more saturated obscuration. Do not use an airbrush or spray paint of any sort.

Perform the painting operation in two steps. The first time, paint out just 1 or 2 mm of the far edge. If you are fortunate, this narrow ribbon will largely cure the edge problems. Star test the mirror again. If the image doesn't improve, extend the painted zone inward. At the end of such a procedure, the ugly appearance of the mirror can be shocking. Keep in mind, however, that you have, in effect, trimmed off the poorest part of the aperture. Do not attempt to grind a wider chamfer. Besides the possibility of damaging the coating, you may relieve stress in the edge glass.

The effect of a corrected turned edge can be seen in the star test. How it will improve actual observing is not immediately apparent. Remember, the most objectionable feature of a turned edge is a hazy glow in the vicinity of the bright point image—less than 20 to 40 Airy radii out. Inspection of the MTF chart reveals a quick dip at low spatial frequency. For this 5% wide zone, most of the damage has already occurred by the time 2 to 5% of the maximum spatial frequency of the MTF target has been reached. If the aperture is 200 mm (8 inches), then detail separations less than about 10 to 30 arcseconds are degraded. Masking or painting the turned edge will have the most benefit with richly-detailed large objects, such as the cores of tight globular clusters or planetary disks.

Chapter 12
Chromatic Aberration

Because refraction changes with wavelength, a simple refractive lens focuses light of different wavelengths at different distances behind it. Also, lenses that are not centered or otherwise have a significant thickness of glass far away from the pupil will divert light different distances to the side. These properties result in color errors. The different length of focus is called longitudinal chromatic aberration. The other error spreads point images into a short linear spectrum. It either depends on the distance from the center of the field, in which case it is probably lateral color, or it is fairly uniform over the field, where it is wedge error.

To get an idea of the difficulty involved, hold up a simple one-element lens and look at the transmitted image with a low power eyepiece. Such a lens often appears in toy telescopes. Obviously, chromatic aberration is profound. Every bright object appears to be surrounded by a rainbow-hued glory, and a subtle error is revealed in the overall lack of contrast. The MTF values for wavelength components not in focus are profoundly lowered, and the overall contrast is diminished. Figure 12.1 shows the MTF weighted for the photopic color curve following the sensitivity of the human eye, and Figure 12.2 shows the spot diagram at the design wavelengths. Not surprisingly, reasonable imaging only occurs for fields comprised, of only one pure color.

Early astronomers reduced the importance of simple-lens color error by increasing the aperture ratio to enormous values. This stratagem increased the depth of focus until it encompassed the spread of colors, and they found that performance improved. The 100-mm simple lens spot diagram is shown in Figure 12.3, but the focal length was here lengthened from 1.5 m to 15 m (50 ft). This battle was hard to win, however, because modest increases in aperture had to be accompanied by huge increases in the focal length, and operational difficulties worsened considerably (King 1955; Bell 1922).[1] Nevertheless, refractors of this type were used by scientists during much of the 17th century.

Color errors are not really geometric aberrations in the narrow sense

---

[1] Many 5×24 finders on department-store telescopes are actually stopped-down simple lenses. If a roughly 8-mm stop is visible immediately behind the objective lens, it is not corrected for color. The finder is useless. Discard it.

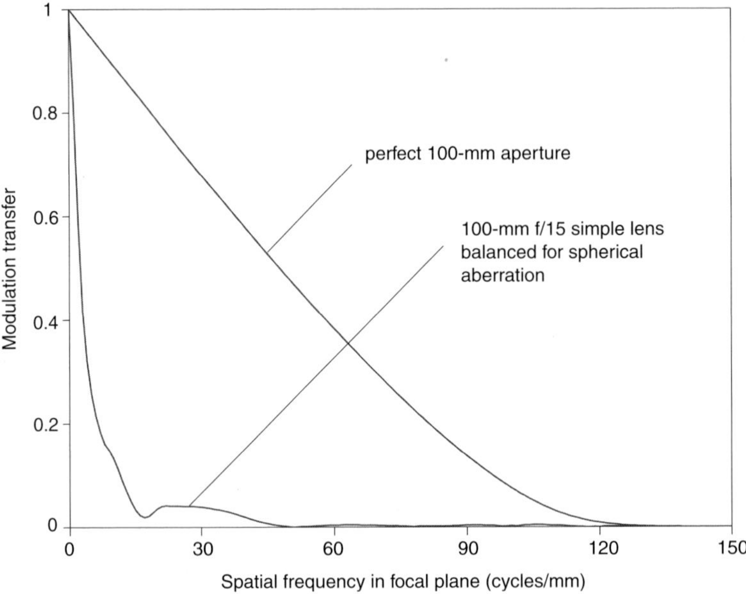

Fig. 12.1 Photopic-weighted MTF of simple 100-mm f/15 lens, with curves balanced for minimum spherical aberration.

that aberrations are departures from the ideal spherical shape of a wavefront. One could easily describe (if not make) an aperture that focuses each color precisely on a much different axial point. In any given color, the wavefront converges spherically. Such an aperture could justifiably be called perfect, yet it would not work as a visual telescope objective.

Another important distinction between wavefront (or "geometrical") aberrations and chromatic aberrations is the absence of interference effects in chromatic aberrations. Different colors do not mutually interfere. For most of the discussions appearing in this chapter, *wave optics and ray optics are identical.* Wave optics still applies to each color separately when used to explain geometrical aberration, but there are no new diffraction effects that derive their origin from the mixture of colors.

Whatever the origin of the optical degradation, color error is objectionable, and one goal of the star test is to make certain that it is as small as possible.

12.1 Dispersion

An ideal lens would focus all colors at the same lens distance, and the performance of lenses would be indistinguishable from mirrors. Unfortunate-

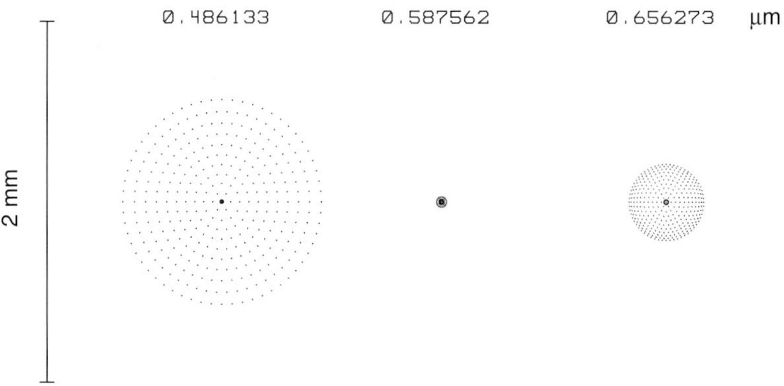

Fig. 12.2 Spot diagram of simple 100-mm f/15 lens. Note that the red and blue images are about 1 mm across.

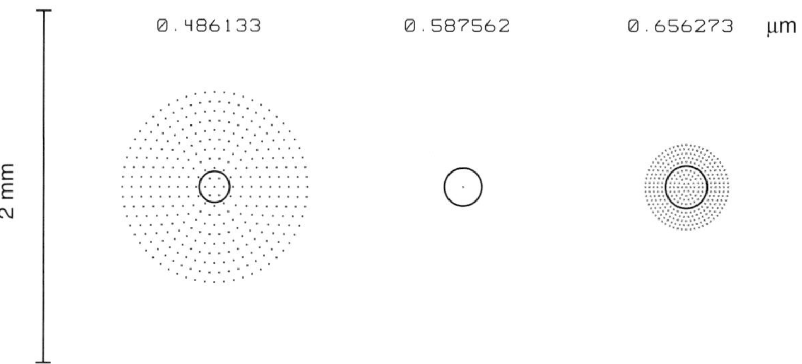

Fig. 12.3 100-mm f/150 simple lens spot diagram. The Airy disk has expanded, even though the image spots have not changed. The much longer eyepiece needed to give the same magnification would make the defocus less objectionable.

ly, all simple lenses are dispersive. Dispersion arises from the inability of all colors of light to move at the same speed in glass. Like the word propagation, its source is botany or agriculture. "Dispersion" originally referred to randomly spreading seed. Its optical meaning refers to the spreading of colors, but such dispersion is anything but random. In fact, if it had not been applied already to color error, dispersion might have been a better word for what is known today as "light scattering."

Red light typically has a higher speed in glass than blue light. Thus, red-light waves outrace blue-light waves while they are traversing the material. The net effect is that after passing through a prism blue light is diverted to a larger angle than red light.

Say we have two prism materials, as in Figure 12.4a, that divert or *refract* light to the indicated angles. The average behavior of the two

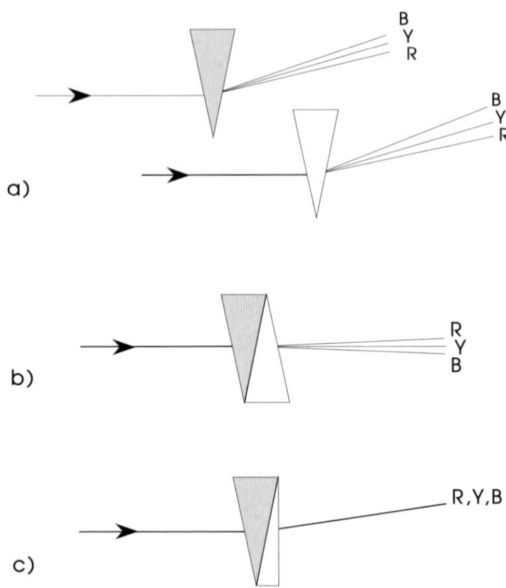

Fig. 12.4 Simplified achromatism: a) "crown" glass (dark prism) showing dispersion of light and "flint" glass (light prism) exhibiting the same average refractive power but twice the dispersion, b) the two glasses combined to generate a "straight-through" rainbow (dispersion without diversion), c) a combination that results in an achromatic prism (diversion but not dispersion).

prisms is the same. Each kicks the yellow wavelength of sodium light to 10°, but the behavior at either end of the spectrum is somewhat different. One glass (call it "crown" glass) diverts red light to 9° and blue light to 11°. The other glass (call it "flint") spreads the spectrum twice as far, to 8° and 12° respectively. This second sample of glass is said to have higher dispersion because of this spreading property.

We can combine these prisms to produce interesting effects. If we invert the flint prism and place it snugly against the crown, as in Figure 12.4b, we have made an approximately plane-parallel window. The combination passes the central wavelength undeflected, but red is pushed up to 1° and blue is bent down to −1°. What we have envisioned here is a straight-through dispersive element, one that spreads light into its color components, but does not refract it on the average. Such a device was once made and marketed as an objective prism by the firm of Merz and Mahler, but its expense prevented wide adoption (King 1955, p. 294). The advantage of such a prism was that cameras could be directly pointed at regions of interest.

Figure 12.4c depicts another clever trick we can perform with these two materials. This time, we divide the flint prism in half, making another prism that only deflects yellow light to 5°. Blue light would exit this half-prism at an angle of 6° and red light at 4°. If we invert this prism as before and place it close to the crown prism, the exit angles of the combination become: yellow, 10°–5° = 5°; red, 9°–4° = 5°; and blue, 11°–6° = 5°.

If one were designing a crystal chandelier, this prism combination would be a disaster. Instead of dividing the light into a profusion of sparkling rainbows, the compound prism keeps the light packed tightly in a white beam. It diverts all colors equally. On the other hand, if one were designing telescopes, it is just what is required. This device has two important characteristics. It bends light and does so colorlessly. What has been demonstrated here are all of the essential elements of an achromatic refracting telescope.

12.2 The Achromatic Lens

Figure 12.5 shows the progression that must take place to get from the achromatic prism to the achromatic lens. If we envision the lens as composed of little prism pieces, and allow the divisions to get finer and finer, we eventually get to a cylindrical lens. It takes very little imagination to rotate the other way and extend the situation to a spherical lens.

Almost as soon as the concept of dispersion was developed, this trick of achromatism was envisioned. All that was left was to find suitable materials. Isaac Newton conducted limited experiments where he compared the dispersive power of different media with their refractive power. His hasty conclusion was that dispersion and refraction were inextricably linked. Thus, dispersion could not be counterbalanced without also eliminating bending of the light beam. He reached this erroneous result by perhaps relying too much on a visceral feeling that dispersion was a property of the light itself. By this argument, materials didn't matter; dispersion was proportional to refraction and thus always existed until lenses were weakened to be no different than windows. Therefore, achromatic refractors were impossible.

Various conflicting statements about Newton's mistake have been made. Bell (1922) said that Newton had never published this result, and King (1955) referenced a section of Newton's *Opticks* in which the great physicist despaired of ever curing chromatic aberration but didn't give his reasons. The confusion can perhaps be reduced by referencing yet another portion of *Opticks*, where Newton said (Book One, Part II, Prop. III, Prob. I, Exper. 8):

> I found moreover, that when Light goes out of Air through several contiguous refracting Mediums as through Water and Glass, and thence goes out again into Air, whether the refracting Superficies be parallel or inclin'd to one another, that Light as often

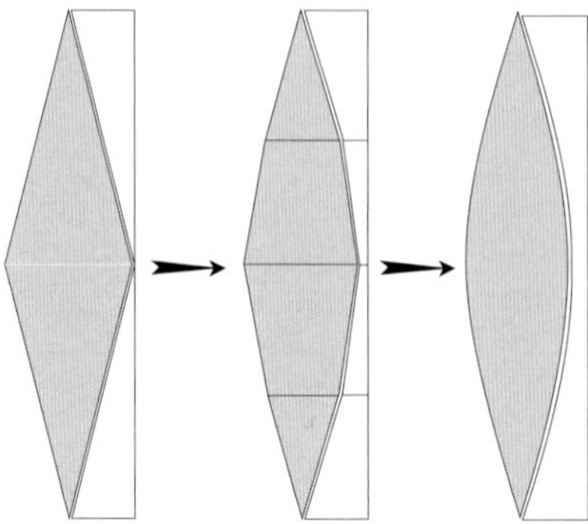

Fig. 12.5 How a prism achromat can be imagined to lead to an achromatic lens.

as by contrary Refractions 'tis so corrected, that it emergeth in Lines parallel to those in which it was incident, continues ever after to be white. But if the emergent Rays be inclined to the incident, the Whiteness of the emerging light will by degrees in passing on from the Place of Emergence, become tinged in its Edges with Colours. This I try'd by refracting Light with Prisms of Glass placed within a Prismatick Vessel of Water.

In other words, conditions leading to the straight-through spectrum of Figure 12.4b do not occur, implying Figure 12.4c is also impossible. He followed this experiment with some speculative theorems elaborating his ideas. In Newton's defense, I don't believe that discrediting the idea of achromatism was the main thrust in this section of *Opticks*, but the experiment was among those interpreted by later readers to be a stronger condemnation than it actually was (Newton 1730).

Partly because of Newton's powerful reputation, opticians gave up on achromatic lenses for 50 years. Then, an English barrister and amateur optical designer named Chester Moor Hall made the first reduced-color lens from two different materials. He kept the lens design hidden, although the trick was reverse-engineered by a lensmaker who happened to be subcontracted by separate opticians to work on both lenses at the same time. The lensmaker didn't realize the importance of the invention and news of it languished until it was redeveloped a generation later by John Dollond. When it appeared to be a profitable development, lawsuits were filed by London opticians saying that Dollond had stolen the idea.

Dollond likely had heard rumors about Hall's lens, but certainly he did enough scientific work himself to be justly credited with reinventing it. Perhaps it would be more accurate to say that Dollond was the first to reduce the achromatic lens to common practice. He was certainly the first to announce it publicly (King 1955, pp. 144–150).

12.3 Residual Chromatic Aberration

Unfortunately, the ideal materials of Figure 12.4 do not exist. Most dispersion in the range of visible wavelengths originates with a resonance in the ultraviolet. In the frequency band of the resonance, the wonderfully transparent materials go opaque. The transparent nature of these materials (called *dielectrics*) in the visible spectrum does not persist for every wavelength. Over much of the spectrum, the materials are content to accept energy at the entrance side and emit it at the exit side with very little loss. The energy brushes over the molecules, disturbing them little. However, at certain frequencies these materials are unusually excited by the incident energy. For wavelengths near this ultraviolet resonance (usually about 100 nm, but strength and bandwidth is altered for different materials), the glass molecules suddenly absorb energy and convert it to heat.

The material behaves like a child's swing. If the swing is pushed once every quarter hour, the pendulum motion does not build. If the swing is pushed three times per second, most pushes are poorly timed and once again, pendulum motion does not build. If the impulses are carefully timed, the energy lost to the swing (or glass) steadily increases. The swing is driven at the resonance frequency. The energy contained in the light wave doesn't float through glass anymore because it must drive the oscillation in the material. Transparency is destroyed (Hecht 1987, p. 63).

The presence of resonances in the ultraviolet causes the refractive index of optical materials over the visible spectrum to increase sharply toward the blue end of the spectrum. Most glasses also have a distant resonance in the infrared caused by molecular vibration, but this resonance affects the slopes less profoundly in the visible band. The exact position of the ultraviolet resonance varies from glass to glass, and so the non-linear dispersion curves are not only stronger and weaker for flint and crown glasses, they have different shapes.

One aim of the lens designer is to choose powers of the lenses in such a way as to cause the dispersion of the "crownlike" lens elements to cancel the opposite dispersion of the "flintlike" lens elements. Because the dispersion tilts differently at the violet end of the spectrum and the number of acceptable materials is limited, the color correction cannot be perfect. This was particularly true back when the achromat was invented, where the primary difference between glasses was the amount of lead oxide added to the mixture. All glass was of the same general family.

Lens designers thus were unable to choose lens combinations so that the dispersions nest like spoons for every color, and this is still true today for common glasses. The differences of refraction toward the blue (caused by the different central wavelengths of the ultraviolet resonances) usually mean that they can choose only two colors with the same focal points. The remainder of the spectrum must go where it will. Each color, naturally, will be paired with another color from the opposite side of the spectrum, but for regular types of glass, the designer can deliberately choose only two.

For most visual achromatic telescopes, the two colors that the designer attempts to bring to a common focus are red (the Fraunhofer C line at 656 nm) and blue (the Fraunhofer F line at 486 nm). The focal point of green is slightly closer to the objective, and the far ends of the spectrum (deep red and violet) are beyond the C-F focus. Violet is farthest away, but that doesn't matter. The human eye is not sensitive to violet except at high brightness, so the defocused halo of violet light is mostly invisible. The residual color spread of achromatic objectives between the two chosen colors is known as the *secondary spectrum.*

Chromatic aberrations scale with size. As the diameter of the objective increases, it must be made at higher focal ratios to squeeze the light between C and F inside the diminishing Airy disks. While small 80-mm lenses can still perform admirably at $f/10$, a conventional achromat six inches in diameter must be made at $f/18.5$ to focus the different colors as well (Sidgwick 1980, p. 67; Rutten and van Venrooij 1988, p. 55). A.E. Conrady is even more conservative, stating that the aperture ratio for an 80-mm lens must be $f/15$ and a 6-inch should be $f/29$ (1957, I p. 201). Since secondary spectrum is about $\frac{1}{2000}f$ for ordinary achromats, we can use Appendix E to show:

$$F = \frac{D[\mathrm{mm}]}{8.8\Delta n}$$

12.1

where Δn is the number of wavelengths defocusing error we are willing to tolerate in the secondary spectrum, D is the aperture diameter, and F is the focal ratio. Following the depth-of-focus discussion in Section 5.1.1, we place Δn at $\frac{1}{2}$ wavelength (i.e., $\pm\frac{1}{4}$ wavelength). The formula becomes $0.23\, D[\mathrm{mm}] = F$, or $f/18$ for the 80-mm aperture and $f/35$ for the 6-inch. This result is more conservative than Conrady's. Nevertheless, I have observed through a 6-inch $f/15$ refractor and was only moderately bothered by the excess color. Perhaps the restriction can be eased to 1 wavelength without too much loss. In this case, it drops to $0.12D[\mathrm{mm}] = F$, which is the same as that reported by Sidgwick. Plotting the wavefront at various colors for a 152-mm $f/15$ achromat in Figure 12.6 (design in Section G.1) demonstrates that the aberration is indeed about 1 wavelength (focus is not placed at the best position for this refractor, but it helps the measurement). Note that most of the color error is pure defocusing error. Figure 12.7 displays

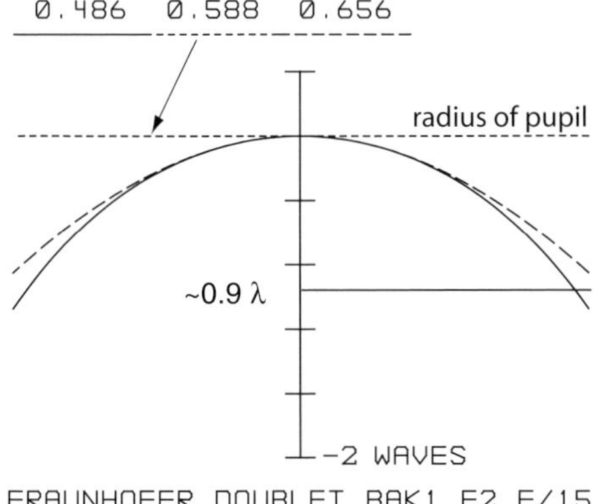

Fig. 12.6 Plotting the wavefront at various colors for a 152-mm f/15 achromat.

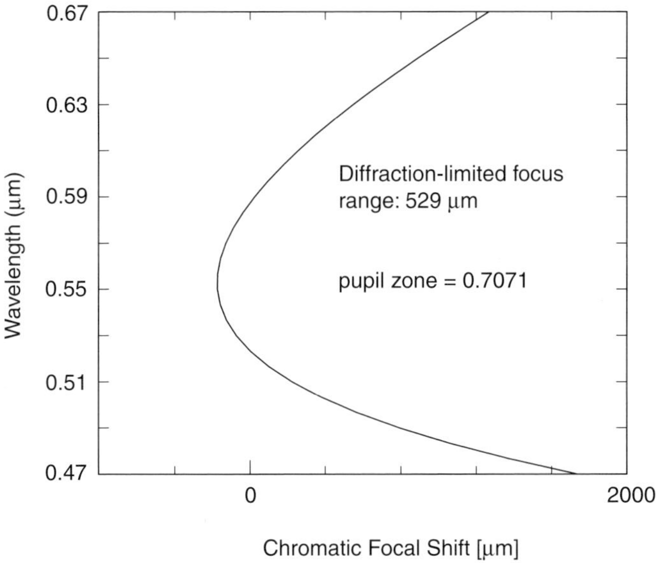

Fig. 12.7 The chromatic focal shift of a 152-mm f/15 Fraunhofer achromatic refractor. The secondary spectrum (between C and F) is smeared out about 0.9 mm along the focus direction where the allowed focal range is 0.53 mm.

the color shift diagram of the same refractor.

This focal-ratio discussion sounds a disturbing note about the so-called perfection of conventional refractors. Perhaps it also says something about the human tolerance for chromatic errors and a change in favored observing targets. Color error is no better in modern achromats than it was in the early days, yet today's amateurs seem to tolerate 100 mm f/10 achromats where they used to tolerate only f/15. The only way of explaining this is to point out that in the 1800s, planetary objects were the prime targets. Deep sky now occupies a greater fraction of the observer's time. Frequent use of lower magnification to view dimmer objects would tolerate lower optical quality much of the time, and the $f/10$ aperture ratio helps achieve wider field with common eyepieces plus a degree of portability that is not easy to dismiss.

Lens designers have found other useful telescopic color corrections besides pulling C-F into common focus. In the early years of astrophotography, special-purpose achromats were corrected for the orthochromatic emulsions then in common use. Orthochromatic plates were most sensitive to blue through near-ultraviolet light, and completely insensitive to red. Photographic telescopes could only be focused by the tedious process of taking actual exposures, but they gave sharper images on the plate than a visually-corrected lens. Modern imaging devices, on the other hand, demand good correction all across the spectrum and particularly in the infrared. The sensitivity range of CCDs is between 350 and 1100 nm.

Since the designer had to take the whole optical system into account, the personal preferences of the observer were figured into the design. Color correction curves appearing in Bell (p. 91) showed that the finest makers of the 19th century favored bringing the F (blue) line into common focus with the deep red at 680 nm (the B line). In older refractors, much of this color-correction shift is presumably caused by the chromatic aberration inherent in the eye and in the eyepieces used at the time (Taylor 1983). Also, with today's faster refractors, a color curve may be designed with the intention of using anti-fringing filters.

12.4 The Apochromat

Achromatism can be compared to tying the spectrum in a knot. The brightest parts of the visual spectrum are deliberately folded into the tightest bundle, with the deep red and the violet ends hanging out like shoelaces. Some of the earliest optical workers (most notably Peter Dollond, son of the achromatic lens developer) tackled the cause of these spectral defects. Dollond could choose from only a handful of glasses. He reasoned that if a flint element were "designed" from a composite of two glasses, then the dispersion of that element could be tuned so that it would nest more closely with the opposite dispersion of the crown element.

Peter Dollond made and marketed such a triple objective, but apparently the lens was designed by trial-and-error. In any case, the glasses of the time were not yet good enough to consistently allow such refinement. Typical crown glasses had pretty much the same dispersion behavior of other crown glasses and the same was true of flints. It was not until later in the century that the physics of dispersion in materials had progressed to the point that unusual dispersion curves could be designed into glasses, usually by the addition of exotic materials to the mixture. In 1892, H. Dennis Taylor produced the first popular apochromatic lens using an abnormal flint for the third element of a composite triplet (King 1955, Ceragioli 2003). Taylor not only corrected secondary spectrum more fully, but folded the violet tail of the spectrum close enough to the visual to allow the blue-sensitive photographic plates of the time to use the same focus as the human eye. He called this lens the "photo-visual" (King 1955).

Using different forms of glass, of which one must be abnormal, allows the lens designer to perhaps put an extra kink in the dispersion curves, which in turn allows the simultaneous focus of three chosen colors. The color spread is knotted yet again. Often, the colors selected for common focus are further separated than the Fraunhofer C and F lines. One such correction brings the lines C, e, and g into common focus; or red, green, and violet (Kingslake 1978, p. 86). The deep-violet and deep-red ends of the spectrum are tucked in closer to the visual focus. Maximum correction of color error represents a trade-off with geometric aberrations, however. Glass combinations that result in three-color focusing at long aperture ratio perform less well at short aperture ratio, since maintaining good off-axis behavior may limit the color correction. Typically, the residual tertiary spectrum has been reduced a factor of five to $\frac{1}{10,000}f$ (Rutten and van Venrooij 1988, p. 54), meaning that we are able to achieve half-wavelength secondary-color focusing at

$$F = \frac{D[\mathrm{mm}]}{20}.$$

12.2

If color were the only consideration, 100-mm telescopes could be made at $f/5$ and 150-mm objectives at $f/7.5$. In practice, there are geometric aberrations that may drive minimum aperture ratio slightly higher.

Tuning of color correction is more dependent on the choice of materials. Apochromat design is a lot more than plunking down some likely materials and twiddling the curves. If the designer chooses the wrong starting materials, some optical goals are completely impossible to achieve. Today, there is another form of apochromatic objective possible, one where the abnormal glass is placed in the crown element. Such a lens is made with a low-dispersion crystal such as calcium fluoride (CaF_2, called "fluorite") or extra-low dispersion glass (so called ED glasses like Ohara S-FPL53). In practice, this type has all but displaced the old composite-flint types. The flint-like glass in the case of fluorites or ED glasses is what

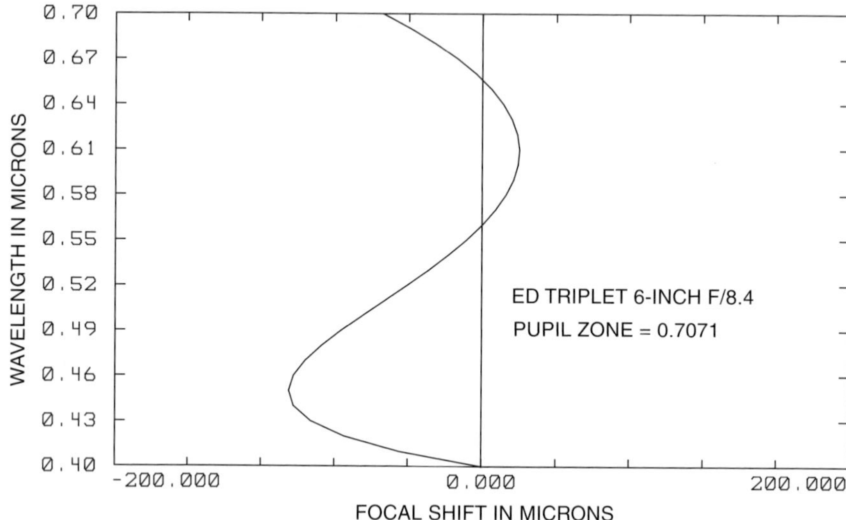

Fig. 12.8 Chromatic focal shift of an example apochromat. This color behavior is plotted at the 71% zone.

was once a normal crown. Figure 12.8 shows the color curve of such a triplet designed in Section G.2. It has three crossings.

The proper behavior of the bright portions of the spectrum is no guarantee that the spectral tails are close to focus. Much depends on what the designer has in mind, and what trade-offs must be considered. Some apochromatic refractors might be designed for purely visual use, minimizing the spread of focus between C and F. These apochromats might bring violet or infrared only a little closer than ordinary doublets. Others may be designed for photographic or digital images without using filters, focusing deep into the blue or infrared with better visual correction than the achromat but not optimal.

Figure 12.9 displays the color curves of a set of 6-inch *f*/10 refractors of varied types. We see that a 6-inch *f*/10 simple lens slashes through the diffraction-limited region, occupying it for a range of only 2 nm. Better is the achromat, which comes in like a comet and hangs briefly in the diffraction-limited zone between 510 and 600 nm. The two apochromatic refractors are notional instruments that might have been designed for old blue-sensitive photographic plates and the modern near-infrared sensitive silicon array of a CMOS or CCD camera. The blue-sensitive apochromat is within the diffraction-limited range between 390 and 775 nm and the near-infrared optimized apochromat is within it from 465 to 1000 nm. Only the emulsion-type photographic apochromat approaches being apochromatic in the visual region, but people wanting to do digital photography would

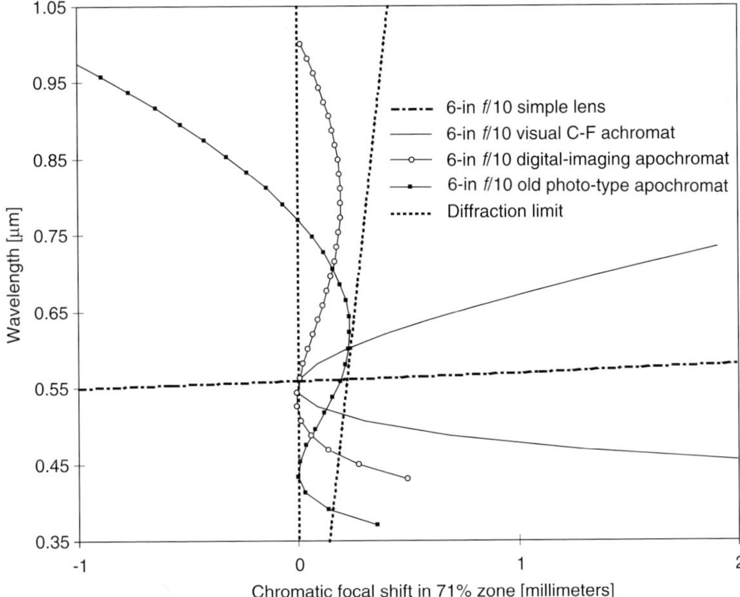

Fig. 12.9 Color focal-shift curves of a number of 6-inch $f/10$ instruments.

probably prefer the near-infrared crossing to the theoretical apochromatism.

A lot is made of the fact that an apochromat should focus three chosen colors to the same point together with certain corrections for geometric aberrations (for a more precise definition, see Buchdahl 1970), but apochromats are designed with more goals in view. They are generally made in lower aperture ratios than old-time refractors, and thus have difficulties with geometric aberrations that older telescopes were never forced to confront. The designer may choose to minimize the complete set of geometric and chromatic aberrations yet never achieve the theoretical condition of apochromatism. Hence, a very good ED refractor having only a fifth of the $\frac{1}{2000}f$ color aberration of a typical achromat may never technically focus three chosen wavelengths within the visual spectral range at the 71% zone, yet still deserve the name "apochromat" because of its colorless and accurate imaging.

Sadly, there are also refractors that use inexpensive fluorocrowns only to acquire the "ED" name commonly associated with apochromats, but they have secondary spectra similar to ordinary achromats.

12.5 Testing Refractors for Geometrical Aberrations

Since each color, in effect, goes through different apparent thicknesses of glass, one would anticipate that the correction for other aberrations might vary over the spectrum. In fact, some aberrations aren't corrected at all. Much depends on how many free parameters the lens designer is allowed to play with.

For example, the variables of a flint-crown doublet are all four of the curves, the separation of the lenses, the position of the aperture stop, and the glass formulations (of which there are hundreds of useful combinations). If allowed to vary these parameters at will, designers can focus two chosen colors simultaneously and adequately correct for coma and spherical aberration over much of the spectrum.

You should test for geometrical aberrations in refractors by using an eyepiece filter. In fact, using such a filter to suppress the polychromatic nature of white light is encouraged even for critical inspection of reflector optics. A very deep green filter is recommended to test the central wavelengths near 560 nm. After you have tested at this color, you may wish to change filters to red or blue to verify that other aberrations are small at the edges of the visual spectrum. Except for limiting the color band, testing refractors for geometric aberration is the same as testing reflectors.

Figure 12.10 shows the aberrations visible in a well-corrected achromat in three colors, while Figure 12.11 shows the same aberrations in a well-corrected apochromat. In both telescopes only a slight amount of spherochromatism is visible. One can see that the inherent geometric aberrations of a properly made refractor are very mild indeed.

12.6 The Star Test for Chromatic Aberration

Different colors do not mutually interfere. Hence, chromatic aberration does not manifest itself in modifications of the diffraction rings, either focused or defocused. Chromatic aberration appears as a different focus position for each color. The shift can be a lateral one, as for a mild prism, or it can be a longitudinal offset. For all real lenses, each color has a slightly different focal length.

To see color effects, *pull off all filters and focus on a white star or an artificial source.* Be sure to filter the flashlight for a nighttime artificial-source test of chromatic aberration (see Section 5.2.4).

Achromat RGB image 33

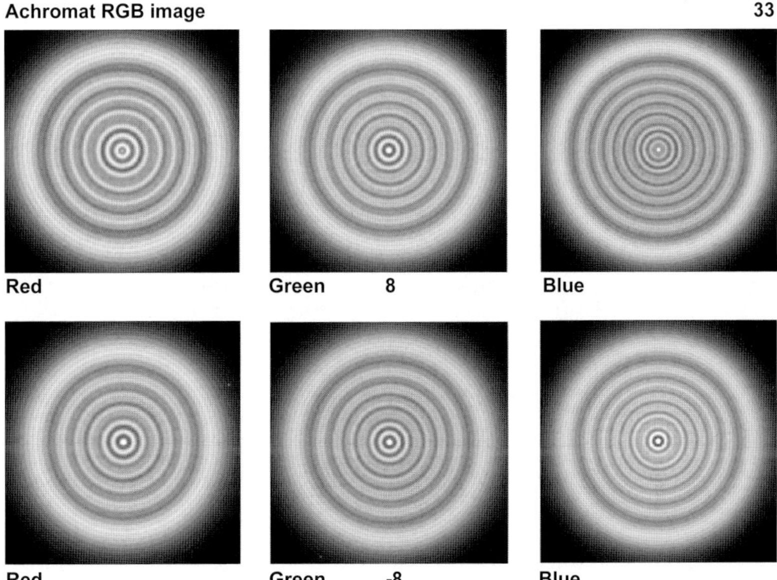

Fig. 12.10 Geometric aberrations of an C-F optimized BAK1-F2 152-mm f/15 achromat in three colors. The 8 wavelength defocusing aberration is measured from the position of best focus in each color, rather than from an absolute position.

12.6.1 Wedge, Assembly Errors, and Atmospheric Spectra

Look for smearing of the focused image into a short linear spectrum, the effect of decentering or wedge in optical components. Either of these problems puts a red fringe on one side of the image and a blue-green fringe on the other. Decentering may also cause other aberrations, depending on the details.

Decentering is a sideways shift of the elements with respect to one another that usually exhibits itself in wedge. It occurs because the optical center of the lens doesn't correspond to its geometric center. Wedge is an extremely shallow prism that is added to the optical system. It results from having an element thicker on one side than the other. After years of trouble-free use, wedge can appear suddenly after a lens is disassembled for cleaning. Makers sometimes cleverly remove the last bit of wedge in their objectives by canceling it between lens elements. Thus, if the crown element has a wedge of 0.04 mm and the flint element has a wedge of 0.03 mm with a maximum tolerance of 0.02 mm, the total wedge can be reduced to 0.01 mm by putting the thick part of the crown next to the thin part of the flint. An unsuspecting owner can turn these elements to yield a total wedge of 0.07 mm, or more than 3 times the tolerance. Upon disassem-

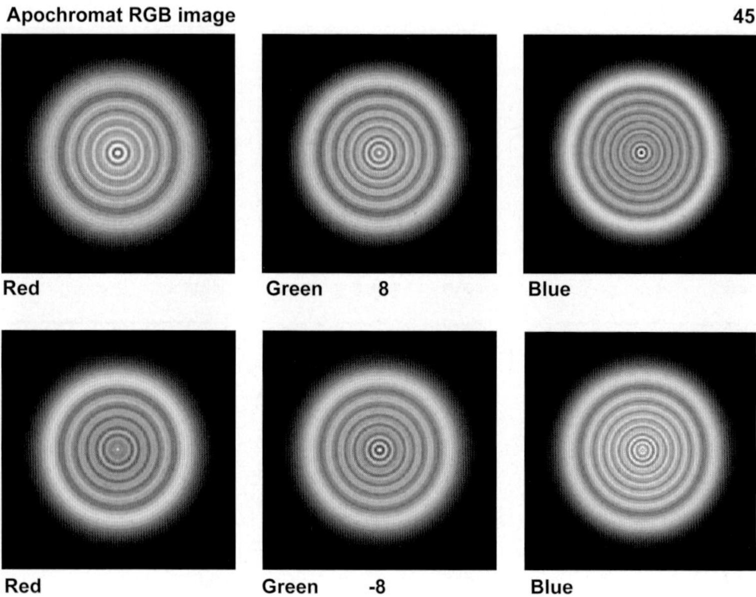

Fig. 12.11 The same aberrations in a well-corrected 152-mm f/8.4 apochromat. Defocusing in each color is measured from the same point.

bling the lens elements, look for alignment marks, either arrows or scratches, on the edges of the disks. These are rare with production lenses, but may be seen with custom or old lenses.

Another accident can also occur when refractor lenses are taken apart for cleaning. Refractors can be so long-lived that even well cared-for instruments will eventually acquire too much internal grime. Occasionally, owners improperly invert the crown element during cleaning. This error sometimes happens even to huge observatory instruments, as is documented in Leslie Peltier's *Starlight Nights* (1965). He had obtained a 12-inch Clark refractor that had a hideous purple glow around star images. Doubting that the Clarks would deliberately release such a poor instrument since it would damage their reputation in professional circles, Peltier speculated that the crown element had been inadvertently flipped during washing in the past. He inverted the element once again, and recovered the fine performance of the original design. Any instrument that is air-spaced may suffer this indignity, regardless of size. I once saw it in an airspaced 2-inch Unitron refractor, but in the case of that miniature doublet, the inversion also damaged the spherical correction.

The direction of spectral dispersion may be vertical. In this case, the atmosphere and not the telescope is at fault. The same spectrum would appear in a similar-sized reflector, and certainly no mechanism exists in a reflector to cause the same dispersion. The presence of a vertical spectrum could either be caused by cooler air puddling at the bottom of the telescope, or (more likely) a slight prismatic effect in the atmosphere itself.

Observers commonly witness a color spread on low-lying planets and bright stars. Any object below 45° elevation is likely to be smeared to a small extent. Such errors are unlikely for artificial sources, but in the sky they are all too prevalent. Rotation of the tube will serve to isolate this error to air effects. Choose a star closer to zenith.

12.6.2 Star Test for Conventional Astronomical Visual Doublets

The design of the crown-flint refractor lens froze into place in the 19th century. Individual makers may have chosen slightly different residual dispersions curves, but all were more-or-less bound by the availability of materials. The quality and homogeneity of materials have improved, but the simplest astronomical refractors are still made using designs that would have been recognized by makers in the 1800s.

The following star test does not apply to modern advanced refractors made from uncommon materials. It is based on a star test done by the author on a 4-inch $f/15$ Alvan Clark refractor built in 1881. This test used Polaris as a target. It is essentially the same as that described by Taylor (in 1891), who probably used similar instruments, and would apply to high-quality doublets even today.

Inside focus, a very pale yellow-green disk with a trace of a magenta fringe is visible. Just beyond focus to the outside, a somewhat surprising red spot appears in the center of a pastel green annulus. This wonderfully tiny crimson point of light is astonishing to someone who has never noticed it before. Taylor says that its origin is in the deep red beyond the C line.

An additional factor contributes to the appearance of this red dot. It approximately coincides with the location for which yellow-green is 1 wavelength out of focus. The diffraction pattern of yellow-green light for this situation looks like the left side of Figure 12.12, which shows an annulus with a hole neatly punched in it. Equation 5.1 gives the focus shift to +1 wavelength defocus as $8F^2\lambda$. The fraction of the focal length is just $8F^2\lambda/f$. For a 4-inch $f/15$ refractor in yellow-green light this ratio amounts to $8(225)(2.2 \times 10^{-5})/60 = 0.00066$.

For conventional doublet refractors, the difference between yellow-green and C (red) focus is about 0.0005 times the focal length. For the same location as the focus in Figure 12.12a, red light slightly beyond the

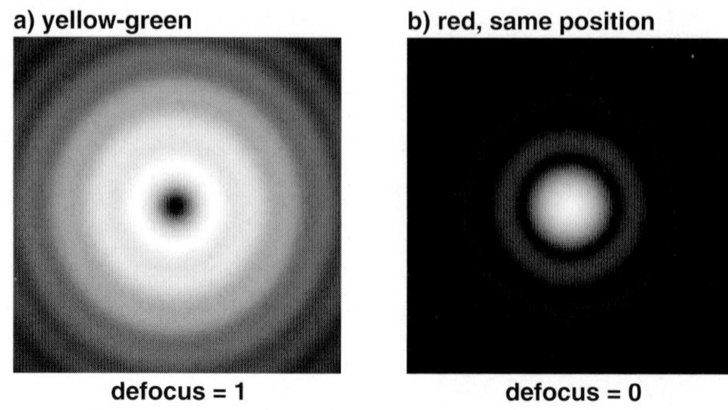

Fig. 12.12 The "red dot" effect just beyond focus in conventional doublet refractors.

C Fraunhofer line is being focused in the right side of the pattern of Figure 12.12b. Thus, a convenient little hole is present in the bright yellow-green diffraction pattern through which the red focus peeks.

Beyond the red spot, the pale greenish disk expands once more. Taylor says that sometimes a green fringe appears on this disk, but it didn't appear or was very weak in the Clark. Farther out, an indefinite blue-violet focus is supposed to form at the center. In my observation, focus is much too strong a word, however. The 4-inch showed a smaller blue fuzzy region superimposed on the green disk that never condensed well enough to form what could be termed a focus.

Taylor's observations were made with a Huygenian eyepiece while the modern views were supplied by a modified Orthoscopic eyepiece. This difference probably accounts for the few changes, together with variations in optical design. Also, the Clark was not coated, and it had the residual coloration of the glass types of its era, slightly affecting comparisons with modern doublets.

A dim in-focus image is mostly colorless. Its most objectionable feature is a watery purple or violet glow that forms around bright objects such as the planet Venus. Taylor said it best when he described a planetary image as a sketch in black and white, where the artist made a last pass dabbing on the far red and violet colors with a sponge. However, the bad effects of this halo should not be exaggerated. It attracts notice on only the most dazzling objects, and even then it seems to interfere only slightly with the ability to discern detail. On conventional refractors smaller than 80 mm, this purple glow is almost unnoticeable. It begins to become intrusive at 4 and 5 inches (100 to 130 millimeters), but only for large instruments does it become objectionable. I saw it in a 6-inch *f*/15 refractor used as a guide telescope for the Schottland 16-inch Schmidt camera (King 1955, p. 370). Saturn had a crisp white-yellow disk surrounded by a bright purple

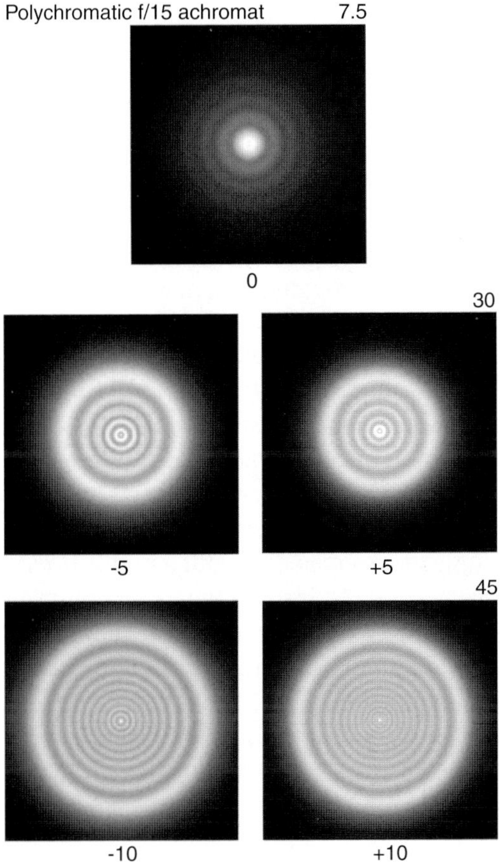

Fig. 12.13 Photopic polychromatic star test of a perfect 6-inch f/15 achromat.

blur.

Incorrectly made modern doublets will show a too-strong halo of red or greenish-blue surrounding the focused image. Too much bright color is certainly a reason for concern. Such behavior is not normal.

Lesser amounts of chromatic aberration can be also be discerned by examining the fringes of the image just inside focus. The wavelength that focuses closest to the lens in a normal doublet is about 550 or 560 nm (yellow-green). If the place where the achromat folds the spectrum is too close to the red end, the inside-focus fringe is blue with perhaps a small green component. If the fold is too close to the blue end, the fringe is a scarlet red instead of magenta (Sidgwick 1980). A grayscale representation of the polychromatic star test is shown in Figure 12.13.

The main point demonstrated by Figure 12.13 is that a perfect achromat star tests in the usual way, if taken far enough from focus (about 10 wavelengths) that inconsequential aberration levels are suppressed. The trouble with achromats is that the focuser must be moved a great distance to achieve this defocusing (for an $f/15$ instrument, Chapter 5 demands nearly a centimeter). Star testers are dismayed by the aberrations visible on trusted and proven objectives when they defocus lower distances than this, and considering how far they must turn their hands, the testers almost always do defocus too little. Measure the protrusion with a ruler and defocus 8 or 10 wavelengths. If you don't see the even disks visible at the bottom of Figure 12.13, investigate further.

12.6.3 Star Test of Apochromats or Advanced Refractors

Coloration is far less noticeable in apochromats. Taylor claimed that the out-of-focus star disks in his photovisual lenses were virtually colorless. I have never inspected a Cooke photo-visual lens, so I cannot provide confirmation. According to Sidgwick (1980, p. 201), a focused photovisual lens had a dazzling yellow-green disk with a purplish-red fringe. It should be pointed out that a weak red fringe appears on point sources even in reflectors, although it is difficult to observe. The diameter of the diffraction disk increases with long wavelengths, and the telescope is unable to pack red light into the small bundle.

My apochromatic refractor had a slight magenta fringe at a short distance inside focus and a green fringe at the same position outside focus. In focus, no color was obvious. This description is consistent with yellow or green focus nearest the objective and red or blue focus slightly farther away. A blurred blue-violet focus formed nowhere. Presumably, violet is folded back near the focus of other colors or it is obscured by brighter wavelengths. Because violet is faint to the eye, it was no longer seen in the background of the other strong colors. The location of the red focus was not far enough behind that of yellow-green to be distinctly noticeable. A polychromatic representation of the design appearing in Section G.2 is shown in Figure 12.14. Note the balanced grayed-out region in the center of both sides of 10 wavelengths defocus. These water-smooth, ring-free regions are a hallmark of apochromatic refractors. Note that they do not appear so flat for ordinary achromats.

The white appearance of a focused star image or the muting of color in views of the planet Venus (in comparison with a conventional doublet) should be a powerful indicator that an apochromat or advanced refractor is properly color-corrected. You should tolerate little spurious color in any modern ED design. Planets should not show easily discerned blurry colors beyond their limbs.

Polychromatic f/8.4 apochromat   10

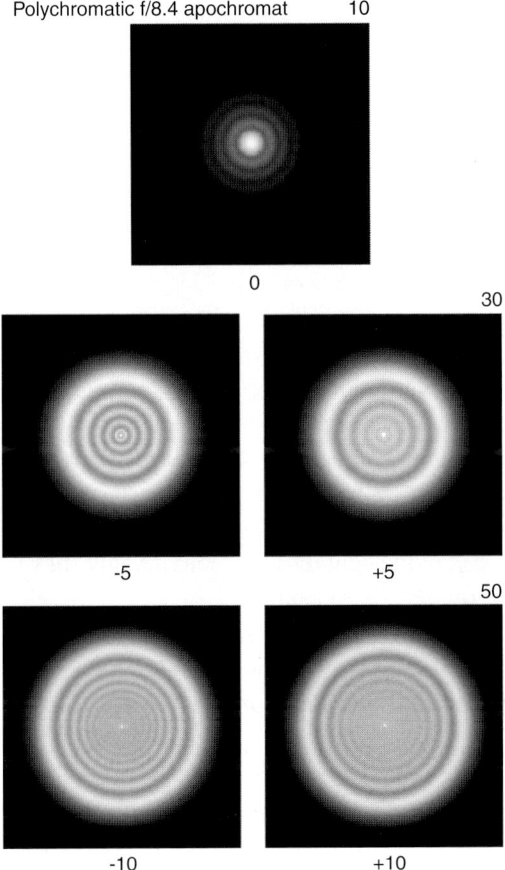

Fig. 12.14 The star test of a perfect 152-mm f/8.4 apochromat. Photopic weighting.

12.6.4 Chromatic Effects in the Eye

Make certain you are not blaming the objective lens for color errors in your own eye. Most people do not realize that their eyes are not achromatic. In fact, the erroneous assumption that the human eye is achromatic led later scientists to question Newton's research on dispersion. In normal daylight vision, the eye-brain system is able to process out much of the color error encountered. Astronomy is no common activity, however, and the processing is subverted during observation. For example, Taylor mentions how the apparent color correction of a telescope was upset with change of magnification and exit pupil.

The answer to this problem is to use the eye at a reduced pupil size. As the magnification of the telescope increases, it illuminates less of the eye's pupil. The eye's color correction improves with smaller pupil size for the same reason that refractors perform better at high focal ratio. Most perceived chromatic aberration is then produced in the telescope instead of the eye. Star test with a shorter focal length eyepiece, but still move the focus in or out the indicated distances.

12.6.5 The Eyepiece

An objective can present a perfectly acceptable image that is destroyed by imperfect achromatism in the eyepiece. One may be unjustly blaming the telescope's objective lens for an error that arises later in the optical train.

Fortunately, the modern general-purpose eyepiece is designed to work passably well even with a steep $f/4$ light cone. Most refractors that we would want to star test operate at $f/8$ and higher. At these mild aperture ratios, compound eyepieces such as Orthoscopics or Plössls work superbly. Not unless they have been improperly made or assembled do they add significant chromatic aberration to the image. Still, they perform best when the star is in the center of the field. Do not make chromatic aberration judgments on decentered stars.

The easiest way of checking your eyepiece is to change to a different one and see if the color error goes away or changes. Also, put the suspect eyepiece in a reflecting telescope and see if the color difficulty is still present.

12.7 The MTFs of Corrected Refractors

Figure 12.1 demonstrated how poorly simple lens refractors transfer contrast from object to image. Let us now look at how well the refractor may perform. Figure 12.15 shows a number of curves. The top curve is the perfect MTF, here weighted by the relative strengths of a 15-color photopic wavelength distribution between 420 and 700 nm. The horizontal axis is normalized to the approximate maximum of the spatial frequency. Of course, the qualifier "approximate" is necessary because no defined maximum for a distribution of wavelengths exists, only an apparent maximum where the transfer is observed to diminish to nothing. The other curves are indexed to this maximum, including the multiplicative factor due to the achromat and apochromat working at two different aperture ratios.

When the $f/8.4$ apochromat that was designed in Section G.2 is weighted photopically, it will generate the next lowest line. At best focus, this telescope has a polychromatic Strehl ratio a little over 0.95, which is hardly a surprise. Apochromats are widely regarded as good telescopes, albeit miniature ones. But there is an interesting fact that can be gleaned

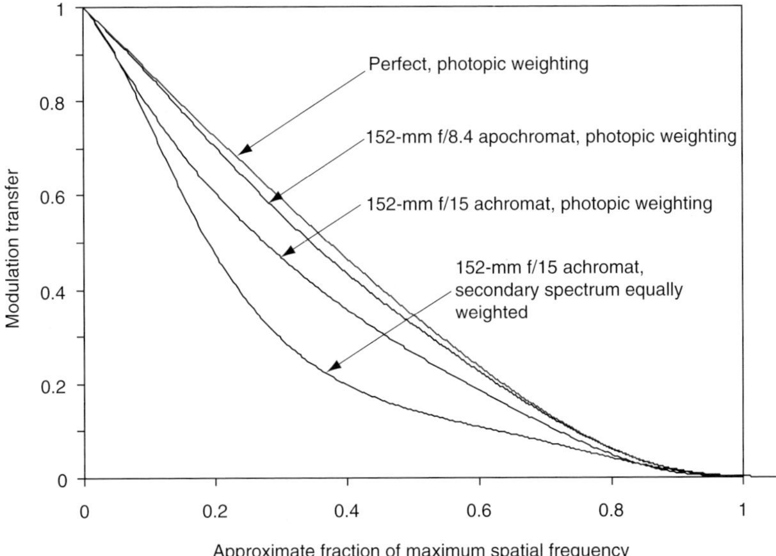

Fig. 12.15 MTFs for real refractors, showing how apochromats are nearly perfect and giving an explanation for why achromats work as well as they do.

from this chart, and that is an explanation of why conventional achromatic doublets work so well.

We saw earlier how an achromat works poorly at the limits of its secondary spectrum, and how various authors have come up with widely-ignored recommendations of how f-number should scale with aperture. Conrady, as you'll recall, recommended $f/29$ for a 6-inch instrument. No one seems to pay attention to this limit, and 6-inch refractors work just fine at $f/15$ and even at $f/10$ if observers are willing to forgive some color. How can this be, when such a bloated ray pattern is visible at the limits of the secondary spectrum in Figure G.1 and such a brief range of diffraction-limited performance is shown by the $f/10$ achromat in Figure 12.9?

Figure 12.15 has a curve depicting the performance of achromats if the wavelength range is equally weighted between the wavelengths limiting the secondary spectrum. The polychromatic Strehl ratio of such a weighting is 0.59, which more or less matched the poor performance anticipated by the spot diagram. However, it also has an MTF curve weighted photopically, and that curve has a diffraction-limited Strehl ratio of 0.81.

12.8 Conclusions and Remedies

About the only optical problems discussed in this chapter that are changeable are those that deal with atmospheric effects or improper assembly of the lens cell. Most other color errors must be handled at the factory. You must be very cautious in blaming a perceived chromatic aberration on the instrument. Personally, I have never seen a quality refractor with grossly improper correction of chromatic aberration, although I have heard of a few cases secondhand. Even department-store models seem to get color correction right although they botch nearly every mechanical feature on the instrument. I have seen fast richest-field refractors or large binoculars that had only marginal color correction, but specialty refractors must be judged to a different standard than the lunar-planetary models. Owners of these telescopes should realize that they have traded increased color error for a wider field. Also, a mishandled chromatic aberration is unlikely to appear isolated from other errors. If improper color appears, other geometric aberrations (especially spherical aberration) are likely to be equally evident.

Testing with a bad eyepiece presents the greatest opportunity for error. Be certain that the suspected chromatic aberration appears in many different eyepieces, preferably not of the same type. Do not test with the Huygenian or Ramsden eyepieces that are often included with inexpensive refractors. If a Barlow is used, make certain that it is achromatic. Because of the high aperture ratios of department-store refractors, some companies have been known to include simple-lens Barlows. It generates an incorrect color correction in the star test. Perform the test at high magnification to avoid color problems in your eye, and be absolutely sure that an unusual coloration in your light source is not shifting the results. Finally, test on multiple occasions before the final assessment is made.

Chapter 13
Roughness

This chapter discusses the diffraction effects caused by quasi-random or asymmetric errors polished into the surface of the glass. The circular rings break up into tiny speckles in focused images, and the out-of-focus disk shows non-circular detail. The reader should take four important points from this chapter:

1. Wavefront roughness usually follows a continuous spectrum at increasingly finer scales, which telescope makers arbitrarily divide into categories such as "dog-biscuit" or "microripple." However, such distinctions are more matters of nomenclature than descriptions of real phenomena.

2. Medium-scale roughness (primary ripple or dog-biscuit) errors are the ones which most severely damage the image because they don't divert light far from the core of the image. The scattered light is hence more condensed and brighter.

3. Roughness errors are difficult to distinguish from turbulence, and careful star testing is required to avoid unfairly judging an instrument. Tolerance for roughness errors must be viewed in the context of likely turbulence errors.

4. Roughness at the scale of microripple is of interest only to makers of specialized instruments who have already reduced other forms of diffracted light to the vanishing point. Microripple of small amplitude has little relevance to general-purpose instruments.

13.1 Roughness Scales and Effects

Harsh methods of polishing and use of fast polishing compounds can lead to random or nonperiodic error that is not a figure of revolution. This so-called "surface roughness" is not generally viewed as a problem on the global scale, the way figuring errors are perceived, but it can be harmful to the image in its own way.

We must carefully define what is meant by surface scale before introducing the deformations of that surface. Think of the aperture as being expanded to the size of the United States. The spherical curvature of the Earth is analogous to the focusing curvature of the wavefront. In the aberration function plots, this curvature has been removed as a noninstructive universal constant. The map has been flattened on the average.

Figuring errors are large-scale deformations. An example is spherical overcorrection, which starts fairly level at the center of the aperture and rises to a peak at the 70% zone. The aberration function then falls rapidly until it reaches the edge. Similarly, the center of the U.S. starts fairly flat in the plains states, rises toward the Appalachian or Rocky Mountain ranges, and then falls rapidly toward the oceans.

No serious mapmaker would suggest that U.S. topography could be completely represented by a simple model of two ridges with a flat area between. Nevertheless, we may usefully describe the coarsest features of the landscape with broad-brush concepts like *continental divide* and leave the narrower details for later. Describing the figuring errors of the optical surface as "spherical aberration" or "zonal defects" is that sort of large-scale reference. Here "scale" refers not to how high the aberration is but how *wide* it is, or rather, how persistent the aberration is over long distances.

To improve the map, the landscape is refined by adding rivers or watersheds. Many of these features extend over areas the size of an entire state. On the mirror, we can decide to measure roughness with ruler divisions of around $\frac{1}{10}$ or $\frac{1}{20}$ of the aperture. These are "medium-scale" roughness errors. We could also map individual mountains or county-sized variations in the terrain and the analogous "small-scale" roughness errors. With sufficient magnification, the map could chart the position of boulders, plowed fields and ditches. Likewise, if we examine the optics on the molecular scale, we see a convoluted surface, but such errors are so much smaller than the wavelength of visual light that they cannot be sensed by ordinary means. The wavefront remains flat after encountering molecule-sized roughness.

Medium-scale roughness errors go by the colorful name "dog-biscuit" and the less-colorful name "primary ripple." Their greatest width scale approximately matches the spaces between grooves on the polishing tool. These channels are always cut or cast into the polisher to provide space for the pitch to spread with pressure and to supply reserves of the finely-suspended polishing abrasive. The grooves are a necessary evil. If they are not cut into the lap, large-scale shape errors are generated that are even worse than moderate amounts of roughness.

Small-scale roughness errors, called "microripple," have spacings of about 1 to 2 mm. The cause of these errors is less obvious than primary ripple, but their origin is probably found in the choice and use of polishing materials. Cerium-oxide polishing compound seems to give rougher results than rouge. Waxy laps yield more rippled surfaces than pure pitch, and paper laps are worse than wax. At the basic level, however, non-uniformities in the glass itself seem to limit the smoothness. Texereau claims that the lap is able to attack the surface of the mirror through a combination of physical and chemical means, and that once begun, such errors are self-sustaining (Twyman 1988, pp. 578–584; Texereau 1984, pp. 88–91).

Certain fast-acting laps deliver a roughness with characteristic dimensions intermediate in scale between classic primary ripple and microripple, descriptively called "lemon-peel" surfaces. This appearance is rare in instruments intended for astronomical use, however. Usually, telescopes are polished on gentler materials.

We might expect the diffraction image from the roughness facets characteristic of primary ripple to be 5 to 20 times larger than the unaberrated image, but that simplified logic does not take into account the accidental correlations that occur when nearby scattering facets act in phase with one another. Antinodal bright areas and nodal dark regions will form. The net effect of mild primary ripple is to blow the scattered light into a knobby glow surrounding the image, which has its greatest brightness within a radius less than 5 times the Airy disk. Such scattered light can be a bad problem because it is condensed enough to easily see.

Let's compare that defect with the likely behavior of microripple. Texereau states that microripple is occasionally as bad as 6 nm on the wavefront and has an average spacing as small as 1 mm (Twyman 1988, p. 580). However, we are fortunate that the slope of each 1 mm facet is seldom correlated with the slopes from nearby facets, so the effective apertures of the facets only statistically combine. This 6 nm case is also the worst one; most wavefronts have microripple below 1 nm ($< \frac{1}{500}$ wavelength). Because of the small size and lack of correlation of the scattering surfaces, the diffraction pattern of scattered light from microripple is a shattered dim glow, quite similar to the aura that occurs with a turned-down edge. Microripple of small amplitude is hard to see with the Foucault knife-edge test. It requires specialized equipment to detect unambiguously.

When roughness is small, as it is for optics, it only affects the diffraction shape of the image in a minor way. It removes light from the focused image and shoves it out into a blotchy halo of small diameter for primary ripple and large diameter for microripple. The missing energy is calculated by seeing how much the central intensity is lessened.

The Strehl ratio of roughness can be calculated from an approximation (Born and Wolf 1980, p. 464). Here i_s represents the Strehl ratio at best focus and σ_{RMS} is the root-mean-square deviation of the wavefront (in wavelengths) as measured from the reference sphere centered on best focus:

$$i_s \approx 1 - (2\pi\sigma_{\mathrm{RMS}})^2 .$$

13.1

For example, a $\frac{1}{14.05}$-wavelength RMS deviation yields a Strehl ratio of 0.8 (the Maréchal tolerance). A $\frac{1}{20}$-wavelength error typical of noticeable primary ripple gives a ratio of 0.9. A severe case of microripple might have a deviation as large as $\frac{1}{100}$ wavelength, so the intensity is reduced only to 0.996. Clearly, microripple has a very different character than primary ripple.

Another approximation to the Strehl ratio has been given by Mahajan (1982):

$$i_s \approx e^{-(2\pi\sigma_{RMS})^2} .$$

13.2

It gives a more accurate number than Equation 13.1 at large aberration amplitudes. We can invert Equation 13.2 for a $i_s = 0.8$ to define a "Mahajan tolerance" of about $\frac{1}{13}$ wavelength RMS. For small roughnesses, however, the difference between these approximations is negligible.

13.2 The Terminology of Roughness

We must distinguish the nomenclature from the reality of surface error. Roughness is typically modeled either as a *continuous* spectrum from large scale to small scale, or as a discontinuous *composite scale* spectrum. The "spectrum" in this case is not composed of light intensities plotted versus colors, but magnitude of roughness versus width scales. A composite scale is something like huge swells of water on which are superimposed tiny wind-blown capillary waves—smooth undulations with wrinkles. The terminology of calling lap-sized roughness "primary ripple" and small roughness "microripple" originates from assuming that two separate causes generate composite-scale roughness.

I have seen mirrors that obviously obeyed this composite-scale model very well. They were covered with smooth, wavy roughness that displayed little of the smaller-scale roughness in between the primary ripple and microripple. More often, though, mirrors appear less and less rough at more diminishing scales, but there is no one scale at which the tester can say the roughness stops (an example appears in Figure 13.1). Such names as "primary ripple" mean less on such surfaces because there is always a scale just below it. We might call it a "not-so-large ripple" scale, followed closely by an "even smaller ripple" scale. The model described below follows this continuous spectrum behavior instead of the composite-scale that

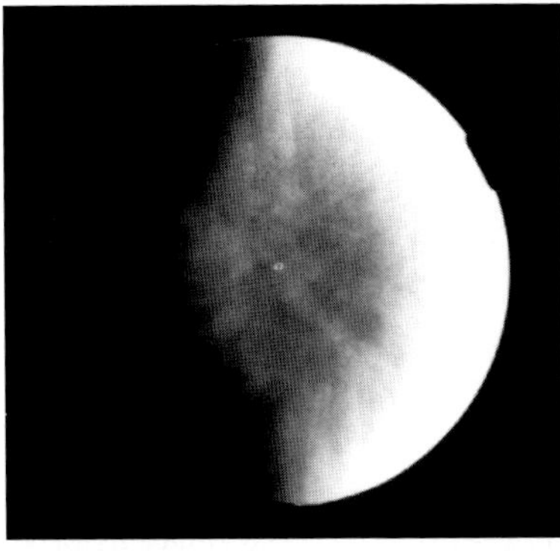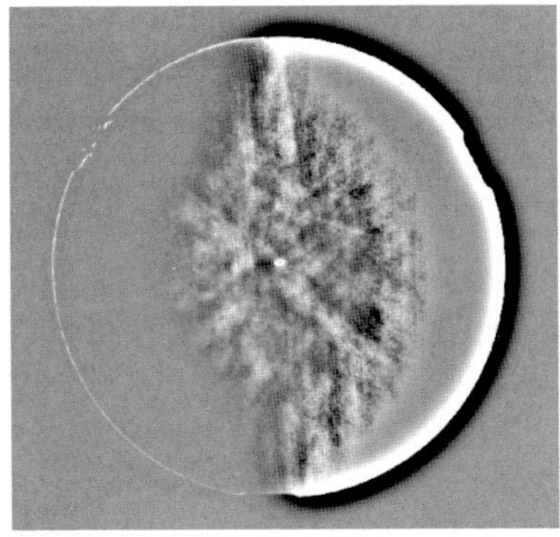

Fig. 13.1 Roughness visible in the central zone of a mirror in the Foucault test. At right: contrast is enhanced by subtracting an unsharp-mask image.

the terminology is based on. The words will continue to be used because they are convenient and firmly lodged in the literature.

Another reason for using the artificial divisions of roughness into medium-scale versus small-scale is that they are typically found in different tests. The Foucault test is good down to a small-scale of roughness intermediate between primary and microripple, but then its sensitivity fails. The investigation of microripple requires a phase-interference test that deliberately suppresses sensitivity to large-scale error by use of a coarse slit. One may incorrectly assume that there is no roughness at in-between scales merely because it has not been seen.

13.3 Medium-Scale Roughness, or Primary Ripple

The appearance of a mirror suffering from primary ripple is shown in the Foucault test photograph of Figure 13.1. The roughness is only apparent in the center of the mirror at this knife setting. Ripple extends into the bright and dark regions, though it is less visible in these areas. The roughness is dimly seen as a random structuring with a set of superimposed grooves beginning in the center and extending outward.

This 6-inch (150-mm) *f*/5 mirror was produced during the manufacturing boom at the last return of periodic comet Halley and is typical of the sloppy fabrication practices of the time. As bad as it looks, the roughness is estimated at only somewhere between $\frac{1}{10}$ or $\frac{1}{20}$ wavelength RMS (peak-to-valley roughness between $\frac{1}{4}$- and $\frac{1}{5}$-wavelength). Spherical aberration amounting to 1 wavelength overcorrection was also found during this bench test. Obviously, the spherical aberration was the worse failing.

Your eye also suffers from medium-scale roughness. Take aluminum foil and perforate it with a pin. Hold the foil about 8 to 15 cm in front of your eye and look through the pinhole at a frosted incandescent light bulb. Try to focus your eye on the lamp, not the pinhole, and cover the other eye. If you have punched the right size hole in the foil, you should see a mottled disk that roughly approximates the out-of-focus patterns seen in this book. The outside ring is perhaps the only one clearly delineated. The appearance may be cleared up slightly by placing a colored filter between the lamp and the pinhole.

As you blink, horizontal lines appear briefly on the defocused disk. You can see that some of the details change with every blink. They are probably caused by variations of the moisture thickness on your cornea. Depending on how bright the light is, you may also see some dim radial spikes outside the disk. These spikes may be caused by diffraction from the non-circular iris opening or streaks in the roughness.

The roughness is visible as coarseness in the expanded disk. This coarseness does not vary from blink to blink. The aberration is many wavelengths high, so the appearance of individual rings is obscured and confused. When I was 19, I had a fleck of metal removed from my cornea, and the evidence of the trauma can still be seen in the out-of-focus image. So roughness can originate from corneal defects, surface roughness in the eye lens, and non-uniformities in the refractive index of the eye or the transparent parts of the eye itself.

The human eye is not even close to diffraction-limited. An eye with a 3-mm iris opening (typical during daylight) can theoretically resolve lines separated by 0.6 arcminutes, but a person who resolves lines only 1 arcminute apart is deemed to have excellent vision. How can we test telescopes to the diffraction limit through such an imperfect aperture?

In fact, the answer to this seeming paradox is quite simple. Angular errors in the instrument are magnified until they are bigger than errors in the eye. Once the separation between the finest possible details have been magnified beyond 5 arcminutes (i.e., ⅙ the diameter of the Moon), aberrations in the telescope begin to dominate aberrations in the eye. A 1-inch aperture should resolve lines separated by 0.092 arcminutes, so sufficient image size is reached by 5/0.092 ≅ 50 power/inch (20 power/cm). Somewhere beyond this magnification, even perfect telescopic images begin to become fuzzy.

Ironically, some people boast about telescopes that can "withstand more than 100 power/inch" (40 per cm). What they don't realize is that they're not bragging about the telescopes. They are inadvertently admitting the poor quality of their own visual acuity. When using extremely high magnifications beyond 100/inch, the diffraction disk appears bigger than

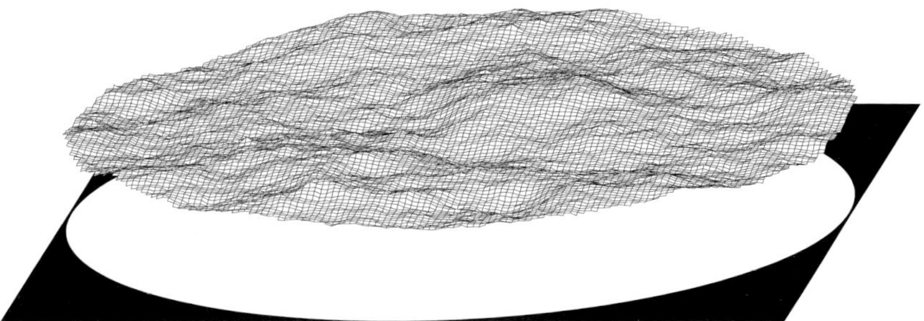

Fig. 13.2 The aberration function of medium-scale roughness, also called "primary ripple" or "dog-biscuit."

two-thirds the angular diameter of the full Moon.

13.3.1 The Aberration Function of Medium-Scale Roughness

To generate rough wavefronts, the fractal model described in Chapter 7 on turbulence is again used, with certain modifications.

The first change is to suppress midpoint deviations for two iterations. This step ensures that the surfaces so generated will be uncorrelated at distances greater than $\frac{1}{8}$ to $\frac{1}{4}$ of the aperture. One doesn't expect that medium-scale roughness will persist over long distances, and by not allowing the surface to deviate until it is divided into a grid of 16 squares, this correlation scale is achieved. Only 16 points out of almost 13,000 are artificially clamped to zero, but the entire character of the surface is changed.

The other modification is to avoid quenching the roughness. Figure 13.1 shows fine detail over scales smaller than the tool spacing (presumably about $\frac{1}{8}$ of the diameter). In the case of turbulence it was desirable to suppress the deviation at small scale because no mechanism existed to produce it. Turbulence cells have a quasi-period of about 10 cm. Roughness in the glass, because it originates from many causes at a number of scales, will be modeled here to behave as a self-similar fractal.

Figure 13.2 shows an example aberration function of medium-scale roughness. The streaks visible in Figure 13.1 are not represented in this algorithm. One expects such grooves to diffract light into low-contrast spikes at right angles to their extent (as a spider vane does). We must not overemphasize their importance, however. The visual perception system tries to create order in what we see. It is especially fond of straight lines and often creates a line where only a hint of one is actually present. Orion's belt, for example, is curved a great deal. The eye imposes linearity because it *prefers* linearity.

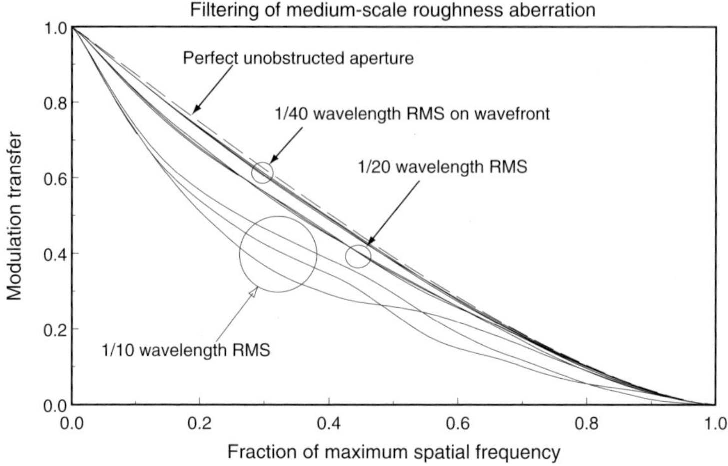

Fig. 13.3 Filtration caused by primary ripple. Three amounts are shown: 0.1, 0.05, and 0.025 wavelength RMS wavefront deviation.

13.3.2 Filtering Effects of Medium-Scale Roughness

Because the scale of roughness is much smaller than the whole aperture, one expects a brisk drop at low spatial frequencies, a condition similar to turned edge. The MTF thereafter should remain a fairly fixed fraction of the perfect MTF. Thus, the average degradation drops from unity to a constant at about the correlation length (Schroeder 1987, p. 208). We see in Figure 13.3 that the previous guess of a correlation length of ¼ to ⅛ of the aperture is a good one. The sagging of the curves seems to have reached a steady fraction of the perfect aperture's MTF at about that range.

Roughness aberration is nonsymmetric, so three curves are plotted for target patterns with bars oriented up-down, left-right, and at a 45° angle. Because these curves represent a single realization of the rough surface rather than an average over many such surfaces, the MTF wiggles somewhat. These curves are examples of the variations that can be expected from changes in the MTF-target orientation or from slightly different surfaces.

The degradation is severe for 0.1 wavelength RMS wavefronts, but it improves rapidly for smaller roughnesses. The quality is acceptable for wavefront roughnesses less than 0.05 wavelength RMS. Some manufacturers guarantee optics that smooth, but they always give the specification on the surface rather than the wavefront. Read claims carefully.

Also shown is the smooth wavefront with RMS deviation of only ¹⁄₄₀ wavelength (¹⁄₈₀ wavelength on a mirror surface). If any reasonable care is taken, all astronomical optics can be made this smooth. Global wavefront aberrations are difficult to reduce below $\frac{1}{28}$ wavelength RMS ($\frac{1}{8}$ wavelength peak-to-valley), but optics can readily be smoothed until the wavefront roughness is less than $\frac{1}{40}$ wavelength RMS deviation.

13.3.3 Star Test on Medium-Scale Roughness

Two focus runs appear in Figures 13.4 and 13.5. The first is an image sequence of an otherwise perfect $\frac{1}{40}$-wavelength RMS roughness wavefront. Even though the wavefront is very good, the roughness is detectable in the out-of-focus images. Figure 13.5 doubles the aberration and uses a different fractally-derived wavefront. This $\frac{1}{20}$-wavelength RMS aperture is acceptable in the MTF chart, yet it seriously distorts out-of-focus images. Fortunately, it seems to tuck away the messiness visible out-of-focus to yield a fairly crisp pattern while in focus. If this pattern were turbulence, it would be at least a 9 on the 1–10 Pickering seeing scale.

We go approximately 8 wavelengths on either side of focus (Table 5.1). If the wavefront has primary ripple close to $\frac{1}{40}$ wavelength RMS, the effects of roughness are very delicate and hard to detect. At $\frac{1}{20}$ wavelength RMS (about the limit of what you should tolerate), you will see it all too plainly.

Often, you must test for roughness alongside some spherical aberration. Roughness is easier to see on the soft-edged side of focus. The dim outer portions of the disk flare into a twisted, asterisk-like pattern. Don't concentrate on the roughness until you have determined that spherical aberration is acceptable. Spherical correction errors are much more damaging to high resolution images than roughness errors because their surface scale is so large.

13.3.4 Roughness and Turbulence

Turbulence closely resembles roughness, so it interferes strongly with the star test for that aberration. Thus, roughness is nearly impossible to check under real skies using an actual star. Nights where the air is absolutely still are so rare that they will never coincide with a deliberately-planned star test. Besides, star tests are the last thing the observer wants to do on nights of exceptional steadiness.

An artificial source is often crucial to test for roughness. Also, testers cannot check for primary ripple just anytime and anywhere. They must try for a serendipitous combination of time and place that results in a tranquil testing path. Maybe the necessary conditions will occur during a night when the temperature is not dropping too rapidly. The likeliest good tests are conducted on a windless evening or a very early morning over grass.

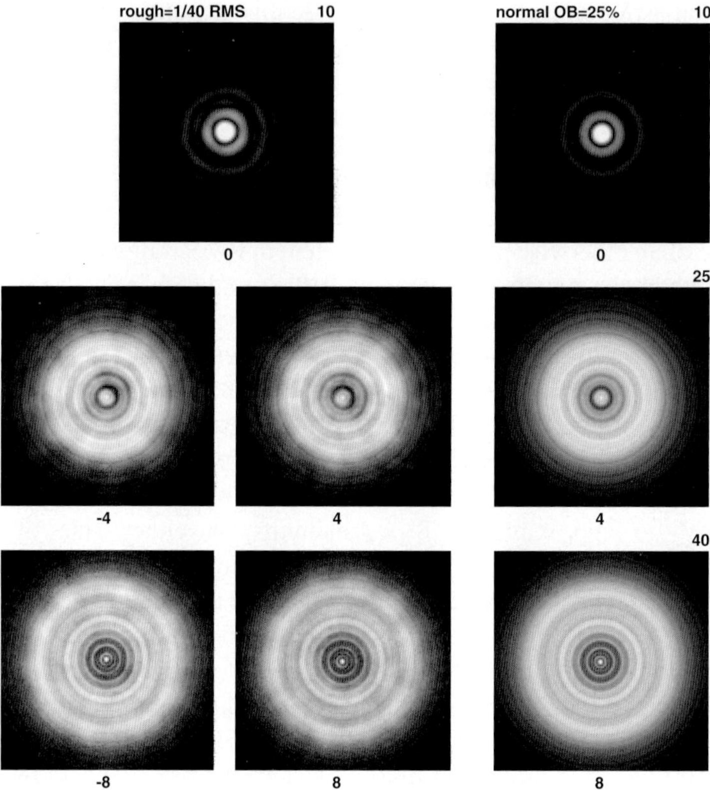

Fig. 13.4 Medium-scale roughness of ¹⁄₄₀ wavelength RMS with defocusing aberration from −8 to +8 wavelengths. Obstruction is 25% and the perfect image is seen in the right column.

The artificial source test goes best when the flashlight and sphere are set up in the bright period after sunset but before twilight has ended. Use the Sun to illuminate the sphere in the early morning. If these times are difficult to arrange, test for roughness with the artificial source at night. (See Hufnagel 1993, pp. 6–12.)

Also, you must be realistic in your expectations. Roughness is only important if it is a significant fraction of the similar turbulence aberration. Thus, roughness is judged to be objectionable only if it appears during the nights that turbulence affects the telescope least. If the image is flickering a small amount when one sets up the test, the result is not useless.

See if the fixed roughness pattern is visible even against the relatively light turbulence in front of an artificial source. If you cannot discern a fixed roughness pattern under these excellent conditions, then you may be assured that during actual use the roughness is affecting the image little.

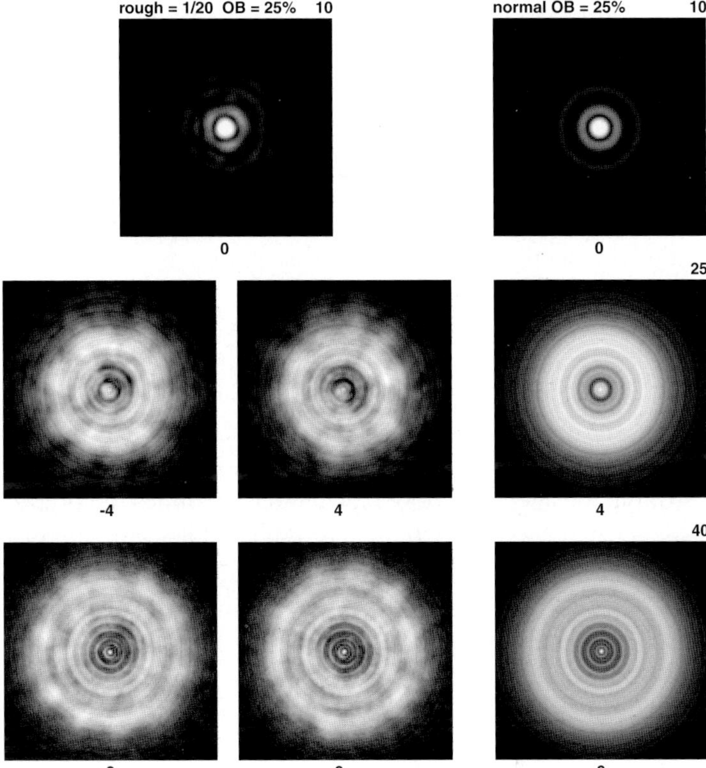

Fig. 13.5 Another primary ripple wavefront, this time having a statistical deviation of $1/20$ wavelength RMS, is focused from −8 to +8 wavelengths. The smooth wavefront is to the right.

The scale must also slide a little to accommodate local conditions. If seeing is abysmal during 99% of the nights at your location, perhaps roughness is less important. In locations with good seeing, roughness standards must be tighter.

If roughness is still grossly objectionable even after taking these mitigating conditions into account, the optics must be refigured.

13.4 Small-Scale Roughness, or Microripple

The original concern over microripple stems from efforts early in the 20th century to observe the solar atmosphere all of the time, not just during total solar eclipses. The solar corona is a thin, high-temperature gas that extends out several solar diameters. Observations during eclipses were excellent, but resembled infrequent snapshots. Scientists wanted a method to view the inner corona daily and to monitor its changes. The corona is brighter than the full Moon, but it can be lost next to the hellish intensity of the Sun.

To make a telescope capable of blocking out parasitic scattered light requires more advanced methods than the empirical baffling recipes generally used by astronomers. André Couder stated that the intensity of the corona 5 arcminutes from an occluded solar image is only about 1 millionth as strong as the light intensity streaming from the unblocked Sun. Even nonuniformities in the glass of his coronagraph lens scattered light only a little less intense. He estimated the scattering from the atmosphere, even on a clear mountaintop, was about half as intense as the corona, and the scattering from the most carefully cleaned lens was equally bright (quoted by Twyman 1988, p. 585). Clearly, contrast was already suffering badly from unavoidable effects, and little room was left for scattering by microripple. The severe $\frac{1}{100}$ wavelength RMS microripple example mentioned above by Texereau would be unacceptable. It scatters 0.4% of the energy striking the aperture, more than a thousand times too bright.

Texereau also describes a finely polished surface with microripple of approximately 0.05 nm RMS deviation, or about $\frac{1}{11,000}$ wavelength. This value inserted into Equation 13.1 results in an intensity reduced $(2\pi/(11,000))^2$ from 1, or only 3×10^{-7}. This tiny amount of scattered light would be even less intense by the time it was considered at an angle of 5 arcminutes. Such a surface is sufficiently smooth to be used in a coronagraph (Texereau 1984, p. 88).

Optics exhibiting the primary ripple of Figure 13.1 demand little or no remediation of the relatively subtle effect of microripple. In fact, the fractal model of medium-scale roughness automatically includes a moderate amount of microripple (about $\frac{1}{1,000}$ wavelength RMS), but the presence of coarser-scale roughness dominates the lesser scale.

Nevertheless, we want to investigate what happens when primary ripple is stripped away and all that is left is small-scale roughness. Microripple is often blamed for low contrast. Can this mysterious roughness scale be responsible for so many optical worries?

13.4.1 The Aberration Function of Small-Scale Roughness

The fractal algorithm was not used in the microripple model, because we want to remove medium-scale features and concentrate on small-scale effects alone. A pseudo-random number generator was used to assign heights to the 128×128 pupil grid. Because asymmetry was expected, tending toward flat tops with sharper grooves, a square root was taken of this starting surface, and the result was normalized. The surface is shown in Figure 13.6.

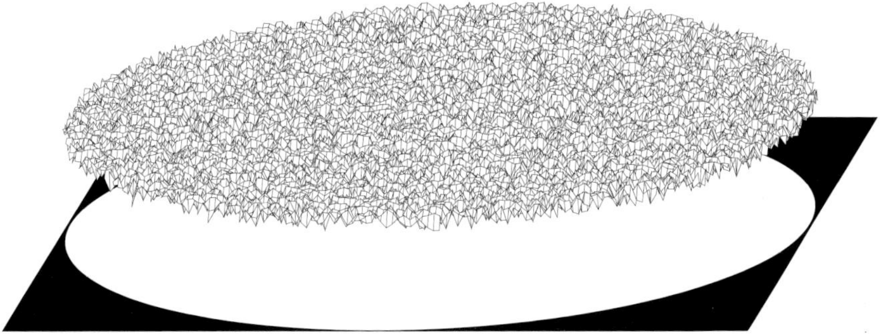

Fig. 13.6 A modeled microripple surface with amplitude expanded greatly.

13.4.2 Filtering of Small-Scale Roughness

Consider what was said above concerning correlation length. Because the microripple has a very tiny correlation length (perhaps less than a millimeter or so), we should expect a precipitous drop in the modulation transfer function, followed by a constant degradation. These effects are illustrated in Figure 13.7. MTF drops quickly and then remains a more or less constant fraction of the perfect value.

In fact, microripple as large as $\frac{1}{10}$ wavelength RMS is an unlikely event, even though it appears on the graph. It is shown only to make the fast initial drop more apparent. Texereau gave a worst-case amount of only $\frac{1}{100}$ wavelength. We can see by the behavior of the MTF graphs that $\frac{1}{100}$ *wavelength of microripple would be indistinguishable from a perfect aperture in most dark-field observing situations.* Calculated images differed little from perfection, so no star test diagrams of microripple appear here.

13.4.3 The Great Unknown

So why do we hear all sorts of warnings about the debilitating effects of microripple? Microripple is one of those myths of telescope making that feeds on folk wisdom and hearsay. Part of the trouble is the difficulty of measuring it. Texereau describes a technique (originated by Lyot) that requires an attenuating phase plate, a specialized device that delays and weakens the propagation of the unscattered wavefront. Since the strength of microripple cannot be quantified easily, it gets the blame for any unidentified optical difficulty.

Also, published descriptions of the construction of unusual instruments such as coronagraphs (where microripple *does* matter) tend to frighten readers. People believe that microripple may affect more prosaic forms of ordinary observing. The light diffracted from a slightly turned far edge is vastly stronger than that coming from microripple. Nearly every

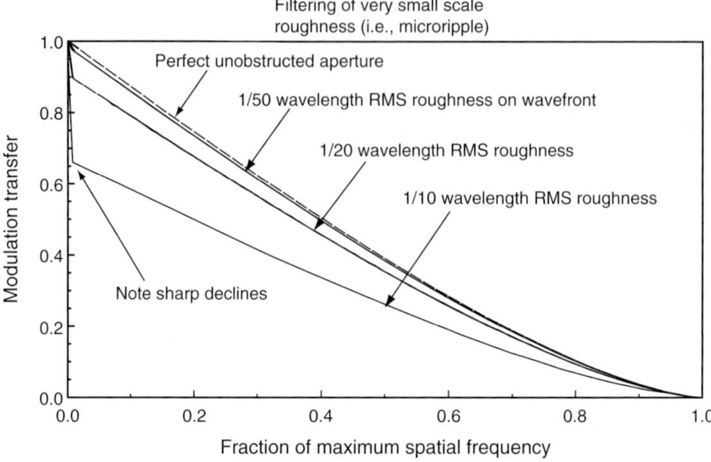

Fig. 13.7 Various MTF curves characteristic of microripple. The aberrations appearing here are exaggerated to show the shape.

telescope with an unmasked edge has much more diffuse light than the scattering caused by the microripple roughness.

Extended objects suffer far greater degradation from obstructions like spider vanes, mirror clips, and tiny screws projecting from the side of the secondary holder than from microripple. For example, spiders deflect more light than any likely case of small-scale roughness. A spider with 4 vanes 0.5 mm thick on a 200-mm aperture diffracts 0.5% of the incident energy, slightly more than Texereau's $\frac{1}{100}$-wavelength worst case microripple.

A root-mean-square microripple error of 0.5 nm (about $\frac{1}{1000}$-wavelength) results in a Strehl-ratio decrease of 0.00004. If this missing light is distributed over the field of interest, it results in a signal-to-background ratio on extended objects that can be as bad as 25000. This value is well above the 1000 maximum defined in Section 9.4 for dirty optics.

For most dark-field observing, microripple is not very harmful. For example, 1 mm width-scale roughness scatters light into a halo about 100 arcseconds wide. Maybe 10% of that energy will cover a 20-arcsecond image of Mars. The rest of the stray light is beyond the planet's limb where it doesn't contaminate the image. Thus, the signal-to-background ratio improves to 2.5×10^{6}, a very large value indeed. An analogy to this signal-to-background ratio is the darkening of the image of the Sun as viewed through a safe solar filter. If the signal were the unfiltered solar image, the noise caused by $\frac{1}{1000}$-wavelength microripple would be about as strong as the filtered Sun. If you require an instrument capable of discerning dim

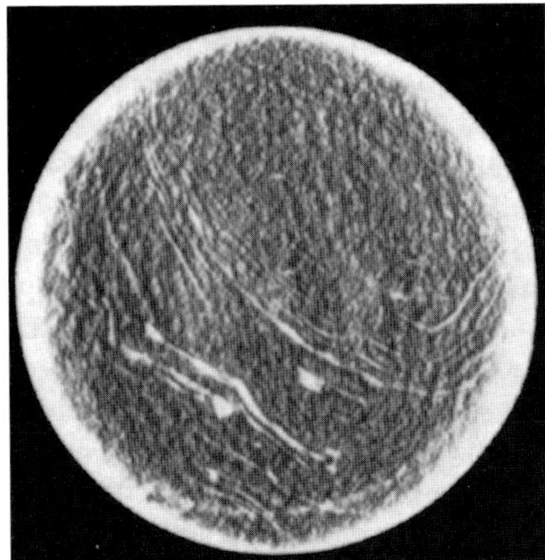

Fig. 13.8 A testing technique originated by Lyot reveals microripple and veins of differing hardness on the glass. (From How to *Make a Telescope* by Jean Texereau, copyright 1984 by Willmann-Bell, Inc. and used with permission.)

details next to a bright source of interference, you may well need to worry about microripple. Small-scale, low-amplitude roughness is not a threat for most observers.

Chapter 14
Astigmatism

Astigmatism is the tendency of an objective to focus at two distances. Each focal position is stretched at right angles to each other. An average focus is found between them, with the image forming a crosslike pattern. Because of the stretching at the angles of the cross, resulting in lack of a unique focus, contrast of high-spatial frequency information is lessened. This chapter will discuss a number of points:

1. Astigmatism has many separate causes, and most can be cured at moderate or no expense.

2. Unambiguously locating the source of astigmatism takes careful logic (and to be honest, a certain amount of luck).

3. Small amounts of astigmatism are detected by gently rocking the focuser about the position of best focus.

4. The way to determine whether the telescope has too much astigmatism is to try to detect non-circularity of the image at a calibrated focus setting.

5. Astigmatism is frequently caused by a simple curvature in the diagonal mirror, either those inside Newtonian telescopes or diagonals at the eyepiece end. A sensitive test exists for these.

14.1 Astigmatism in Eyes and Telescope Optics

Astigmatism means *not stigmatic,* or *not focusing to a spot.* In the widest possible definition, all of the nonsymmetric aberrations discussed in this book are astigmatic. The symmetric aberrations may produce fuzzy spots, but at least they are round.

This type of aberration was originally defined by Ludwig von Seidel and modified by Frits Zernike (Born and Wolf 1980, p. 470). Astigmatism divides focus into two stubby lines separated by a short distance along the direction of eyepiece travel. The lines are at right angles to one another. A sort of "best focus" is seen at the halfway point between them, where the illumination pattern takes the shape of a cross.

Astigmatism is frequently corrected in human vision.[1] The *cylinder*

---

[1] More common problems fixed in eyeglasses are the *spherical* corrections, which are modifications of the focusing power of the eyes. "Spherical" to an optometrist refers to the $A_2\rho^2$ focusing term of the expansion of Equation 10.1. The words "spherical aberration correction" used by telescope makers designate the $A_4\rho^4$ term of that expansion.

numbers in an eyeglass prescription are adjustments for astigmatism. This aberration is usually caused by a deformation of the cornea, the transparent outer covering of the eye. Astigmatism could result from a tighter "stretch" of the cornea in one direction than in the other, as well as a misshapen eyeball. The aberration can be temporarily corrected in some cases by pulling one corner of the eyelid.

For example, my right eye has a cylinder power of 2 diopters[2] at an axis of 90° and my left eye has a power of 2.5 diopters at 88°. Because I also have significant focusing correction, I see point sources stretched up in one eye and flattened in the other. The approximate 90° axis is probably no accident; the stretch direction may be related to the muscle attachment points of my eye or to the shape of the orbital cavities in my head.

The simplest example of an astigmatic lens is a cylinder with no curvature in one direction and significant curvature in the other. Truncated chords of cylinders are often used in reading magnifiers. By putting the magnifier right against the page, one can suppress the poor focusing properties of the lens and stretch tiny letters in one direction.

Truly cylindrical lenses or mirrors are almost never used on purpose and never happen accidentally. The most common form is a hybrid of spherical and cylindrical shapes. In fact, one can subtract the spherical focusing power of a lens entirely and be left with a purely astigmatic form.

Such a lens is modeled as part of a torus, a doughnut-shaped surface. The particular form that is most interesting here has a hole in it with exactly the same diameter as the torus' cross section. A drawing of such a doughnut appears in Figure 14.1, with half of the torus chopped away to reveal the interesting region. Consider the tiny circle drawn on the inside of the hole. When lifted away from the rest of the torus, that circular piece has the same surface shape as an astigmatic lens with no focusing power (the other side of the lens is flat). The curve goes into the paper along the up-down direction and out of the paper to the left or right. Imagine looking down on a saddle shape, with the left-right directions toward the pommel and seat, and up-down toward either stirrup.

Astigmatic surfaces can be modeled as combinations of purely spherical surfaces and toroidal surfaces. If we move the tiny circle of Figure 14.1 to the outside of the doughnut, its surface no longer resembles a saddle. Instead, it looks like one of these hybrid surfaces. It has lower astigmatic power than before, with some spherical curvature mixed in. In the astigmatism of a real lens, the converging wavefront doesn't truly have

---

[2] A diopter is the reciprocal of the focal length in meters. The typical 2-meter focal length of a small Schmidt-Cassegrain telescope produces a focusing power of only 0.5 diopter.

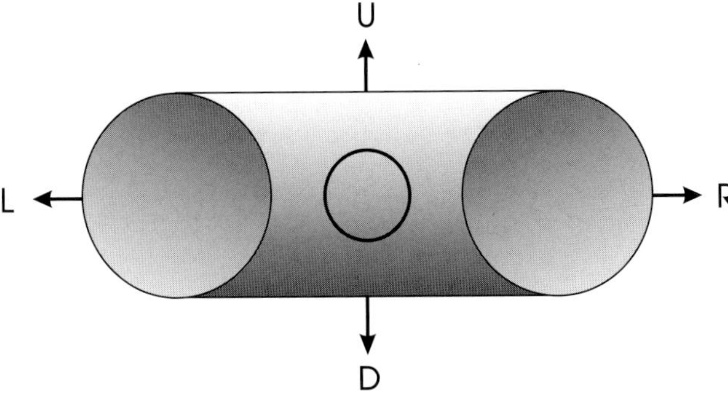

Fig. 14.1 A tiny circle on the inside of a torus is the purest form of the astigmatic deformation.

a saddle shape. The shape is produced only after the spherical focusing component is subtracted away.

14.2 Causes of Astigmatism

Astigmatism in telescopes is a symptom of several optical problems, some of which are listed here:

1. Misalignment (often mixed with coma).

2. Poor support of the optic's weight, causing pressure directed along an axis (the "potato chip" sag).

3. A supposedly flat diagonal or right-angle prism that is actually slightly spherical (a surprisingly common problem).

4. A slight cylindrical deformation ground or polished into the glass (a very tiny amount is found in most mirrors and lenses).

5. Poorly annealed glass (often appears in "porthole" mirrors or other thick pieces that were made with another application in mind).

6. Uncompensated astigmatism in the observer's eye.

The most disastrous of these causes are number 4, true astigmatism ground or polished into the surface, and number 5, unrelieved stresses contained in the glass itself. Saving money by using undocumented glass for the mirror substrate is a risk that many makers are willing to accept, and they often get away with it. The astigmatism that is derived from the other sources, however, is no less objectionable. Luckily, these causes are easily repaired.

In the days when all mirrors were small and thick compared with their diameter, astigmatism was comparatively rare. Even halfhearted adherence to the optical shop practice of rotating the tool with respect to the mir-

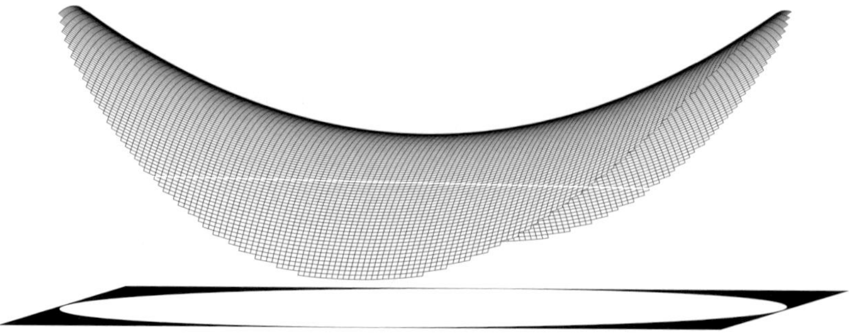

Fig. 14.2 The saddle-shaped aberration function of astigmatism just after it has passed through the aperture.

ror and inverting which piece was on top usually kept astigmatism dormant. Only when stresses were frozen into the mirror disk itself (cause #5) did astigmatism appear.[3]

Cause #4 is more common now that large, thin mirrors are being figured. Even careful opticians can polish cylindrical curves into such a flexible surface. If conditions that result in astigmatism appear, they are usually a function of how the mirror was supported during grinding or polishing and are therefore persistent. Astigmatism, if it appears in such mirrors, is usually severe.

Discovering the cause of observed astigmatism to determine possible strategies for its elimination takes methodical but straightforward detective work. The process will be described beginning with Section 14.5.

14.3 Aberration Function of Astigmatism

The term added to the wavefront aberration function to account for astigmatism alone is (Born and Wolf 1980, p. 470)

$$W_{\text{astig}}(\rho, \theta) = A_2^{\text{astig}} \rho^2 \left(\cos^2\theta - \frac{1}{2} \right). \qquad \textbf{14.1}$$

Here A_2^{astig} is the coefficient giving the amplitude of the astigmatism, ρ is the distance from the axis, and θ is the angle from the axis of astigmatism. The constant ½ subtracted from the $\cos^2\theta$ term is the needed focus shift to produce a Zernike best-focus aberration. Below, the astigmatism axis is conveniently placed at $0°$, but it can occur at any angle. W_{astig} is graphed in Figure 14.2.

---

[3] If the disk itself was at fault, Russell Porter gave a succinct, though final, solution: "Seek out a good hard, solid hydrant. Hurl the mirror as fiercely as possible at said hydrant. Walk home." (Ingalls 1976).

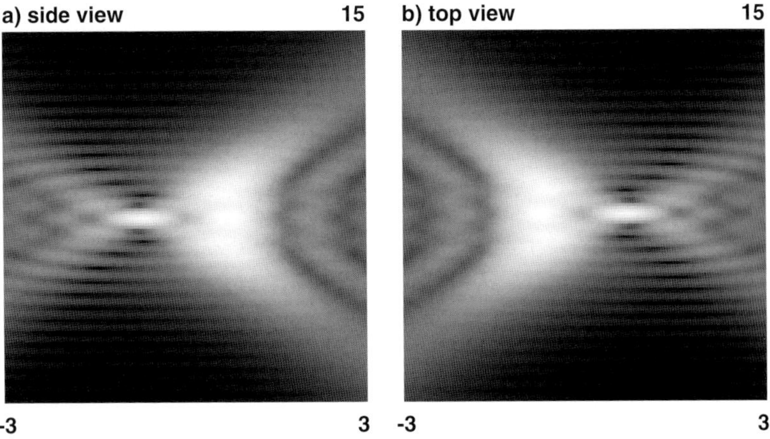

Fig. 14.3 Two slice patterns depicting 1.5 wavelengths of astigmatism. The up-down dimension of each frame is 15 angle units (1.22 = Airy radius). Defocus is between −3 and 3 wavelengths. Best focus is at the center.

For example, if A_2^{astig} is ¼ wavelength, then W_{astig} goes from $+\frac{1}{8}$ wavelength to $-\frac{1}{8}$ wavelength. This definition is slightly different from the usual one for primary astigmatism. Coefficient A_2^{astig} is the total peak-to-valley aberration and is the number that compares most closely with the Rayleigh tolerance.

Equation 14.1 has two interesting features. The first is the way the astigmatism scales with distance from the axis. It has the same power, ρ^2, as defocusing aberration, so one may interpret astigmatism as a pathological form of defocusing. Also, at a defocusing aberration of $A_2 = \pm(1/2)A_2^{\mathrm{astig}}$, the aberration becomes flat along one line, with the line for the "+" value oriented at right angles to the line for the "−" value. This fact demonstrates that an optimum focal point exists for each axis of the mirror and that these optimum foci are on either side of conventional focus. The actual best focus position is a compromise between these axes.

The division of focal regions is best seen in Figure 14.3, where the same longitudinal slice through focus is viewed first from the side and then from the top. The views are mirror opposites of one another. The approximate lines of brightest focus are the three-lobed wing structures, only one of which appears in each diagram. The little bright lozenge that appears fore or aft of the wing is just the other wing viewed end-on. If these slices could be perceived in three dimensions, one would see a pair of illuminated regions vaguely resembling boomerangs at right angles to each other, overlapping nose-to-nose.

As the astigmatism gets worse, the bright region looks less like a wing

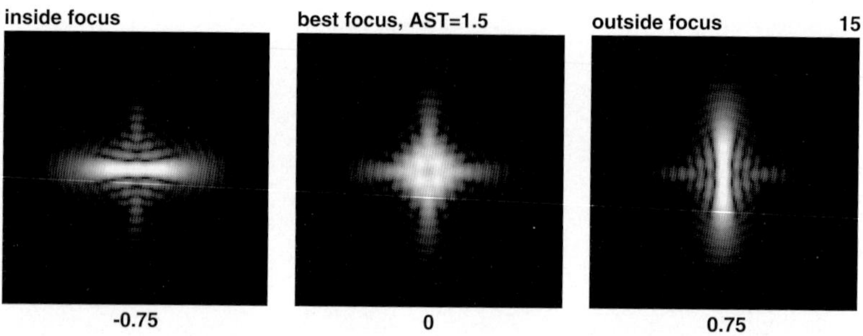

Fig. 14.4 Strong astigmatism (1.5 wavelengths total aberration): inside focus, best focus, outside focus.

and more like a long bar. The two bars separate and move apart. In the diagram, the bright regions separate and become more like lines.

The best focus diffraction pattern pinches the rings into two bars for small aberrations (see Figure 14.4). For larger amounts (many wavelengths) of astigmatism, best focus (if any focus could be called "best") resembles a square-woven basket. At 1.5 wavelengths of aberration, an on-axis dip in intensity or a nodal minimum exists. By defocusing to ±0.75 wavelengths, we see the wings of Figure 14.3 as they would appear face-on in the eyepiece.

Another interesting feature of Equation 14.1 is that the toroidal shape is approximated by a parabolic radial dependence. Like spherical aberration, there are higher order terms to this expansion as well, but they are customarily ignored.

14.4 Filtering of Astigmatism

Since astigmatism is not circularly symmetric, the transfer function depends on how the MTF bar target is oriented. Hence, Figure 14.5 shows each aberration amount as two lines, with all intermediate MTF curves approximately between them. Clearly, astigmatism is profoundly harmful at 1.5 wavelengths. The 0.8 Strehl ratio occurs at a little less than ⅜ wavelength peak-to-valley.

Because it affects the aperture on the same scale as defocusing, astigmatism is primarily an intermediate-to-high spatial frequency error. It affects the transfer of contrast only mildly at low spatial frequencies. The transfer function of astigmatism doesn't fall as quickly with increased spatial frequency as the MTF for primary spherical aberration or obstruction. Look at its performance at middle spatial frequency, however. The lowest curve even for the marginally acceptable ⅜-wavelength range degrades the

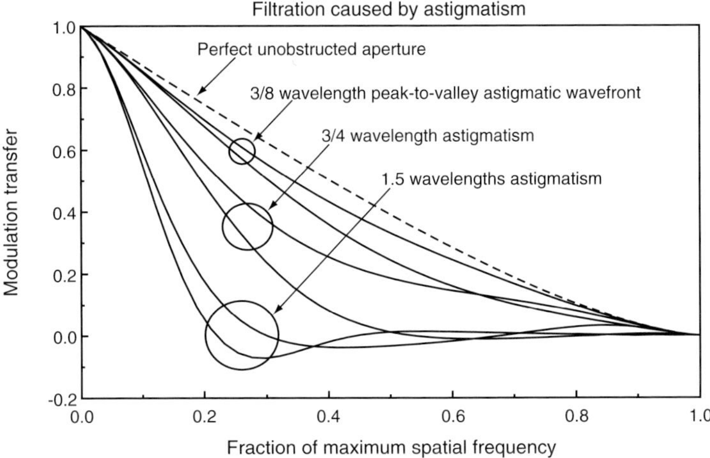

Fig. 14.5 Modulation transfer functions for astigmatism at best focus. Three amounts are shown: ⅜ wavelength, ¾ wavelength, 1.5 wavelengths. An aberration of ⅜ wavelength is just outside the Maréchal limit (0.8 Strehl ratio). In each pair, the lower curve is for MTF bars oriented along the axes of astigmatism. The higher curve is for bars at 45°.

MTF to an average of 70% the perfect value.

The degradation bottoms out on the lower curve of the ¾-wavelength MTF at about 0.5 of the maximum spatial frequency. For a 200-mm aperture, details with spacing of about 1.2 arcseconds and less would be the most severely damaged. Because it attacks high spatial frequencies, astigmatism reserves its worst behavior for lunar-planetary observation. This aberration may partially account for the poor reputation of thin-mirror Newtonians for high-magnification observation.

14.5 Star-Test Patterns

Star-test focus runs appear in the next two figures. The aberration amount in Figure 14.6 is ¾ wavelength. It is shown as it would appear with 20% obstruction. Figure 14.7 shows effects of ⅜ wavelength of unobstructed astigmatism. The signature of astigmatism is the oppositely directed oval appearance on either side of focus. Recall that these patterns are calculated with an astigmatism axis aligned with the square. In real observing, astigmatism is not required to sit upright. It can be seen at any angle. As defocus is increased, the pattern becomes less elliptical. Defocusing is kept small in these figures because astigmatism most severely affects high spatial frequencies and is most readily detected close to focus.

In fact, the most useful method to detect astigmatism is to focus a dimmer star and then rock the eyepiece back and forth across focus. The astigmatism stretches the image first one way and then the other, and the aberration is immediately apparent. A focused image is supposed to appear as a cross, but you may have trouble seeing the diffraction disk at all in the telescopes for which astigmatism is likely. More common are apertures that just show a hint of the pattern, as in Figure 14.7.

If ellipticity is seen, turn to Table 5.1 and determine how far you must move the eyepiece to defocus 4 wavelengths. For a focal ratio of *f*/6, the amount is only 0.025 inches or 0.63 mm. Defocus this far and carefully inspect the image of a dim star. If the pattern is distinctly elliptical, the telescope has too much astigmatism. Ideally, astigmatism is difficult to detect at 2 wavelengths defocus.

However, do not be surprised if most telescopes suffer from a trace amount of this aberration, if only because of the ever-present force of gravity. Few instruments have none.

14.6 Identification in Newtonian Reflectors

Just because the instrument suffers from astigmatism doesn't necessarily mean that the problem is ground into the glass. Finding astigmatic error is easy; identifying its source is more difficult.

First of all, determine if your own eye is causing the astigmatism. Simply using your other eye to look through the telescope often does not help (see my prescription above), so that's not the way to tell. You can try rotating your head with respect to the eyepiece and see if the astigmatic axis follows, but changes caused by such slight rotations are difficult to perceive.

The best way to determine if the problem is in your eye is to increase the magnification. You will make the astigmatism easier to see if it is contained in the telescope, and diminish its strength if it is contained in your eye. Few eyes are so bad that they will show strong astigmatism if the exit pupil of the telescope is set below 1 mm. The bundle of light exiting the telescope decreases in cross-sectional area with higher power, and the illuminated area of misshapen cornea is reduced significantly. As the exit pupil approaches a pinhole, your eyes perform better.

At this point, spin the eyepiece. The astigmatism axis should remain fixed. If it follows the rotation, the eyepiece is astigmatic.

The next thing to investigate is the angle along which astigmatism seems to compress or stretch the image. In refractors or Schmidt-Cassegrains, the angle is obvious. For Newtonians, you can perform the same trick used in Section 6.5.1. By defocusing a long way and poking your

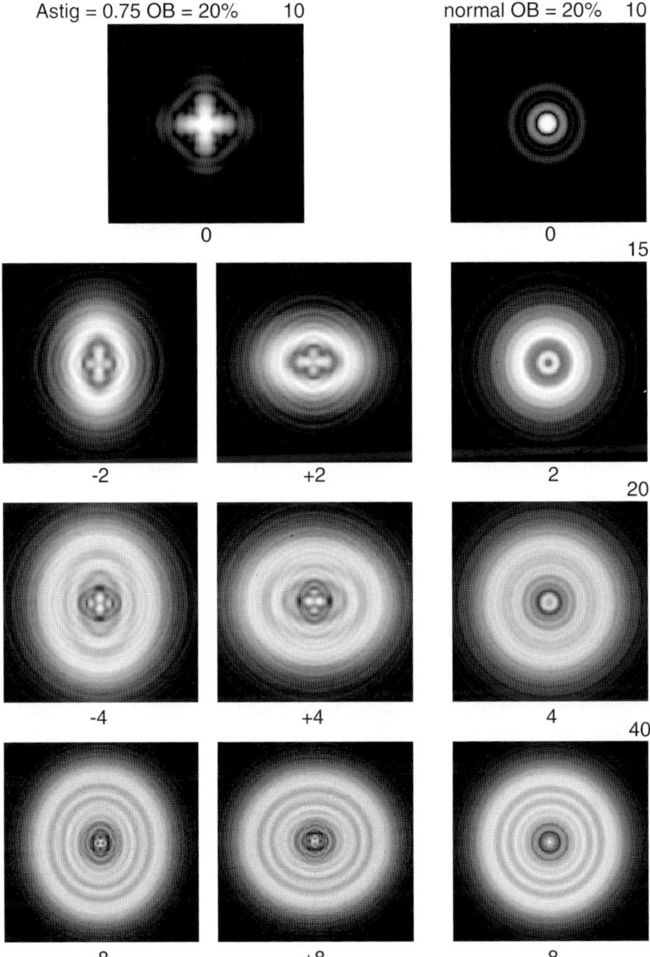

Fig. 14.6 Focus run of ¾-wavelength astigmatism. The normal aperture is in the right column. Obstruction is 20%. Ellipticity is still visible at 8-wavelengths defocus.

hand halfway into the optical path, you can determine the direction of stretching. Be suspicious of the main mirror cell if the astigmatism axis is oriented along the horizon or elevation direction (that is, in the gravity axis). Suspect the diagonal cell or the diagonal itself if the astigmatism seems to be aligned with respect to the tube (in the diagonal-tilt axis).

Unfortunately, these two situations sometimes occur together, especially for large altazimuth Newtonians with a level eyepiece. Therefore, we must change the deformation for one mirror or the other. The first thing to try (especially if you're doing a level-telescope artificial source test) is to point the telescope at a star near zenith and see if the astigmatism is still

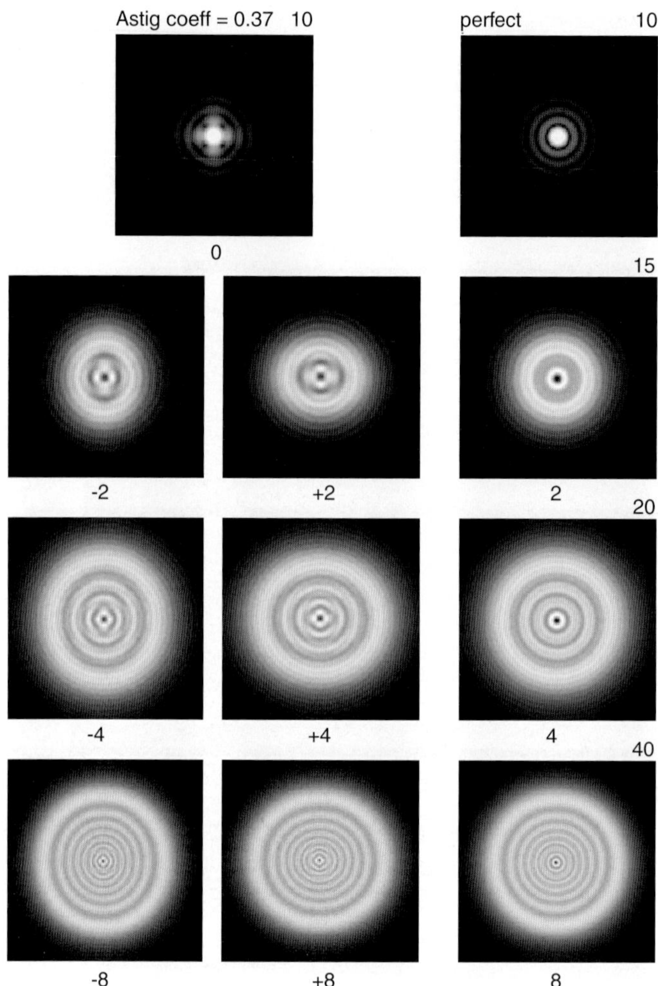

Fig. 14.7 Unobstructed aperture with ⅜-wavelength total astigmatism. The right column depicts a perfect aperture. Stretching is visible at 4 wavelengths defocus, but is hard to see at 8 wavelengths.

visible. If the main mirror is riding on a single edge point, this step will redistribute the weight of the mirror and hence its compression. Perhaps the pattern will still be bad, but at least it will be different. It might, for example, transform into a three-sided pattern. If the image is unaffected, shake the telescope slightly and try again.

If no change is seen, main mirror warping may yet be the problem. In the case of a mirror glued into its cell, one adhesive pad may be unduly straining the mirror, and the tension might have little to do with the mirror's weight.

The worst possibility is that the mirror actually has an anomalous curvature in the glass. To determine if that problem afflicts your telescope, try a 120° rotation of the mirror. Realign the telescope and test again on a star near zenith. The error should also rotate if it is contained in the mirror.

The image rotates. Ask yourself if the mirror performed well in the past, and only recently has performed poorly. If so, the weight of evidence seems to indicate that the mounting has failed or the mirror has become chipped or cracked.

Disassemble the whole mirror cell, inspect the mirror, and carefully reassemble the cell. Make certain that the mirror is adequately held but is nowhere pinched. For mirrors glued to a flat metal plate, make certain that you do not overly tighten this plate into place.

If the telescope is still astigmatic after these corrective measures or if it has always performed badly, then consider the possibility that the aberration is either ground into the glass or the mirror substrate material was poorly annealed.

The image doesn't rotate. When a 120° rotation does not affect the axis of astigmatism, then your attention must turn to the diagonal. Again, ask yourself if it performed well up to a recent date, then suddenly turned bad. Or have images always been marginal, and you are just now investigating with the star-test techniques presented here?

If the loss of quality has been abrupt, then the diagonal mounting is probably at fault. Take it out and look it over. Have you recently taken it apart for cleaning? Have you recently adjusted it?

Many Newtonian diagonal holders use cotton wadding behind the diagonal. Often, too much cotton stuffs the holder and causes tension on the mirror. Use only enough cotton behind the diagonal to hold it into place. Some diagonal holders have a split-cylinder construction. During reassembly, carefully widen the slit enough to give the diagonal mirror room. Make sure that a screw is not protruding from the base to strike the mirror in its back side, and be certain that the tip of the diagonal does not touch the base.

If all these mounting changes do not help, perhaps the diagonal has a spherical curvature polished into its surface. Because this sphere is at a 45° angle, the effect on the image turns to astigmatism.

You can check this poor performance by removing the diagonal holder and rigging it on a separate support (a photographic tripod is ideal). First, perform a straight star test on a refractor larger than the minor axis of the diagonal using an artificial source. Verify that the image is nicely

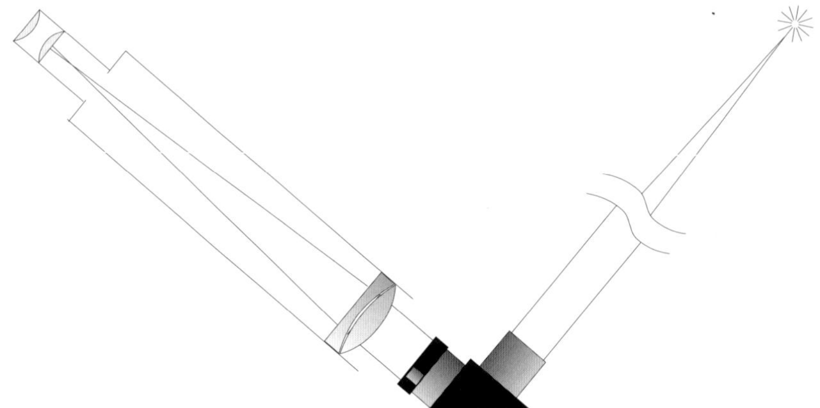

Fig. 14.8 A sensitive test of diagonal mirrors and star diagonals.

circular. Then, do the star test again, looking at approximately 45° through the diagonal (see Figure 14.8).[4] If the diagonal has curvature, you will see that the refractor has developed a sudden case of astigmatism.

You can also check the operation of star diagonals or right-angle prisms using this auxiliary telescope technique. You may rely on any mirror satisfying such a check because the test demands equal path length over the whole surface of the right-angle bend. This condition is much harsher than actual operation demands (see Section 14.8 below concerning spherical deformation in diagonals). In normal use the converging light cone for any point of the image leaves much of the surface dark, but during the test the whole path is equally lit.

14.7 Refractors or Schmidt-Cassegrains

Except for instructions concerning the diagonal, the comments and instructions dealing with Newtonian reflectors above are still useful. First, remove the right-angle prism or mirror if you are using one. Many such devices have poor optical quality, and the problem may be cured by this simple expedient.

If astigmatism is seen in a refractor, one possible cause is poor alignment. Enough surfaces are present in such telescopes that designers, if they so choose, can strip coma away from off-axis images. Only astigmatism is left. Try to realign using instructions in Section 6.5.2.

The secondary is not strongly tilted in Cassegrain-style reflectors, as it was for the Newtonians described above. Thus, a spherical error in the

---

[4] Pointing this combination can be frustrating, but you will eventually succeed.

secondary will not show itself as astigmatism. Astigmatism must come from misshapen or severely strained optics. One likely source is the secondary holder itself. Some secondary mirrors aren't sufficiently isolated from the adjustment plate at their bases. Distortions in that plate can be transmitted to the mirror and show up as astigmatism (or worse) in the image.

The secondary mirror of a Schmidt-Cassegrain adjusts by the turning of set screws with a hex-head wrench. Because of the mechanical advantage of such a tool, you can unknowingly put huge tensions (500 pounds or more) on the secondary cell. The secondary is the only mirror to adjust in Schmidt-Cassegrains, so if you can't eliminate astigmatism by reducing forces in the secondary holder, you must return the instrument to the maker for servicing.

14.8 Spherical Deformation of Diagonal Mirrors

What is the tolerance for spherical deformation of a mirror set at 45 degrees, either for directing the beam out of the instrument (e.g., Newtonian mirrors) or making the viewing position comfortable (a star diagonal on a refractor or Cassegrain)? It depends on the amount of diagonal illuminated

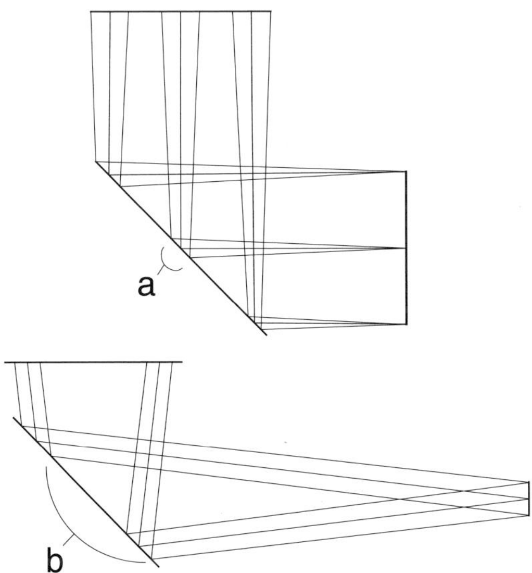

Fig. 14.9 Two situations requiring much different spherical curvature tolerances. The top figure depicts a typical situation for a star diagonal used with Cassegrains or refractors. The spot a) reflecting light from a single image point is much smaller than a spot b) derived from an f/4.3 Newtonian reflector.

by a single point on the object. The minor axis of such a spot is much easier to calculate than the diagonal size of Appendix C. It merely is the distance from the focal plane to the diagonal T divided by the aperture ratio of the telescope F,

$$D_{\text{spot}} = \frac{T}{F}.$$ **14.2**

For the example star diagonal on the top of Figure 14.9, T ranges from about 30 mm to 60 mm and $F = 12$. Thus D_{spot} is from 2.5 mm to 5 mm, with a representative average of 4 mm. For the 10-inch $f/4.3$ Newtonian diagonal of the bottom drawing, T is 175 mm and D_{spot} is 41 mm. It is not difficult to see that the Newtonian diagonal can tolerate less curvature than the star diagonal. There is a bit of mythology that says something like "as the beam approaches focus, optics can be less accurate." This is usually followed by an incorrect explanation that an error close to the eyepiece has less opportunity to divert a ray as far as an error far away. In fact, wavefront error occurring anywhere along the optical path has an equal effect on the image. It's just that the errors from a single spot are applied to smaller parts of the diagonal, allowing the whole surface to deviate more.

Looking at the deformation to the wavefront (see Appendix C), the radius of curvature of the diagonal can be calculated that will result in a Strehl ratio of 0.8 at 560 nm:

$$R_{diag} = 842 D_{\text{spot}}^2 \ [mm] \ .$$ **14.3**

For the two examples above, the star diagonal used at $f/12$ can tolerate a radius of curvature of 13500 mm, and the $f/4.3$ Newtonian diagonal a radius of curvature of 1.4×10^6 mm. Of course, a maker would never want to push the tolerance right against the limiting Strehl ratio, so a safe limit at twice those radii of curvature yields a Strehl of 0.95. It is interesting (and useful, if a diagonal must be ordered) to figure out what the radii listed here come out to in terms of perpendicular surface flatness over the whole diagonal's surface. In the case of the 20% obstructed 10-inch $f/4.3$, the maximum spherical error tolerable on the diagonal is 0.8 wavelength. In the case of 13500 mm radius of curvature over the long axis of the 25x35-mm star diagonal, we find that the deviation is as much as 20 wavelengths! In each case, however, the error on the wavefront of an astigmatic beam is ⅜ wavelength peak-to-valley, together with trifling amounts of other aberrations.

For a non-spherical error, the situation is more complicated, primarily because the wavefront is not able to be refocused. In this case, if the peak-to-valley surface error over each beam spot is confined to $\frac{1}{5.4}\,\lambda$, it is well within the Strehl = 0.8 tolerance, and by 0.1 λ, it is superb.

14.9 Remedies

Nothing can be done about astigmatism in the glass except to refigure or replace the optics. Perhaps the maker will agree to refigure them for you, since the optics either left the factory with astigmatism or developed it later because of improperly annealed glass.

Luckily, astigmatism is most often the result of improperly assembled mirror cells and secondary holders. Increasingly, mirrors are held in cells by the technique of gluing them to a plate instead of the more benign method of holding them loosely by clips. The thin plates buckle and transmit their warped shape to the optics through forces applied by the adhesive pad.

We tend to judge everything by familiar uses, and in this case what we know is tightening bolts in, for example, an automobile. Everyday vibrations would quickly shake an automobile to pieces if the screws weren't tight. However, telescopes are delicate. The worst vibrations they are likely to encounter is a gentle, infrequent jiggling during transport. They don't turn at 2000 RPM like an internal combustion engine but shift position perhaps 40 times in a night. They are scientific instruments. Tighten screws firmly but gently.

Chapter 15
Accumulated Optical Problems

15.1 Breaking the Camel's Back

What you have been shown thus far is how the individual aberrations and transmission variations can affect the image. More important, however, is the way minor problems add up. "The wobbly stack" of Figure 3.1 shows how many errors accumulate as a collection of filters. Even if each filter is relatively unimportant, the total filtration could render the image fuzzy and indistinct.

The concept of modulation transfer outlined in Section 3.3 presented a single standard around which we could rally. By defining optical quality as the ability of the system to preserve contrast in an image, many disparate optical problems were compared on an equal footing.

As each optical problem was discussed, it was assumed that the single error was the only difficulty. Of course, this is nonsense. Like wolves, optical problems travel in packs. The expression "maximum tolerable" was used in earlier chapters to indicate the worst amount of a single error that can be endured. Unfortunately, any further loss of imaging quality, whether it was derived from that particular error or not, causes the MTF curve to sag even lower.

For an aperture troubled by acceptable amounts of several errors, let's calculate the ability to preserve contrast as each successive optical problem is piled on. This will show, as no individual aberration section could, the deterioration of optical quality as an accumulation of small weights. Eventually, the modulation transfer function collapses under the load. Ironically, we will see that the inadequacy of imaging performance in our example is not caused by any one error on the glass. Instead, poor imaging generally results from the summed effects of several errors, including poor telescope alignment and the deterioration of atmospheric conditions.

The optical problems of Table 15.1 are sequentially added to a perfect aperture. The example pupil function is not unusual and may even be considered better than normal. The total errors in the first column do not add up straightforwardly, so the root-mean-square deviation is shown as it accumulates in the last column. The amounts of cell pinching and misalignment seem excessive, but these errors do not compare well with the ¼-wavelength Rayleigh criterion. Because they are more limited in area

Table 15.1
Aggregate errors in wavelengths

	Total of each	Peak-to-valley error	Cumulative RMS error
25% Obstruction	–	–	–
Undercorrection	0.20	0.20	0.054
Cell pinching	0.29	0.40	0.077
Misalignment	0.30	0.53	0.094
Turbulence	0.27	0.70	0.117

than correction mistakes, they have to have a higher total value to result in the same degradation. All of these errors are about equally bad, and none would significantly damage optical quality by itself.

Note that this aperture is not affected by a turned edge, zones, astigmatism, or surface roughness. In fact, it would bench test very nicely, with only ⅕-wavelength undercorrection error. Warping is an important aberration here, so we might think of this aperture as one of the thin-mirror Newtonians common today. The value of misalignment is consistent. These fast instruments are difficult to keep collimated and many of them routinely are used in a state of poor alignment. A misalignment aberration of 0.3 wavelength, when appearing alone, still reduces the contrast less than ¼ wavelength of spherical aberration. The RMS deviation in the last column points out how some of the aberrations cancel others. For example, the misalignment appearing alone would affect the RMS deviation by a little less than 0.07 wavelengths.

The stacked MTF curves appear in Figure 15.1. Only one curve for each asymmetrical aberration is shown, all for the same orientation of the target bars. Each optical error degrades the image slightly, some seeming to be stronger at some spatial frequencies than others. In fact, the particular mix appearing in this example appears to be complementary. The spherical aberration wipes out contrast at lower spatial frequencies (around 0.2 maximum) while the misalignment acts more strongly at medium spatial frequencies (0.5 maximum). The boxes represent an otherwise perfect aperture that has a defocusing aberration of 0.4 wavelength. This single aberration follows the lower-limit envelope fairly well.

Surprisingly, this diagram represents the typical operating condition for a large astronomical telescope (i.e., one that is seeing-limited more often than limited by deficiencies of fabrication). The choice of aberration amounts in the example has been especially generous as to air turbulence. According to material presented earlier, the chosen amount of turbulence would rate a high 8 to low 9 on Pickering's "seeing" scale.

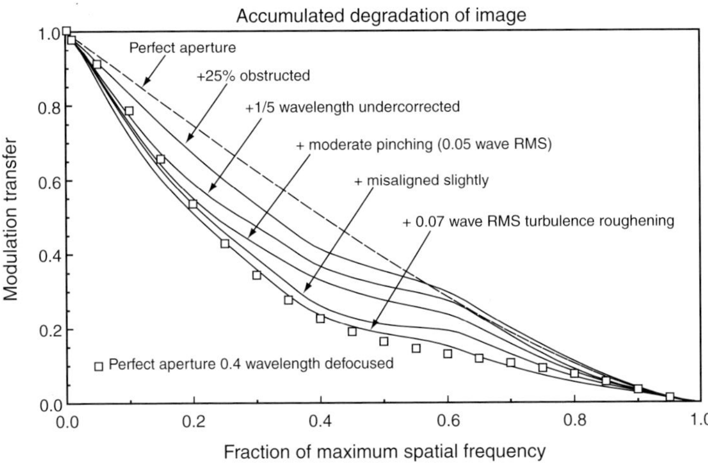

Fig. 15.1 Filtering of a realistic aperture. The curves show the filtration as aberrations are successively added.

Star tests are shown in Figure 15.2 as these optical problems added one-by-one. Each difficulty is not too much worse than the frame above, but the bottom row is considerably poorer than the first row.

Peter Ceravolo has made a well-calibrated set of 6-inch (150-mm) *f*/8 mirrors having peak-to-valley correction errors of 1, ½, ¼, and ¹⁄₁₀ wavelength. When he set them up side-by-side and had observers rate them, he noticed that people had no trouble telling the 1 and ½-wavelength correction errors from the better pair, but had a hard time distinguishing the ¼-wavelength mirror and the nearly perfect one (Ceravolo *et al.* 1992).

Based on the calculated example above, we can speculate on the reasons for this failure to discriminate a nearly perfect mirror from a barely acceptable one. Ceravolo's telescopes were probably less troubled by misalignment, pinched optics, or obstruction than by turbulence. Terence Dickinson rated seeing at 7 out of 10 in one such session. When we consult the turbulence material in Chapter 7, we can estimate that "seeing" of 7 induces an aberration somewhere near 0.10 wavelength RMS. Recall that ¼ wavelength of correction error is about 0.075 wavelength RMS. The lack of a visible performance difference between the better mirrors could have occurred because the contrast degradation was dominated by the seeing. Not until the spherical aberration became big enough to compare to the turbulence-generated roughness was the difference easy to perceive.

Figure 15.3 shows focused examples of these correction errors with the turbulence aberration added (assuming no other aberrations contribute). A 20% obstruction is reasonable for a 150-mm *f*/8 Newtonian.

Place this figure at some distance from your eyes and try to perceive

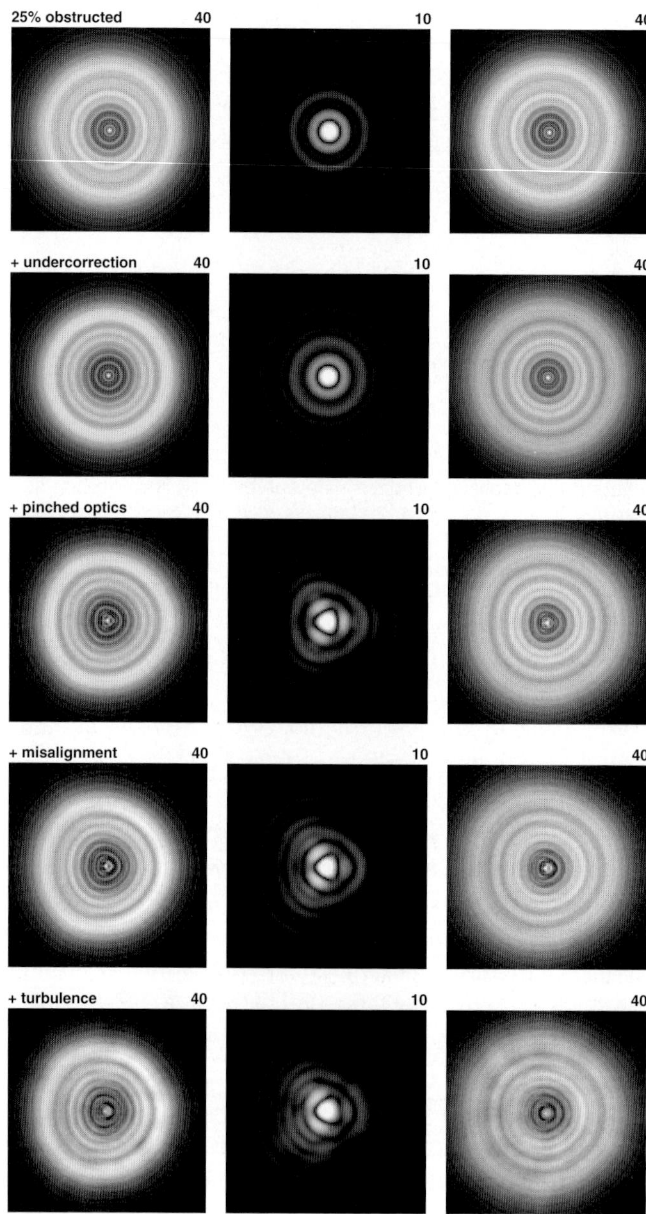

Fig. 15.2 Perfect aperture as each aberration is added in Table 15.1

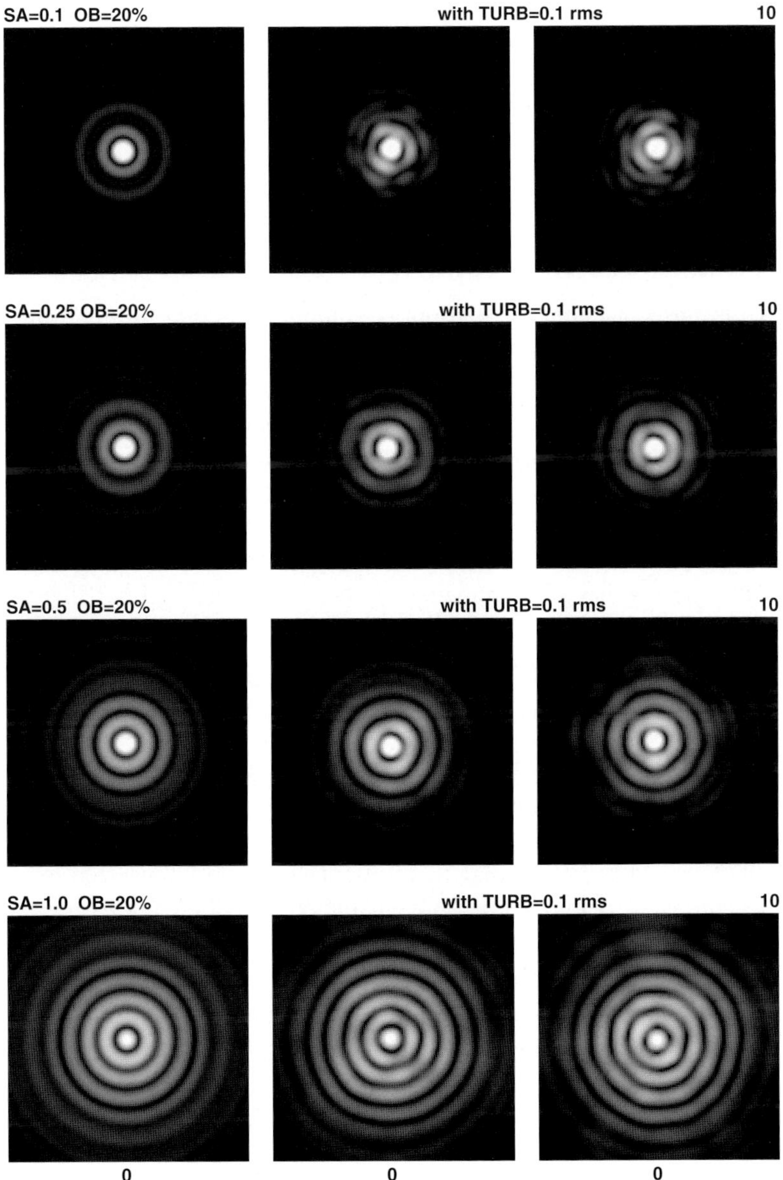

Fig. 15.3 A model of Ceravolo's four mirrors if they are used in the presence of atmospheric turbulence with effect on the wavefront amounting to 0.1 wavelength RMS. Obstruction was arbitrarily chosen at 20%. The aberrations caused by surface errors alone are in the left column. The other columns are examples of fractally-derived wavefronts superimposed on the correction errors.

the difference between $\frac{1}{10}$-wavelength and $\frac{1}{4}$-wavelength image frames. Keep in mind that the turbulence patterns continuously change. The modulation transfer curves appear in Figure 15.4. Turbulence is sufficient to corrupt the excellent optics of the $\frac{1}{10}$-wavelength mirror and it dominates the correction error of the $\frac{1}{4}$-wavelength mirror. Only the lower quality mirrors are so poor that spherical aberration overpowers turbulence.

15.2 Fixing the Telescope

In the absence of hard information, people tend to concentrate on one of the telescope's suspected optical difficulties and blame it for everything. Mirror-making enthusiasts, for example, go to extraordinary lengths to figure ultra-precise optical surfaces. They know optical fabrication, so they see all error in terms of improperly shaped surfaces. People who assemble their telescopes from prefabricated parts deal exclusively with telescope design. Some zealously reduce the size of the secondary obstruction. Others attempt to reduce the spider obstruction or at least mask it by bending the vanes. A few become specialists in the arcane tricks of baffling or try to seal the tubes of their reflectors with optical windows. Many suspect their telescope isn't adequately collimated and focus their primary attention on alignment.

The lesson of Figure 15.1 is that no optical problem is all-important nor can any problem be neglected entirely. Each difficulty deserves an appropriate response, with the important word being *appropriate*. Each suspected error deserves some attention, but no error must be emphasized to the exclusion of others. More to the point, no problem should be emphasized so much that its cure *damages* good operating characteristics of the telescope.

Also, we must think about the type of observing as well as the errors. Specialized observing situations exist where a little spider diffraction, microripple, or a few flecks of dust make a difference. In other cases, they don't matter much at all. Deal only with real threats.

If you mask your mirror to reduce a turned edge, be sure also to carefully baffle the instrument as well. After all, the image doesn't care about the source of the spurious light. One of the most neglected steps is a careful baffling of the final focuser tube, either by a series of shallow rings or by threading. (One clever manufacturer achieves this baffling by fitting a coiled spring inside the tube.)

Pay attention to everything visible from the inside of the focuser. Can a bright star just out of the field of view contaminate the image with internal reflections? Is your Newtonian tube long enough to prevent distant street lights from adding glow to the image? A good way of checking the

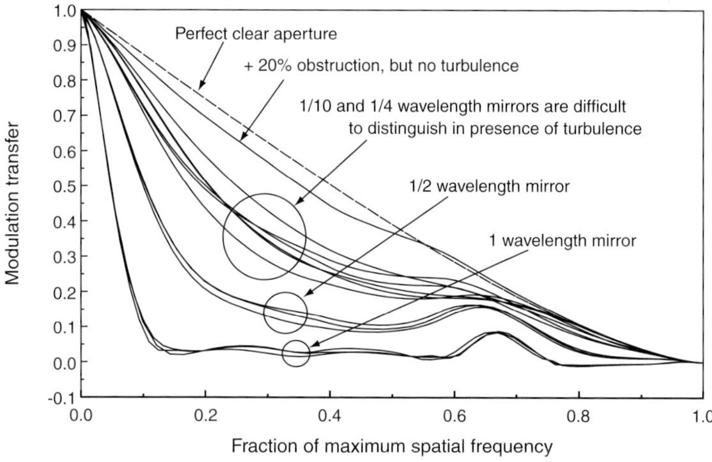

Fig. 15.4 The MTF curves of spherical aberration coupled with turbulence-induced roughness of 0.1 wavelength RMS.

baffling in the daytime is to put your lowest power eyepiece in the telescope and—backing up a foot or so—look at the immediate area around the exit pupil. If there are any strongly lit regions out here, perhaps a ring of grass visible around the mirror, or a section of the focuser tube well-lit from the end of the tube, there will be a problem at night too. A dark-adapted eye will allow the light surrounding the exit pupil into the eye at moderate to high power.

If each of the degradations appearing in Figure 15.1 is cut in half, you can derive an enormously improved image. Certainly, alignment tops the list of offenders, but if the telescope is aligned, consider the job only partly finished. Secondary size is important, but some of the small sizes suggested in the literature (10% to 15% of the aperture) can be too demanding. By reducing the secondary obstruction to 20% or below, you have climbed well up on the curve of diminishing returns. The important thing is that you have addressed each problem in turn. If each MTF degradation is boosted slightly, the net gain can be appreciable.

A solution to a lower MTF that may not be obvious to some readers—and to others may seem like cheating—is to obtain a larger telescope. Recall that the maximum spatial frequency (in units of cycles/angle) is D/λ. Hence, if Figure 15.1 depicts the response of a 400 mm (16 inch) reflector, we can see that it delivers contrast about as well as a *perfect* unobstructed aperture one-third to half its size. That means it's behaving about as well as the finest 6-inch unobstructed telescope.[1] In fact, a 16-inch reflector possessing only those aberrations depicted in Figure 15.1 would be judged an excellent telescope *for its size*. Therefore, if we are troubled by the performance of a small telescope, we can accept similar degradations in a larger aperture and still come out ahead (Zmek 1993). Using such a steamroller method to cure errors may not be subtle, but it's effective. Anything that results in more image information is legitimate. Keep in mind the system design discussion of Section 3.4.4. We must view a telescope as design trade-off of aperture, light-gathering power, and quality with cost, weight, and transportability.

Many issues having nothing to do with design or optical quality affect the performance of the telescope. For example, ask yourself if you have properly dealt with obvious things like obtaining a smoothly operating focuser before you ever consider obscure items like minimizing obstruction. Some instruments are so shaky that one touch on the knob sends the image into wild gyrations. Such telescopes cannot be focused using intuitive hand-eye coordination. Other telescopes are solid enough, but have focusers so tight and hard to turn that an observer is forced to wrestle with them. Some focusers are lubricated with a particularly heavy grease that stiffens in the cold.

No amount of optical perfection can improve a telescope that cannot be focused. No matter how much you have reduced the obstruction, carefully aligned the optics, or baffled the tube, the image of a poorly focused instrument is still substandard. Figure 15.1 demonstrates this principle in a backhanded manner. The little boxes were intended to show how simple defocusing mimics the aggregate curve, but the implied message is that defocusing *alone* is enough to destroy the image.

Telescopes are a mature technology. Severe modifications are probably mistakes. Yet a careful tweak here and there, as long as it's not excessive, can ensure that your telescope operates as well as it possibly can. It will probably even work better than you would have believed before you started.

15.3 Errors on the Glass

Errors polished into the glass are permanent. Things like spherical aberration, turned-down edge, or ground-in astigmatism cannot be adjusted out or disappear with remounting, nor can the user just wait, as with atmospheric or cooling effects. The telescope will never be able to perform adequately.

---

[1] Differences caused by the atmospheric turbulence scale, changes in eyepiece performance at lower focal ratios, and brighter images mean that performance won't be precisely duplicated.

Glass errors demand sober thought. Say you have inexpensively obtained a fast, thin-mirror altazimuth reflector, a "light bucket" having low to medium magnifications as the primary purpose. You cannot possibly expect crisp Airy disks and clearly defined diffraction rings. Few would want to pay for the optician's time needed to obtain such perfection in a large instrument, and few living under typical skies could often make use of such perfection. An optician cannot give an optical surface a great deal of individual attention without charging an enormous amount of money for the service. The time required to figure mirrors to the diffraction limit increases explosively with aperture or low focal ratio. Large and fast mirrors have both in abundance. This large telescope will not give razor-sharp planetary views, but it should at least perform well on the objects it was meant to observe.

If you have obtained a general-purpose telescope of moderate focal ratio and moderate aperture, however, you have a right to expect reasonable performance whenever seeing allows it. Before complaining to a manufacturer, star test the telescope again under different conditions. Make absolutely certain that the telescope is aligned and that the error doesn't originate with the eyepiece. Determine which optical component produced the error. If at all possible, try to see a star test on a good instrument before you judge a telescope as bad. Either stop down the offending instrument with an off-axis mask or perform the star test on a small or slow telescope likelier to give a nearly perfect result. Don't fully trust spherical aberration estimates gathered under rapidly changing temperature conditions. And finally, be certain that you have gone through the effort to quantify the defocus using the tables in Chapter 5. Unless you have put in the effort to test the telescope fully by using all the methods discussed in this book, you hardly have the right to complain.

If, at the end of all these checks, you are still convinced that the telescope has unacceptable optics, contact the manufacturer. Don't waste time discussing your optical suspicions unless you are very sure. Telescope makers can draw on reserves of confusing jargon unavailable to you. Unless you are very knowledgeable, such terminology will soon have you gasping like a landed fish. In your complaint, merely indicate that the telescope is bad. Clearly and carefully explain that the instrument does not focus to a tight spot and say that you are not pleased with the product.

Becoming angry in your dealings with makers serves no purpose and may be contrary to your interests. Always follow phone calls with written correspondence. Few manufacturers knowingly offer poorly made telescopes. Most will work with you until you receive satisfaction.

15.4 The Myth of the Complex Telescope

On the internet there has arisen a persistent urban legend. I have seen it repeated at least for 10 years. I do not know where it originated, so I will quote no secondary references. No doubt they were just passing along what they thought was good information. That myth says that star testing is only good for "simple" systems like Newtonian reflectors. On more "complex" systems, goes the legend, one can see remarkable asymmetries between outside- and inside-focus images in systems that test almost perfectly using other methods.

I first saw this mistake in connection with the commercial Schmidt-Cassegrain. Someone claimed that properly-made S-C telescopes shouldn't star-test well because they are complex. In even the simplest Schmidt-Cassegrain (all spherical mirrors where the polynomial describing the corrector plate is limited to fourth order), we see the nearly ideal behavior in the ZEMAX model of Figure 15.5. Any residuals are due to very slight spherochromatism (see Section G.5), but any circularly-symmetric error that is obvious at 8 to 12 wavelengths of defocus is caused by excessive fabrication mistakes.

When either a simple or complex system is finished processing the light, the thing that results is a focused electromagnetic beam. The influence of details of the system is no longer important; what is left is a free wave propagating in space. The star test senses deviations from spherical convergence (called *stigmatism*, Born and Wolf 1999, p. 159); either the beam converges toward a stellar point or it does not. The test neither knows nor is sensitive to the complexity of what processed that beam.

This myth is most persistent for consumer-quality, unretouched, aluminized-spot Maksutov-Cassegrain telescopes. In Section 11.1.1, we saw how Maksutov-Cassegrains that are left spherical and for which the secondary is aluminized right on the back surface of the corrector have an unavoidable amount of residual higher-order spherical aberration. This aberration may be lessened by polishing a conic correction into one surface or by removing the aluminized spot from the secondary and making it another surface entirely. The only way it is completely removed is by zonal retouching, an expensive procedure and an unnecessary one.

Remember what Welford had to say about sensitivity in Chapter 5? Sharp zonal errors are detectable up to $\frac{1}{60}$ wavelength, and the edge zone exhibited by the aluminized-spot spherical Maksutov-Cassegrain is a good example of a zonal error. However, the sensitivity is variable with defocus. The star test is most sensitive when defocused only a trifle and insensitive when defocused too far, so the amount of aberration it is capable of detecting is a variable with defocus. The key point to remember is that you should defocus far enough (consult Section 5.1 and Chapter 11) to detect a serious and complaint-worthy aberration and not detect an inconsiderable one. It is nice to defocus less to see your residual aberration better, but it is wrong to forget that short defocus lengths are too sensitive. You may well end up complaining bitterly and unfairly to the manufacturer about a residual aberration of no harm. It is not a fault of the star test to be so sensitive, it is the fault of the user to be so inattentive.

I have a 100-mm aluminized-spot Maksutov-Cassegrain that displays a slight diminishment of the outer zone inside focus coupled with the outer

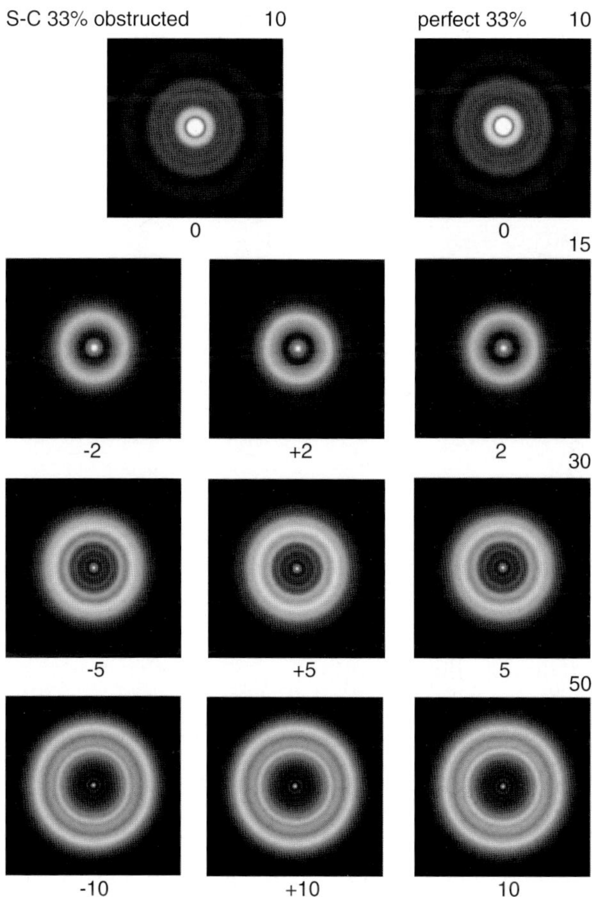

Fig. 15.5 A perfectly-made Schmidt-Cassegrain telescope of the simplified type favored by commercial makers. Images generated with a photopic distribution rendered as it would appear to the receptors of the eye. Contrast has been manipulated to emphasize faint detail in the focused disks.

zone's brightening outside focus. These are the classic symptoms of higher-order spherical aberration visible in Figure 11.2. In fact, observation of the real star test shows asymmetries at 10 wavelengths defocus to be much less visible than indicated by these figures (which are based on linear shadings somewhat more contrasty than real life). In fact, if I had not been looking for them, they would have escaped notice.

The other instruments that have contributed to the myth of complex telescopes, I believe, are refractors. In Chapter 12, I commented on the non-interference between separate colors of light, and the way that different colors should be first filtered out to test individually for geometric aberrations and then considered in concert to test for different color foci. Of course, even interferometers can only test one color at a time. Luckily, you can separate the colors out by filtration, and if the star test is good at the geometric-aberration stage, a moment's glance of an unfiltered telescope is all you need to tell if the foci of the different colors are sufficiently close.

15.5 Testing Other Telescopes

Once you become familiar with star testing, nothing prevents you from evaluating every instrument you come across. Such practice will help you to develop an eye for different aberrations. Sooner or later, an alert star tester will see every type of optical difficulty discussed in this book and even some that are not. You are encouraged in this effort to broaden your experience.

Keep the test results to yourself, though. Considerations of courtesy aside, such opinions could be wrong. You generally know nothing of the history of the instrument. You don't know if it is cool or warm. You have not had a chance to align it first, so you need to mentally subtract a significant alignment error from the pattern. Furthermore, one-shot tests are anecdotal and do not allow for follow-up testing. Your goal in performing the star test on other instruments should be experience and practice for yourself. Remember, you were told to test your own telescope again and again. What makes you think you can evaluate someone else's from a glance in the eyepiece?

If you are asked for your opinion, give it along with a comment about the considerable uncertainties. Don't present the result as a pronouncement from heaven. Carefully explain how you developed the evaluation. Finish with a comment that the owner need to follow up with his or her own testing to verify your quick assessment. If others don't know the star test, teach it to them. After all, I hope you realize by now that star testing is not that mysterious.

15.6 When Everything Goes Right

I don't want to leave you with a fatalistic interpretation of the star test. Once taught how to evaluate telescopes, people tend to be overly critical of their instruments. Nothing pleases them. It seems that the illusion of reality has been stripped away—and with it the wonder.

Here is an example of this loss of illusion: Years ago, a universal flaw in pre-digital motion pictures was pointed out to me. Just before the end of a reel, markers in the form of punched holes appear twice as an aid to projectionists. If punched in the negative, the holes are inverted to dark spots. Because most films are compressed horizontally, these marks appear as flattened ovals, one at about 10 seconds and another just before the reel changes. The second projector is always started at a point when the frame is dark or abruptly changes brightness.

I had never seen this tiny flaw before and I never fail to see it now. I almost wish I had never been told. This place where the bones of the technology poke through invariably jolts me from the film's comforting illusion and reminds me that I am just watching a movie.

Similarly, once optical phenomena are familiar, you will see them everywhere. Glasses wearers will be unable to walk in drizzle at night without noticing the interference bands in a refracted sparkle of light on the lenses. Looking upward into a clear blue sky, you will occasionally notice floaters in your eye surrounded by tiny diffraction rings. You'll see the ominous signature of chromatic aberration in every rainbow.

I hope that you will not become unnaturally sensitive to the flaws in your telescope. I know individuals who own telescopes having abominable optics, yet they continually conduct productive and frequent observing sessions. I have known other owners who complained about telescopes that were little removed from perfection. They hardly ever spent time under the stars but were always adjusting and modifying their light-starved instruments. It seems that the attitude of the telescope user is the final filter in the wobbly stack and often becomes the worst form of degradation. The star test enables more realistic images. It is not meant to turn a happy observer into a sad one or to spoil the glorious illusion that a telescope produces.

I remember slipping into the hyper-critical mode one time when giving a group of elementary students a view of Saturn. I was grousing about the image when one of the students accused me of having a hidden photograph inside the telescope. What I saw as substandard, she saw as an image so good that she thought I was cheating.

Astronomical telescopes can weave delightful images, and I would serve readers badly if I were to leave them with a sense of disappointment to spoil the magic of starlight. I want to describe something I saw with my own eyes when everything went right—when much of the filtration dropped away in the "wobbly stack" of Figure 3.1.

It was one of those rare times when the temperature had been virtually constant all day. One evening, my observing group set up a 16-inch $f/5.6$ Newtonian telescope. It had a 3-inch thickness mirror that often had trouble cooling down, but that whole day it had been near ambient temperature. The evening was wonderfully steady. It was one of those infrequent nights when there seemed to be no upper limit to the useful magnification. We aligned the instrument and turned to Jupiter, then at about 45° elevation.

The shadow of a moon was crossing Jupiter's disk. It stood out crisp and distinct against the planet's brilliance. The gray disk of the moon itself was clearly visible as it transited the planet. So many whorls and crenelations were visible on the surface, that neither hours of sketching nor my limited artistry could have captured them. I wasn't looking through a telescope so much as I was being projected beyond one. I had passed through the eyepiece.

I have had similar experiences rarely, but often enough to make it all worthwhile. Once I saw Cassini's division visible on the whole lit circle of Saturn's rings. The crepe ring was easily visible; it looked filmy against the darkness. Another time, at 350 power, I saw globular cluster M15 resolved clear across the core, each star visible as a tiny spark of light.

I wonder how many of the billions of human beings who have ever walked the earth have seen these things. I feel fortunate and humbled to have done so myself.

On these special nights, I did not see the telescope as a filter. I believed the image was real. And that is the point of all of this labor. We learn to judge the magnitude of optical errors to help the instrument fulfill its purpose. We do so for those brief moments when we can forget that we are looking through a telescope—when we can feel the quiet majesty of the sky.

Appendix A
Other Tests

The 1930s were boom years for amateur telescope making in the United States, primarily through the popularization efforts of individuals such as Russell Porter and Albert Ingalls. Estimates have been made that up to 250,000 telescopes were begun during the years before World War II. In a way, it was the only option. The Depression was still a strong factor, and established optical shops specialized in making expensive refractors or contracting for professional instruments. If you wanted a reasonably priced telescope in those times, you had to make it yourself.

Soon, a few amateurs turned "pro," making instruments of the type other amateurs wanted and could afford to buy. Slowly, as number of such manufacturers increased, the character of amateur astronomy changed. No longer was grinding glass a rite of passage for entrance into the astronomical world. The advent of commercial Schmidt-Cassegrain catadioptric telescopes around 1970 completed the transformation. Telescopes are now consumer items.

In the old days, nearly everyone had some familiarity (if not expertise) with the knife-edge test and a general acquaintance with the concepts behind testing during fabrication. Few modern amateur astronomers have been directly involved in making their own instruments. Amateurs have heard about shop tests but have only looked through testers at a convention display. Most amateur telescope makers today deal with the mechanical design, electronics, and construction of their telescopes, not the optics.

Thus, much of the material below will be new to many people. It may seem that I deal curtly and summarily with the various optical shop tests appearing here. This abruptness is not my intention. The tests below can be entertaining and give new insights into optical quality. Each one is so different from the rest that it offers a fresh perspective into the aberrations on the wavefront. Most of my criticisms are not about the tests themselves, but their improper use.

However, I do want to point out that these tests are not recommended if all you want to know is whether or not your telescope is a good one. They are all interesting, and some are phenomenally sensitive, but they aren't the most direct and painless path to knowing if your telescope can work well. The star test is that path.

For the purpose of completeness, these tests are described below, along with brief lists of difficulties for novices, or unanticipated expenses. The reasons for not recommending them are varied, but most of them boil down to the following:

1. Most of these tests are useful during *fabrication* of individual optical pieces, not *evaluation* of finished telescope systems.

2. Many of these tests involve the purchase or construction of accessory hardware, some of which can be very expensive.

3. Often, they require complicated procedures, data reduction, or difficult theoretical knowledge.

4. Most are oriented toward one type of surface and require multiple tests or additional optics if they are to be applied to the whole instrument.

Readers are encouraged to find out more about one or more of the following testing techniques. Each one of them could fill (and perhaps deserves) its own book, just as this one has been devoted to the star test. You can spend a lifetime discovering the details concerning any of them.

A.1   The Foucault Test

Jean Bernard Léon Foucault was an all-purpose scientist. He is best known for demonstrating the rotation of the earth by precession of the axis of a pendulum and measuring the speed of light. He also took the first daguerrotype photograph of sunspots, thus initiating astrophotography. Foucault made an early metal-on-glass telescope mirror. Finally, he invented a sensitive test of optics at their center of curvature.

Imagine a reflective sphere 6 meters in diameter. Clearly, light radiating from a point at the center of that sphere and diverging outward would strike every portion of the sphere at the same time. It would then be reflected and converge to be perfectly imaged as a point on the source that emitted it originally.

Although it would be a beautiful thing to behold, a reflective sphere 6 meters in diameter has few uses. If only one tiny portion of that sphere were silvered, the reflective part would be a weakly concave area. The non-reflective portions could then be trimmed off. Let's say the remainder is a 10-inch diameter (250-mm) *f*/6 (1500-mm) mirror that we wish to test.

Some method of reading the optical quality must be devised before it can be said that a true test is being done. The first problem is that a source of light at the center of the sphere, an illuminated pinhole for example, is imaged back on top of itself. The imaged light is unavailable to the optician. Foucault solved this problem by offsetting the source slightly. The image point in this case is found across the center of curvature at a distance about equal to the offset. As long as the distance between the pinhole and image is kept small, the test only is slightly affected by astigmatism.

The second problem is coming up with a method of probing the image point. One could inspect it with an eyepiece and thus have a variant of the star test, but this method was already known at the time of Foucault. The pinhole inspection procedure does not readily yield numbers useful to plan the next polishing step. Inspection with an eyepiece also demands an extremely small source; in the 250-mm mirror above, the pinhole should be less than 16μm (0.0006 inches) across. This restriction was especially severe in Foucault's time because portable light sources were based on flame and were hence quite weak, diffuse, and difficult to direct through a pinhole.

Foucault's clever solution was to introduce an occluding edge into the beam near the imaging point. This test goes by the popular name "Foucault's knife-edge test" although the method does not depend very much on the sharpness of the blade. The knife is slowly introduced into the reflected beam near focus. In the simplest configuration, the tester's eye is placed close to the knife and peers past it at the mirror. If the knife is between the mirror and the image point, the dark shadow of the knife appears to move onto the mirror from the same direction as the knife. Outside focus, the darkness appears to cross the mirror from the reverse side.

One need not use a pinhole source at all. A short slit can serve equally well as long as the knife is parallel to its image. This way, the illumination can be increased hundreds of times without increasing the power output of the lamp.

The knife is moved toward and away from the mirror until focus is located. If the focused point is struck precisely, the shadow on the mirror does not exhibit the direction of knife motion. A knife setting for which non-directional behavior is found proves the mirror must be part of a sphere. It dims evenly before darkening entirely. The Foucault test of a spherical mirror is a true *null test*—i.e., the reflection blanks out. See left side of Figure A.1.

If any high or low places on the mirror exist, no mirror-knife separation can be found at which the mirror darkens uniformly. The test superficially resembles the shadows cast from a lamp shining from the side of the mirror opposite the knife. The hill shown on the right side of Figure A.1 acts as a protrusion on the apparently flat surface. One side is very bright, and the other side is very dark. The elevated region is shown perched on a

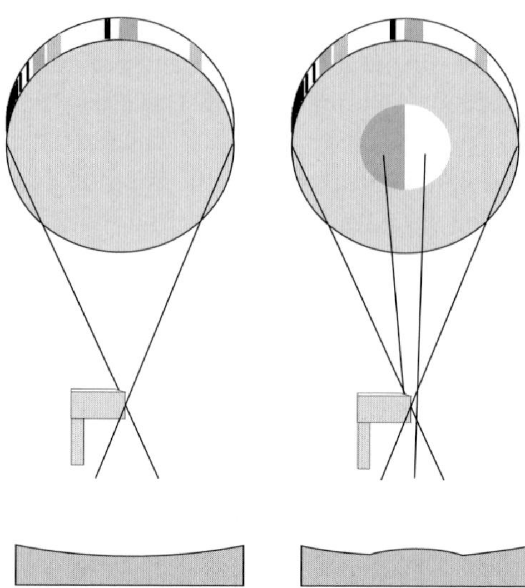

Fig. A.1 Two Foucault test setups, one is perfect and the other shows a slight hill in the mirror's profile (the aberration is exaggerated).

neutral-gray plane. Of course, this side-lighting concept should be viewed as a convenient delusion. The "hills" cast no shadows, and the entire appearance of the display can be changed merely by moving the knife along the axis of the mirror. In the figure, you can pull the knife back until the center bump is uniformly gray and appears to be sitting in a huge cupped depression.

Foucault's genius was that he did not rest once he found this sensitive test for spherical mirrors. He modified his new test to figure paraboloids used as the primary optical element of Newtonian telescopes. The problem with testing paraboloids near their center of curvature is that a parabola is not a circle and a single center of curvature is not defined on its surface. You can squint your eyes and convince yourself that the shallow dish is a sphere to first order, but a perfect paraboloid shouldn't null. The method he worked out is seldom used today (it involved null-testing prolate ellipsoids increasingly near a paraboloid), but may easily be modified to the modern procedure.

The behavior of light rays near the center of curvature of a paraboloid can be straightforwardly calculated. They converge along the horn-shaped caustic defined by an overcorrected surface. The caustic is a region in which ray optics strictly breaks down, but for the purpose of the purely geometric Foucault test, one pretends that it does not.

Jean Texereau drew a very informative diagram of this behavior, which is reprinted in Figure A.2 (Texereau 1984). The caustic appears best

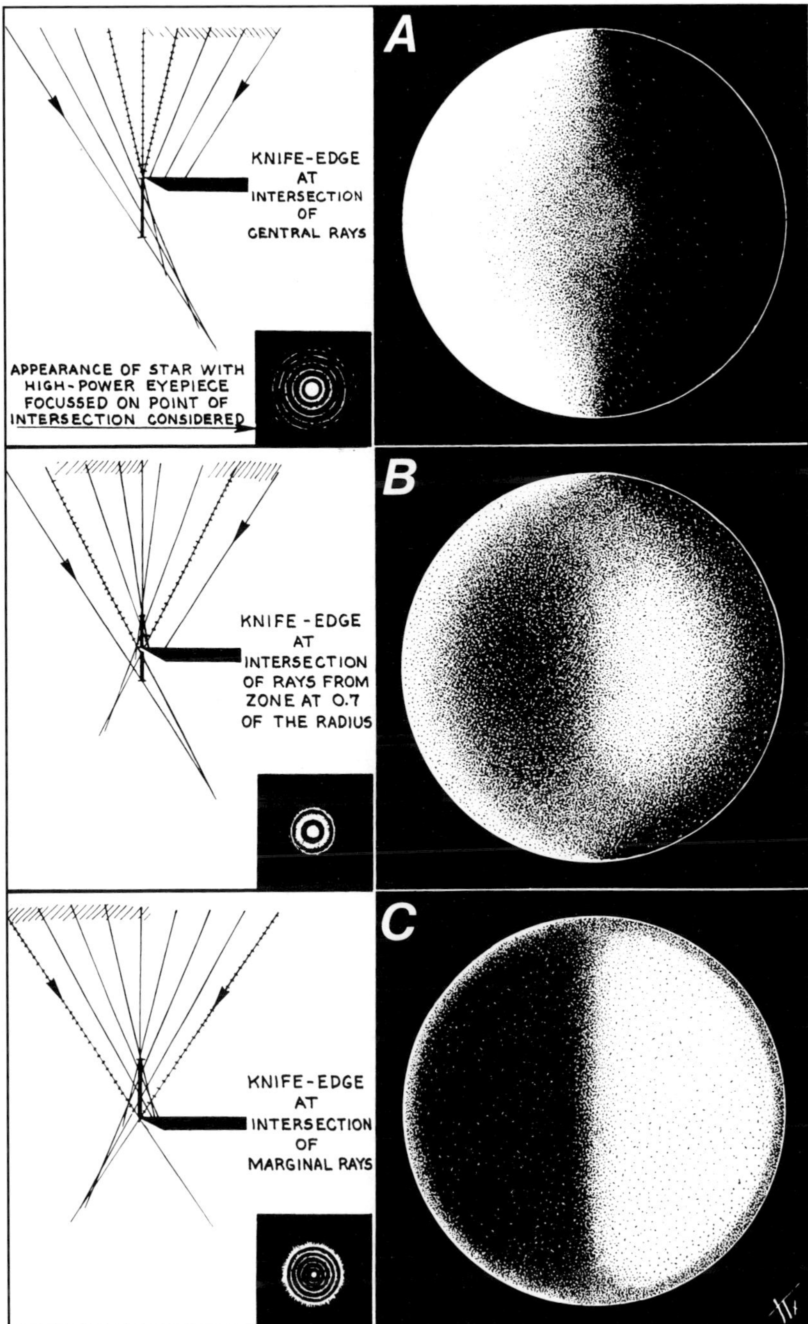

Fig. A.2 A drawing of three knife locations and the resulting appearance of a parabolic mirror in the Foucault test. From *How to Make a Telescope* by Jean Texereau, Copyright 1984 by Willmann-Bell, Inc.

in the ray diagram at the bottom left, where part of it is blocked by the knife. The appearance of the Foucault test at various knife positions is depicted on the right side by stippled drawings. The length of this ray-crossing region (which appears as a black bar at the tip of the knife) is related to the correction of the mirror. A sphere at its center of curvature has a ray-crossing length of zero. For a paraboloid it is where R is the approximate radius of curvature of the mirror and D is its diameter. *LA* means *longitudinal aberration,* the stretch of the black bar in the diagram. For a hyperboloid, *LA* is a larger number, and for a prolate ellipsoid it is smaller:

$$LA = \frac{D^2}{4R}. \qquad\qquad \textbf{A.1}$$

If an optical worker measures the shift between situations A and C of Figure A.2 and keeps changing the shape of the mirror until that shift is at or a little less than $D^2/4R$, then the paraboloid is being closely approximated. Previously, mirror making had been the province of high art and not a small amount of guesswork. In its simplest form, the Foucault test had reduced the procedure of mirror testing to finding the centers of curvature of the center zone (A) and the edge zone (C) and subtracting them. (See Suiter 1988.)

Examples of these situations photographed on a real mirror appear in Figure A.3. This figure also shows a tiny central button zone. Because of its location in the shadow of the diagonal, it would not harm the images.

The contours seemingly shown by the imaginary side lamp have apparent amplitudes of something like 3 to 6 mm for a 150-mm *f*/8 paraboloid. Since this paraboloid departs from a sphere by about ⅛ wavelength, we can calculate a synthetic magnification of errors that amounts to about 4 mm/0.00007 mm or 60,000 times.

This development, coupled with the technology for depositing metal films on glass, set the stage for the huge reflectors of the 20th century. Foucault is the godfather of the massive instruments we use today.

Foucault's knife-edge test is sensitive and proven, but it is not recommended as a final evaluation, for several reasons:

1. The test requires some practice. One must be skilled in setting up, aligning, and interpreting it.

2. It does not allow testing of the convex elements of compound optical systems without expensive additional hardware. Except for an auto-collimation test against a huge optical flat, one must disassemble the telescope and test individual pieces, some of which may not be easy

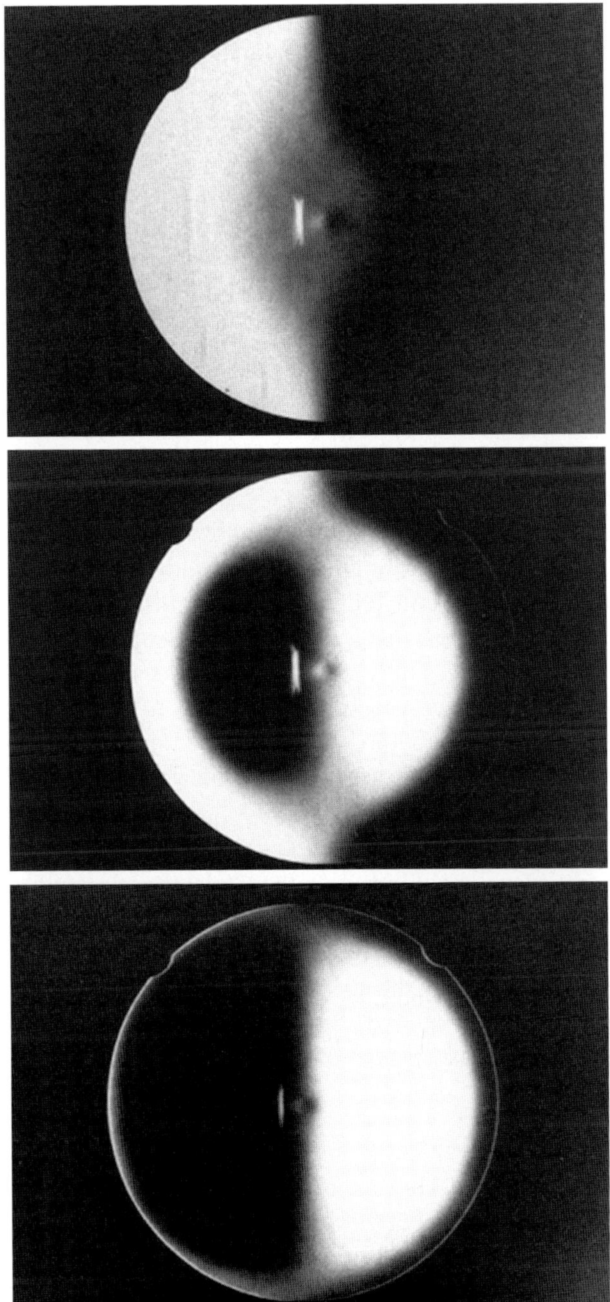

Fig. A.3 Photographs of the three cases of Figure A.2 on a paraboloidal mirror. Bright streaks are spurious reflections of the slit. (Photographs by William Herbert of Columbus, Ohio.)

> to remove.

3. In its more sophisticated forms, it requires a tiresome and easily bungled mathematical reduction procedure. Computers can help with this calculation, but such computations can be mishandled at the input stage.

4. It requires that an elaborate knife-edge tester be constructed. Simpler testers were adequate when people only tested long-focus Newtonian mirrors, but fast instruments give little room for error. Very good motion platforms—called kinematic stages—and compact source/knife assemblies must be built or purchased.

5. In the form most amateurs use, it is insensitive to sideways motion of the center of curvature. A consequence or advantage, depending on your point of view, is that it is insensitive to odd orders in the aberration expansions.

Variants of the Foucault test include the caustic test and the wire test. All suffer more or less from these same difficulties.

One other use of the knife-edge test would be applicable to all forms of optics if it were more convenient. Using a knife edge at the focus of a star recovers the conditions that led to the gray, flat appearance of a sphere at the center of curvature. This method is easiest for owners of an excellent clock drive and a heavy, unshakable mounting, because they can follow the brightest stars. Some method for gradually introducing the knife into the focused beam is also necessary because this variant of the Foucault test is extremely touchy.

Since few telescopes are likely to perfectly null, those wanting to use a knife edge at focus should provide their testing setup with a method to measure the length of focus shift from a situation resembling A in Figure A.2 to situation C. In a typical test, this focus shift should be below $100\,\mu m$ (0.004 inch). Using the artificial star described in Section 5.2 eases the mounting, clock drive, and illumination problems, but some sort of measurement screw must still be placed on the focuser.

A.2   The Hartmann Test

This test, developed by J. Hartmann around the year 1900, is used most often to check the surface of very large observatory instruments. A screen is centered over the objective lens or mirror. Carefully sized and placed holes are cut into this screen. Then the telescope is pointed at a distant star or artificial source at the center of curvature.

Two photographic plates or electronic focal-plane arrays are then exposed, one inside of focus and the other outside. They are then carefully

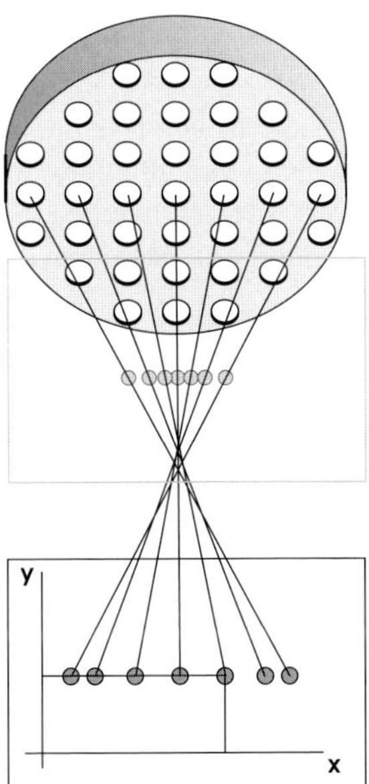

Fig. A.4 Configuration of the Hartmann test. The rectangles represent photographic plates with the intersection of dot-pair lines hovering between them. The measurement coordinates are indicated on one plate.

measured. If the test is successful, the images of the holes are identifiable, resolved, and not so large that their positions are uncertain. Two pictures aren't really necessary when the positions of the holes over the mirror and the amount of defocus are very accurately known, but caution dictates a second longitudinal position.

Finally, the corresponding dots in the two pictures are connected mathematically. Once the intersections of the dot pairs are known, they are entered into expressions that convert longitudinal aberrations to wavefront error, provided that the surface is deformed smoothly. See Figure A.4 for a diagram depicting the manner in which the Hartmann test is used to measure aberration. A two- or three-hole version of the Hartmann arrangement is occasionally used today as a focusing aid (Suiter 1987).

This method is not recommended for first-time testers for the following reasons:

1. The longitudinal offsets are difficult to set in such a manner that the spots are both sensitive to error and their centers accurately located.

2. The two images must be held precisely perpendicular to the optical axis and each other during exposure.

3. Distances between the spots must be accurately determined, either with a photographic-plate measuring engine or by counting pixels

with known pitch and centroiding them.

4. The mathematical reductions are of about the same complexity as advanced versions of the Foucault test, but because of the number of connected dots, many more calculations must be done. This data analysis may be a very tiresome chore. (See Danjon and Couder 1935 for an early reduction procedure that led to the method appearing in Texereau 1984.)

Lately one of the electronic camera manufacturing companies has introduced a Hartmann test utility, but I have not inspected it. The computer relieves much of the drudgery but problem number 1 would still make this test very difficult to use.

A.3   Resolution of Double Stars

While determining the resolving power of a telescope is not a complete optical test, many people treat it as such. Thus, it deserves mention here. This method of evaluating telescope images became popular during the 19th century, when double stars were the subject of very active research. Observers who were primarily interested in the clean separation of two stars began to judge the performance of their telescopes entirely by this characteristic.

Certainly, a telescope that fails to show double stars close to the expected resolution is displaying one of the symptoms of poor optics, but other types of equally bothersome optical difficulties do not betray themselves this way. Spherical aberration of ¼ wavelength insignificantly damages the telescope's ability to split stars. (See the encircled energy plot in Figure 10.6.) However, on planetary detail requiring only moderate resolution, optics with correction errors present distinctly fuzzy images.

Figure A.5a displays the various criteria of resolution. The first is called the Rayleigh criterion, which is not to be confused with the ¼-wavelength Rayleigh limit of wavefront error. The Rayleigh resolution criterion is met when the separation of the two objects is precisely at the radius of the theoretical Airy disk. In other words, the second star is placed on the valley between the first star's central disk and its first diffraction ring.

The degree to which the Rayleigh criterion divides the stars varies with details of the diffracting aperture. Different obstructions and aberrations result in differing depths of the "saddle" between the stars. For a perfect circular aperture with no obstruction and no aberrations, the dip between the stars is to about 70% of the brightest intensity.

The second criterion was stated by double star observer W.R. Dawes in 1867 (Sidgwick 1980). It applies only to unobstructed apertures divid-

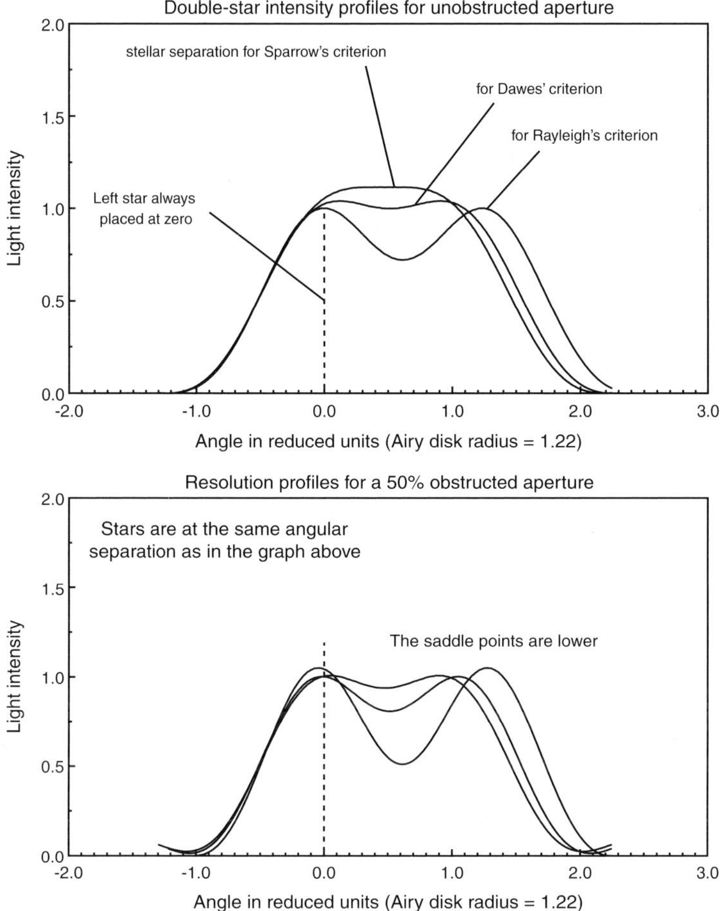

Fig. A.5 Resolution of equal-brightness stars in two instruments: Top a) a perfect circular aperture; Bottom b) the same stars as seen in the same instrument with a 50% central obstruction. The huge secondary mirror actually improves resolution.

ing equal stars. In the Dawes criterion, the separation is a little less than 85% of the separation defined in the Rayleigh criterion. Figure A.5a shows it as the intensity curve with the small drop between the stars. This drop is only about $\frac{1}{30}$ magnitude less than the maximum intensity. When doubles are this tight, their lack of roundness contributes as much or more to distinguishing duality than the intensity drop between them.

The third and narrowest criterion is called Sparrow's, which is defined as that separation which results in a flat isthmus between the stars. The Sparrow criterion is adjusted for obstructed and aberrated apertures until it always gives that flat region between the stars. Thus, it is always uniquely defined and always delivers the same behavior, but its exact separation varies with details of the aperture.

Sparrow's criterion for unobstructed apertures has stars separated by about 92% of Dawes' criterion or about 77% of Rayleigh's separation. In the perfect, unobstructed aperture's MTF chart of Figure 3.5, the inverse of the Sparrow criterion is found at a spatial frequency off the graph at $1.06 S_{max}$. This placement doesn't mean the resolution is illusory. Some double-star observers have even exceeded this value. Stars are point objects, while the MTF target consists of bars. Points can be resolved by using the oval shape of the image as the only discriminator.

Figure A.5b depicts one of the most disturbing aspects of using double stars as test objects. The summed diffraction patterns of two stars seen in a 50% obstructed aperture is calculated with the same stellar separations as were used in the unobstructed aperture of Figure A.5a. For all three curves, equal separations deliver stronger dips in intensity between the stars. To recover the behaviors of Figure A.5a, the stars must be pushed closer together. In short, the 50% obstructed aperture resolves better.

Double-star resolution doesn't tell us much about other types of observing performance,[1] particularly in that case where the 50% blocked aperture will fail badly. Planetary images are greatly degraded by such severe obstruction. Blocking the aperture kicks a large fraction of the light outside the central spot into distant portions of the point image. Resolution of a nearby star is little affected, because the star is close to the center of the image and the light diffracted by the obstruction is largely beyond it.

If the desired object were a small, low-contrast crater on the Moon, the whole image is the sum of all light scattered from light patches around the crater as well as the image of the crater. Much of this spurious light is joined together to fog the interesting image. The scattering from any single diffuse spot is not enough to seriously damage the image by itself, but the combined effect of all of them worsens contrast considerably.

Blocking the aperture by a 50% central obstruction helps resolve some double stars, but the same obstruction leads to severe image degradation of planetary detail. You can easily verify this result yourself. Look at the Moon or a planet some night with and without a large paper obstruction in front of the aperture. Unless the telescope's obstruction is high already, the image produced by the artificially blocked aperture will appear much worse.

When using the resolution of double stars as the sole criterion of opti-

---

[1] Dawes himself was well aware of this difficulty, even though his name is attached firmly to the misuse of this test as an indicator of optical quality.

cal quality, the astronomer is demanding that the judgement favor high spatial frequencies. In the case of obstructed apertures, the spatial frequency response has been robbed of some of its intermediate frequency strength to achieve better contrast at higher frequencies.

Double stars as test objects present other difficulties. Variability is associated with the sky and with the stars themselves. Ideally, one should use equal-brightness white stars separated at the diffraction limit and demand that they be high in the sky. Few stars conveniently arrange themselves this way. Usually, the test must be done with unequal brightness stars separated by a distance close to your instrument's Dawes or Rayleigh criterion but not precisely at it. One of the stars may be tinged with blue or red, and they may be at a low altitude, which produces unfortunate atmospheric dispersion into rainbow-like spectra. "Seeing" will constrict the number of nights on which double star tests can be attempted, particularly for large instruments.

Artificial double stars can be used to test resolution, but they cannot cure the basic inadequacies of the evaluation technique. Because very high magnification must be used, ground turbulence must be low. By bringing the source close to the telescope, one writer avoided the problem of turbulence. In this case, the source was only about 10 meters away (Maurer 1991). However, this a close a distance makes such a test an unreliable check for aberrations (see Chapter 5).

Atmospheric problems also trouble the star test, but because the stars are considered singly, you have more freedom to choose one at high altitude. Since the star test involves inspecting the much larger defocused stellar disk, it does not always require that seeing be perfect.

For the above reasons, the double star resolution test is not recommended as an all-purpose test. In brief summation:

1. Resolution tests are interesting to those who are concerned with double stars. Resolution is of little use as a general indicator of telescope performance because it favors high spatial frequencies.

2. Suitable target stars are difficult to find. Artificial sources have their own pitfalls.

3. Seeing must be superb before the double-star resolution test yields interesting results.

4. Visual discrimination of saddle-point dip is difficult. The test is not precise.

A.4 Geometric Ronchi Test

A test is attractive if it doesn't require a great deal of data reduction or in-

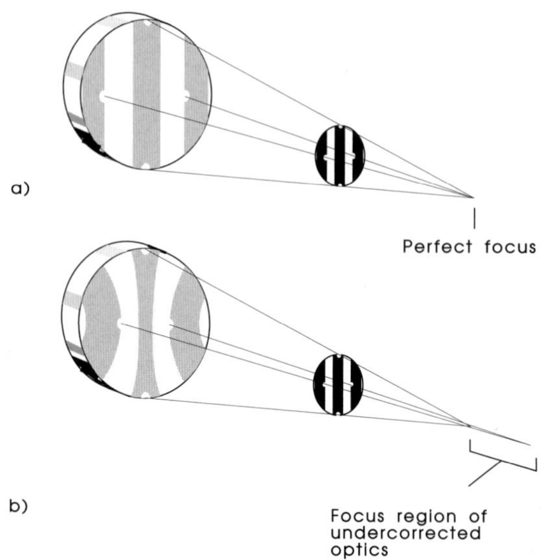

Fig. A.6 A Ronchi ruling is placed near the focus of a star and examined without an eyepiece. a) If the rays are all directed to a common focus, the optics are perfect. b) If the lines distort, they indicate aberrations. The Ronchi grating need not be confined to the tiny circle. Any stripes outside the illuminated portion of the light cone do not contribute.

terpretation. One such test involves placing a coarse periodic grating of dark and transparent bars near stellar focus of the instrument. An example frequency of such a grating is 100 lines/inch (4 lines/mm), with a corresponding period of 0.01 inch (0.25 mm). This method was investigated by Vasco Ronchi around 1923 and thus carries the name of *Ronchi test*.

Interpretation, at least when we assume that light is comprised of rays, is simple. Supposedly, light has originated from a point source very far away (like a star). If the star is correctly imaged, it should be focused to a single point. Starlight passes through a grid of straight, opaque bars, which remove the light rays that strike them. The perceived pattern on the mirror is just the shadows of those vertical bars.

If the optics have aberrations associated with them, the light from different zonal radii is not diverted toward a single focus. In the case of spherical aberration, various axis-crossing points are distributed near the region of the caustic. Because the grating lies at various distances from these points, the spacings of the grating projected on the aperture seem to vary from radius to radius. In the case of undercorrection, this effect is portrayed in Figure A.6.

If the grating is 100 lines/inch and the aperture has a focal ratio of $f/8$, the 2.5 periods (or "lines") shown in the illuminated part of the grating

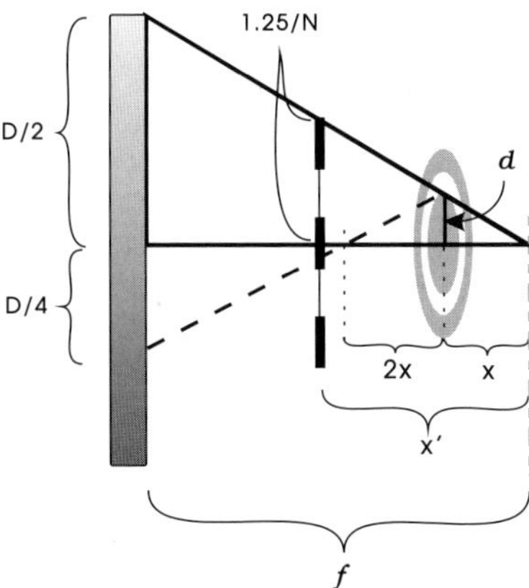

Fig. A.7 A similar-triangles setup used to estimate the Ronchi curvatures. The particular crossing points chosen here would be characteristic of overcorrection, but the derivation in the text would not be affected if they were reversed.

would be about 0.2 inch inside focus and only 0.025 inch in diameter.

It seems as if the Ronchi test is the final answer, a clear and unambiguous null test. Unfortunately, it doesn't always work that way.

A 10-inch *f*/6 mirror came to my attention in 1980. In spite of the fact that it gave soft images, it passed the Ronchi test using a grating of 100 lines/inch. It failed the star test in a stunning manner, however. When removed from the tube and tested with a more elaborate variation of the Foucault test, the mirror showed a ½-wavelength undercorrection.

Still, it had passed the Ronchi test at focus. Something was wrong, either with one of the tests or the mirror. Consider the aberrated Ronchi drawing of Figure A.6 again. The distortion of the pattern seen against the mirror is caused by the change in focus distance as measured from the grating. As depicted, the focus difference (or the longitudinal aberration) is about a third of the distance from the grating to the average point of focus, causing severely distorted Ronchi lines projected on the mirror. The density of lines at the center of the aperture is much higher than at the edge. One might call this curvature a "33% distortion." If the region of focus is a much smaller fraction of the distance to the grating, distortions can be appreciably lower. For optics of conventional focal ratios and diameters, this low-distortion condition is true much of the time.

One can easily calculate the length of the focus region for optics diverting light to the edge of the diffraction disk. If we draw the similar triangles of Figure A.7, the overall triangle has edges of height *D*/2 (half the aperture) and base *f*. The triangle outside of focus has height *d* and base *x*, with *d* being the radius of the diffraction disk. The radius of the Airy disk is (1.22)(wavelength)(focal ratio), or $1.22\lambda F$. We can use similar triangles to show that $x = 2.44\lambda F^2$. Because we must regard a finite-sized region to estimate the number of lines showing, the density of lines at the edge is compared with the density of lines at a point roughly halfway out the radius of the mirror. The dashed line, because it has only half the slope, can cross over at about twice the distance x from the point of focus.[2]

If light from the edge is deflected to cross the axis at more than $3x$ from light at the 50%-radius zone, it begins to miss the diffraction disk entirely and signs of severe optical degradation become visible.

The vertical dashed line depicts a grating located between the aperture and the region of focus. It has the same 2.5 lines as Figure A.6, measured for the outside zone. Above the center, 1.25 lines are showing (a "line" is a complete on-off cycle, or a period). The height of this little triangle is therefore $1.25/N$, where N is the number of lines per inch (or mm) of the grating. The base length of this triangle is left unknown and is called x'. The big triangle is the same as before. Using similar triangles again, one can show that $x' = 2.5(F/N)$.

To calculate the distortion, we estimate the tolerable focus shift divided by the distance to the grating, or

$$\text{Distortion} = \frac{3x}{x'} = 3\frac{2.44\lambda F^2}{2.5(F/N)} = \frac{3(2.44)\lambda FN}{2.5}. \qquad \textbf{A.2}$$

Merely by identifying 2.5 as the number of lines of the grating that are intercepted (called n),

$$\text{Distortion} \cong \frac{7.5\lambda FN}{n}. \qquad \textbf{A.3}$$

For $f/6$ optics with a 100 line/inch grating, taking as the wavelength the yellow-green color that the human eye likes best, the maximum tolerable distortion with 2.5 lines showing is about 0.04.

Anyone who has ever used one of these gratings knows that the sharpness of Figure A.6 presents an unrealistic view of what is happening. The actual shadows are fuzzy and indistinct. A 4% bowing of the lines in the presence of this fuzziness is nearly impossible to see. Figure A.8 depicts the actual pattern that results from a ¼ wavelength of undercorrection, still without the blurring at the edges of the lines.[3]

An arbitrary cutoff can be imposed. This limit is somewhat artificial,

---

[2] The Airy disk in this diagram is greatly exaggerated in size. Hence, the crossing point of the dotted line appears nowhere near $2x$ from focus. However, in real situations with very small Airy disks, an inner crossing distance twice that of the edge is an excellent approximation.

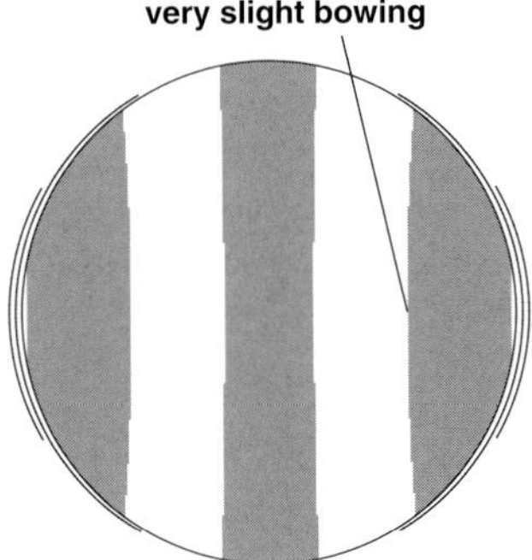

Fig. A.8 The Ronchi distortion expected of a ¼-wavelength undercorrected f/6 telescope when a 100 line/inch grating is placed near focus and the telescope is directed toward a star. The thin exterior lines are the borders of the first- and second-order interference.

but let's set it conservatively at 8%. If we can't set up a test with at least an 8% maximum bowing of the lines, we are not testing the optics with anything close to the kind of precision necessary. One suspects the actual cutoff should be higher, but for the purpose of argument, let's give the geometric Ronchi test the benefit of doubt.

Examination of Equation A.3 reveals how the Ronchi test could recover enough sensitivity to barely achieve the limit. If the grating were moved closer to focus, reducing the number of lines seen on the aperture to 1.25, this mirror will have lines that bow 8%. If the number of lines is increased to 200/inch, the pattern will also distort 8%.

These solutions contain inherent problems. Although the geometric Ronchi test is derived with a ray approximation in mind, the real world doesn't care about the designer's thinking. It follows wave physics. Interference between the gaps on the grating result in superimposed offset artifacts called *diffraction-order images*. These side images make the edges of the bars less certain, obscure behavior at the true edge, and generally muddle the view. They grow less important as fewer and fewer lines are strongly illuminated, but they can't be eliminated entirely because the outer portions of the image remain dimly lit.

White-light diffraction expresses itself as a blur at the edge of the lines, so if a line is expanded by moving the grating closer (the first solution), the blur is increased as well. Increasing the frequency of the lines (the second solution) doubles the angles at which the higher-order images appear, making the edges even more difficult to see. The brute force meth-

---

[3] This pattern does not use the approximation in Equation A.3 but is more accurately calculated using an adaptation of the method of Prugna (1991) to include correction error.

ods of either moving the grating to show fewer lines or using a finer grating are limited in their ability to improve sensitivity.

Again consulting Equation A.3, distortion increases when testing optics of high focal ratio. At *f*/12, we have reached the 8% arbitrary limit with no other changes. The test has not suddenly become more sensitive, but the tolerances are wider with slow optics. For high focal ratios, the Ronchi test can indeed discriminate between bad and good systems.

An additional method doubles sensitivity. A Ronchi null test is conducted not on a distant star but a point source at the focus. Light exits the instrument in the reverse direction, bounces against a full-aperture optical flat, and travels back through the instrument in the reverse direction. It is intercepted near focus by a Ronchi grating. This *autocollimation mode* doubles the aberration because the optics have been traversed twice. A well-known Schmidt-Cassegrain manufacturer tests telescopes in this fashion. Because the aberrations are doubled, and the *f*/10 focal ratio is high to begin with, we can see that this manufacturer's test is an adequately sensitive one. The problem for ordinary individuals is the same as it is for all autocollimation tests—huge optical flats are expensive.

What about a Ronchi test at the center of curvature? Several amateur authors have suggested that the geometric Ronchi test can be successful in a test of paraboloids at the center of curvature (Mobsby 1974; Terebizh 1990; Prugna 1991 and Schultz 1980). The problem with most of these tests is that they never calculate the sensitivity of their methods. The way of properly showing sensitivity is to compute the shape of a pattern for both perfect optics and optics which display a ¼ wavelength of correction error. One must demonstrate that the two patterns are sufficiently different that a distinction can be made.

In Figure A.9, an example pattern of a perfect 16-inch *f*/4.5 mirror is calculated together with the pattern of the same mirror if it were a full ½ wavelength undercorrected. Both are calculated as viewed in a 150 line/inch grating. The Ronchi screen is set at slightly different locations that present similar appearances near the mirror's center. One pattern is for a mirror better than any optical surface has ever been made. The other pattern is for a mirror of little or marginal usefulness in an astronomical instrument. Figure A.10 is a photograph of precisely this situation. The photograph shows, as no theoretical argument ever could, the difficulties of the geometric Ronchi test at the center of curvature.

The problem with the Ronchi test conducted at the center of curvature is that it is swamped by the overcorrection of aspherical mirrors operated far from their natural focus at infinity. The Ronchi screen is so obviously

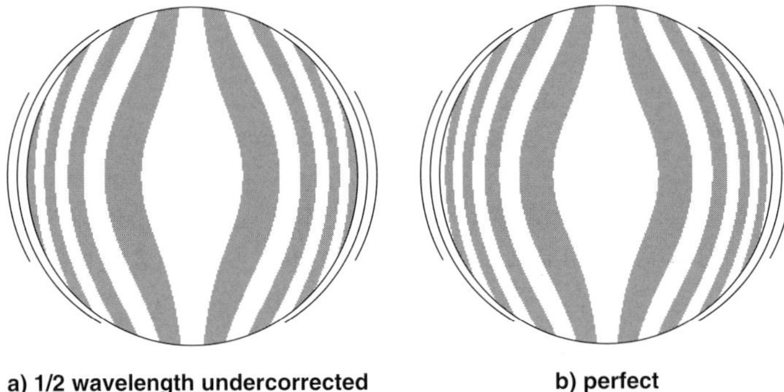

Fig. A.9 Theoretical patterns are generated for a Ronchi ruling of 150 lines/inch placed slightly inside the center focus of a 16-inch f/4.5 mirror: a) if the mirror is ½ wavelength undercorrected; b) if the mirror is perfect. Distances: a) –0.06 inch and b) –0.05 inch.

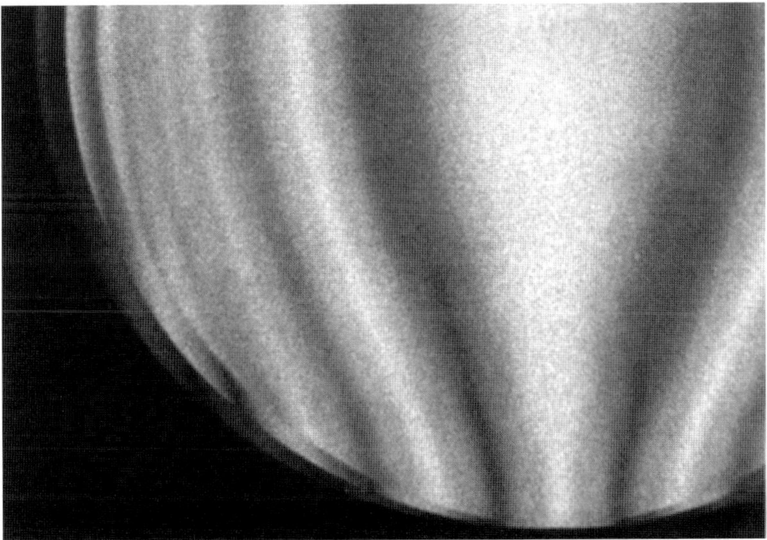

Fig. A.10 A photograph of a Ronchi test on a real mirror as calculated in Figure A.9. Is the mirror perfect or terrible? The mirror was Foucault-tested and found to be ⅛ wavelength undercorrected. The Foucault patterns appear in Figure A.3. (Photograph by William Herbert.)

responding to something that one forgets that the difference between its response to a bad mirror and a good mirror may be slight. The undercorrected prolate spheroid, the perfect paraboloid, and the overcorrected hyperboloid *all* look badly overcorrected at the center of curvature. All but the highest f-numbers show this condition.

For the 16-inch *f*/4.5 mirror at the center of curvature, the length of the blurry focus region is about 0.444 inch (11.3 mm). That's what the Ronchi screen is acting on. However, the length of that region is 0.428 inch if the mirror is undercorrected right at the ¼-wavelength Rayleigh limit or 0.460 inch if it is overcorrected. This difference of 0.016 inch is slight, only a tiny adjustment of those grossly distorted Ronchi patterns.

Sensitivity could be recovered by using a kinematic measurement platform and matching a *number* of theoretical patterns, all the time being careful to record the longitudinal motion of the Ronchi screen (as in Prugna 1991). Unfortunately, the test is almost never undertaken in this manner, probably because it so closely resembles the Foucault test that the user was trying to avoid in the first place.

Professional optical workers have dealt with the sensitivity of the Ronchi test. Cornejo and Malacara (1970) write:

> The Ronchi test is a very powerful test for spherical as well as aspherical mirrors. However, this test is of an accuracy limited by diffraction to a value such that the resulting surface can be used to form images, *but not for wavefronts to be analyzed interferometrically.*

In other words, the test can generate surfaces accurate enough for use in cameras or other coarse imaging devices but not surfaces that are so precise that they can be tested with an interferometer. Such accuracy is demanded in astronomical telescopes.

Ronchi himself commented on the accuracy of the geometric test that carries his name (Ronchi 1964). In an excellent review article he says,

> As long as the grating employed had a very low frequency, like the ones that we had used at first and that had also been used by other authors treating the same argument, the geometric reasoning corresponded quite well with the results of the experiments and measurements; *but at the same time the method did not lead to results as fine as desired.* It was evident that in order to increase this sensitivity it would be necessary to use gratings of the highest frequency possible, but then the results decidedly deviated from those predicted from geometrical reasoning.

Here Ronchi describes his justification for abandoning the geometric test in the 1920s. He goes on to describe techniques to use the overlap between diffraction orders as an interferometric test. The evaluation is made by the complicated interpretation of two equally-aberrated wavefronts somewhat displaced from one another. In this true wave-optics Ronchi test, the separation of the shadows is not determined by geometry but by the interference of light.

The geometrical Ronchi test has some uses, however. It is an excellent way of seeing sharp zones. It can detect seriously defective optics—curvature at stellar focus is often enough to reject an instrument out of hand. It is a good way of testing camera optics or any optics used far from the diffraction limit. But this test has a variable sensitivity that has been unappreciated or ignored by many advocates.

Readers may notice that if a checkerboard grid of squares replaced the Ronchi ruling, the Ronchi test would have a superficial resemblance to the Hartmann test. The corners of the squares would be equivalent to the hole positions. Yet no complaint was made about the Hartmann test's sensitivity. The critical difference is that the Hartmann screen is rigidly fixed on the aperture instead of floating somewhere near the focus. Focal planes are exposed and data are taken with high-precision measuring devices. Systematic errors are lessened by reducing measurements of dot positions from multiple directions. These raw measurements are followed by a sophisticated mathematical reduction procedure.

Many amateurs, particularly those who promote use of the geometric Ronchi test at the center of curvature for aspherical optics, represent the test as a simple comparison of patterns. They have stripped the mathematics away, and with it goes the error-detection ability of the test. The siren song of the geometric Ronchi test is that people can just look at a pattern and avoid the difficulty of measurement. Unfortunately, the measurements contain the sensitivity.

In summation, the geometric Ronchi test is not recommended for telescope evaluators for the following reasons:

1. The sensitivity of the test is variable and depends on the focal ratio of the tested optical system and the frequency of the grating used. Also, results vary depending on whether the test is done at the focus or the center of curvature. Certain combinations are very sensitive; other combinations are fatally insensitive.

2. When conducted at the center of curvature, the geometric Ronchi test is commonly used as a simple comparison of patterns, but the bending of such patterns can differ little from the patterns with unacceptable correction errors.

3. It requires a sufficient acquaintance with the theory of the test so that a "worst case" unacceptable pattern can be calculated. Before declaring tested optics to have passed, one must have an appreciation of what failure looks like. A good place to start is to calculate the anticipated test pattern with and without ¼ wavelength of low-order spherical aberration on the wavefront.

A.5 Interferometry

Many types of interferometers are used to test telescopes. Little purpose is

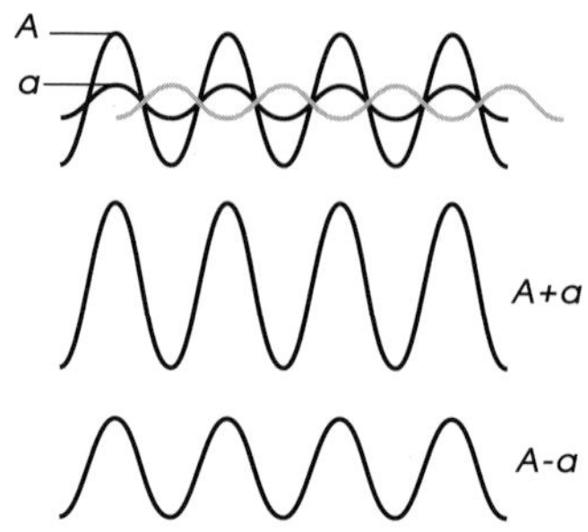

Fig. A.11 The principle of interference.

served in dwelling on each of them here. Most of them require expensive auxiliary equipment, so the likelihood is small that one of these methods will be used by an individual who wants merely to test a single telescope.

A.5.1 How Do Interferometers Work?

Light is a wave, and two waves with the same frequency add together by summing amplitudes A and a. If the waves are moving in opposite directions, the result is a standing wave, like a guitar string. If the waves are moving in the same direction, the result is a wave that can have an amplitude as much as $A + a$ or as little as $A - a$, depending on the respective phases. (See Figure A.11.) This wave is the result of *interference.*

One would seldom expect two waves to line up in direction well enough to see visible, reasonably stationary, interference effects in nature. Indeed, this effect is almost unknown using diffuse white-light sources. The exception can be found in the case of thin films.

With thin films, an incident light beam is reflected twice in quick succession by two very nearly parallel surfaces. The most common way of achieving parallel surfaces in nature is with liquids. Thus, our first experience with thin-film interference is usually found in the colors of a soap bubble. Thin-film interference also occurs in some insect wings, bird feathers, and mollusk shells. It even has a name that was coined before physical understanding was achieved, *iridescence,* after Iris, ancient goddess of the rainbow.

The color in a soap bubble is caused by the reflection from the outside of the bubble interfering with the reflection from the inside. (If you want to try an experiment, the colors are easier to see on a soap film still held in the loop.) The strength of the reflection from the air-to-soapy-water interface is about the same as the strength from the soapy-water-to-air interface, so the soap bubble interference should have high contrast. A color is favored at those frequencies for which one reflection is out of phase with the other by an even multiple of the color's wavelength.

What happens to those colors for which the conditions aren't favorable? A wave contains energy, and that energy doesn't go away when the wave is canceled. If one path has been blocked, it goes in another direction. The light passes straight through.

The reason we see colors at all is because soap bubbles are so thin. If the surfaces had two reflections with a phase delay amounting to 1000 wavelengths of blue light, it would be 999 wavelengths of not-so-blue light, 998 even-less-blue, and so forth. So many frequencies would be favored that the reflected light would be colorless again. The spectral response would resemble a pocket comb, with many selected frequencies jutting up. However, our eyes are not sensitive to such fine structure and we would see only white light.

This is also one of the reasons we do not see bubble colors in ordinary windows. In fact, this color effect is only seen at a certain stage of the ephemeral life of a bubble. A fresh soap film is thick, so one sees many frequencies adding together to a rough approximation of white. As it evaporates, it passes through an iridescent stage where the colors depend on the local thickness. Stripes of red, yellow, green, blue, and purple appear briefly. Then comes a stage when all colors of reflection are equally discouraged. The bubble is colored a pale white or a white tinged with yellow once again.

The bubble is still behaving as a barrier when it appears pale yellow, but soon it will get so thin that the wavelength of light is incapable of sensing the interface at all. If the bubble is long-lived, careful inspection will reveal an area of the film where it is actually invisible. The layer is so thin that it cannot materially impede the forward progress of the wave.

If, instead of using white light, we illuminate a fresh film with one color only, what would we see? As the film dries up, it goes from invisible to visible at any given location, depending on whether that wavelength is favored to reflect. The bubble looks tattered and frayed depending on how fast it is drying at a certain location, and onlookers are able to tell it is still whole only by looking at the edges.

The interference principles exemplified in a soap bubble can help us to understand how interferometers work. If we can generate one perfect wavefront, or at least a known wavefront, we can use it to interfere with the unknown wavefront of our telescopes. If the wavefronts are identical, the interference will be the same over the whole aperture.

A simple example is given to illustrate the ideas behind this principle. Let's say we wish to test a flat used for a diagonal mirror. We set it up in

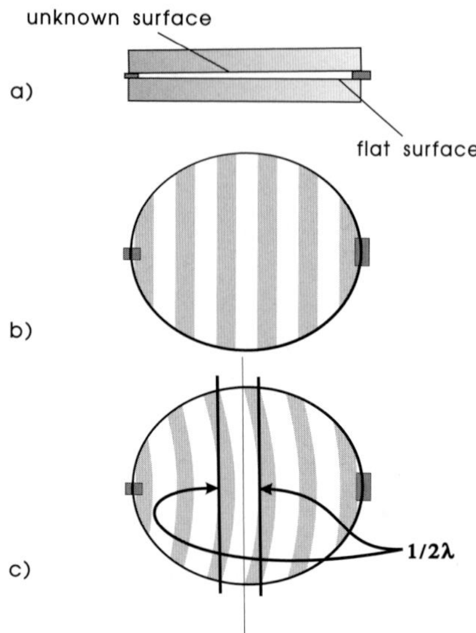

Fig. A.12 Interference using two surfaces in close contact tests the quality of a flat surface.

the configuration of Figure A.12 with a known optical flat, illuminate it from above with fairly pure yellow-green light, and get far enough away in a vertical direction so that perspective does not cause any effects of its own.[4] Here the thin film is not glass or soapy water, but the wedge of trapped air.

If the diagonal is flat and the blocks (actually, thin strips of paper) are of the same thickness, one would detect a uniform brightness depending on the relative phases of the two reflections. Such precision is almost impossible to achieve by luck alone. More likely one will see, as in Figure A.12a, a thin film of air amounting to N wavelengths on one side to $N + n$ wavelengths on the other. If we follow the course of light as it passes through the glass and reflects off the two layers, we see that the two reflections have a phase difference of $2N$ wavelengths on the short side of the wedge and $2N + 2n$ wavelengths on the other side (to within a constant). One can subtract the $2N$ phase difference since it is the same over the whole piece and concentrate on the $2n$. From the discussion with soap bubbles above, the expectation is that light stripes should appear at the locations where the interfering phases appear separated by 0, 1 wave, 2 waves, etc., and dark bars should appear where the phases are separated by ½ waves, ³⁄₂ waves,

---

[4] In practice, such a test is conducted with a mild condensing lens just above the two pieces (Fizeau's interferometer). Such an arrangement concentrates the returning light and rectifies the incident diffuse beam so that the optician can draw close.

$\frac{5}{2}$ wave, etc. Because the phase difference results from reflection, which doubles the distances light propagates, a bar occurs for every increase of a half wavelength in the separation of the plates. Furthermore, the bars should be straight, as in Figure A.12b, because the separation varies only in the direction of offset.

If the diagonal is not flat, however, the bars curve as in Figure A.12c. We may easily view the dark bars (called "fringes") as contours of the tested piece separated by a half wavelength of light. By stretching a dark thread across the test setup, we can even estimate the amount of curvature. This case is out of line by not quite half a fringe so the tested piece is smoothly curved by a little less than $\frac{1}{4}$ wavelength.

This test can easily be generalized to other shapes. If the known surface is a convex paraboloid, a Newtonian telescope maker just needs to test and polish until the bars are straight. The truth is, no one works this way because all of the pieces made would need to have precisely the same focal length (for no good reason). An optician needs a powerful incentive before incurring the expense of a curved testing piece. A camera manufacturer who may have a tight tolerance on thousands of spherical parts has a valid justification to make one. A telescope tester has easier roads to follow.

A.5.2 Interferometers

Most easily interpreted interferometer tests have one feature in common. They always start with a single beam divided into two parts. One becomes the "reference" beam, and the other beam is transmitted through the optics. The reference wavefront must be conditioned into the same converging shape as it would attain had it passed through perfect optics. Finally, the beams must be brought together with only a few wavelengths of tilt. After all of these diversions, the two beams should have roughly the same amplitude in order to provide good contrast between the bright fringes and the dark ones. For example, if the lower surface were aluminized in the example above, the fringes would be very weak. One interfering beam would originate from the transparent glass (maybe $4\% - 5\%$ reflectivity) and the other would bounce strongly off the aluminum coating (92% reflectivity). There are tricks to correct this, but they involve further complications.

Another condition that must be maintained is approximately equal path length. Some, like the point-diffraction interferometer, are inherently common-path (Delvo 1985, Smartt and Strong 1972, Smartt and Steel 1975). Others, like shearing interferometers, are common-path too, but have a more complicated interpretation.

The light source ultimately consists of a molecular or atomic transition, with a finite line width, implying through the Heisenberg indeterminacy principle, a localization of the light beam. A typical stream of electromagnetic energy originating from a single transition is a wave packet about 0.5 m long. Its coherence is further reduced by thermal properties inducing Doppler spreading. If the beam is split, half to enter a compact "perfect" conditioner (where it is immediately returned) and the other half continuing on to the optics, then these two path lengths must not differ by more than its coherence length if the two beams are to successfully interfere. This constraint is eased considerably by using a gas continuous-wave laser with a long coherence length.

Peter Ceravolo described the construction and use of a relatively inexpensive (compared to commercial models) interferometer in *Sky & Telescope* (Ceravolo 1994), so expense is not the problem. The problem is the amount of training and experience it takes to use one. Robert Royce (c. 2000) has written an eye-opening description of the various procedures that must be employed to derive a correct result from a large mirror.

Additional problems occur when interferometers are used to test an asphere. Either null optics must be employed, a huge flat must be used in autocollimator configuration, special reference surfaces must be figured, or a sophisticated computer program must be used to expand the complex iso-contour patterns of one or more interferograms in a series of Zernike polynomials from which the shape of the image at infinity focus may be inferred.

There is an additional subtlety in the type of interferometer used. Interferometers either are one-shot devices, using crest-sampling software like FringeXP (Rowe 2003), or they are phase-shifting interferometers using very expensive hardware and software to increase the number of

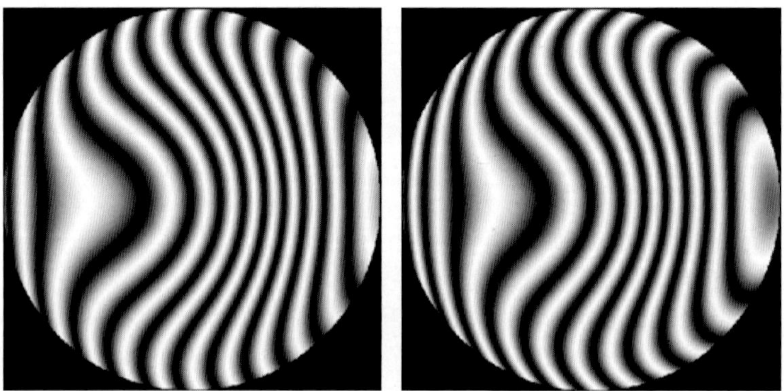

Fig. A.13 Two example computed interferograms are shown as derived from an asphere measured at the center of curvature compared with a spherical reference. The left frame is a 10-inch f/5.6, 1/4 wavelength undercorrected mirror, and the right shows the slight variation for a perfect mirror.

sampled points drastically. The hardware is usually an optical element that can change the optical path length on one side of the interferometer by an adjustable fraction of a wavelength or more. Maximum accuracy in one-shot interferometers is limited by the incomplete sampling of the pupil, the difficulty of identifying the crest of the fringe, image distortion, scaling, and other effects such as the superimposed diffraction patterns from dust motes. These same effects disturb phase-shifting interferometers, but the richness of the sampling of the latter over very many individual interferometer images and the statistical methods to which this information may be subjected makes them much more accurate.

I'm not going to discourage people from trying interferometry, but only warn those observers tempted to try it that it is a whole field to itself and of much more interest to those who make telescopes than to those who merely use them. In particular, interferometry provides direct sensitivity to those types of whole-disk errors for which the knife-edge and Ronchi tests, as typically practiced by amateurs, does not. In short:

1. Interferometry is much easier on small, less deformable, optical surfaces than large optical surfaces for which the test mounting intrudes on test interpretation (as it would for any test method).

2. Interferometry is much more easily interpreted in the "thin film" Fizeau configuration where it can be seen as a lack of straightness in lines or by using null optics.

3. Obtaining contrasty interferometer signals is more complicated when testing already-coated mirrors if the reference surface is not coated.

A.6   The Null Test

This test is actually a subset of many others. Briefly stated, a null test involves a testing configuration in which the perfect result is simple and requires no further reduction. Thus, the Foucault test is emphatically *not* a null test when a paraboloid is tested at the center of curvature, but a knife edge placed at the focus of a distant source or a star *is* a null test. If the optics are well figured, the whole surface will dim uniformly. Likewise, common-path interferometers used at the focus of a star are null tests. If the optics are perfect, the fringes are straight. The Ronchi test at the focus of a star behaves similarly; optics that pass the test show straight shadows of the bars on the aperture. If a full aperture autocollimation flat is used, many of the tests in this chapter become null tests. Even the star test qualifies.

Here, the words "null test" are limited to mean those shop tests for which the spherical aberration of a properly figured surface is undone by putting the opposite spherical aberration in the beam emerging from the source. The Dall null test will be used as an example because it is common in the amateur literature, but there are many null testing devices, some superior to the Dall tester.

The Dall null tester looks like a Foucault tester, except that the source pinhole is behind a small plano-convex lens. This lens puts an equal and opposite amount of anticipated spherical aberration onto the beam that would normally diverge spherically from the pinhole. A good paraboloidal mirror bends the wavefront back into a converging sphere. When the knife enters the image of the pinhole, it does not produce the distorted Foucault patterns of Figures A.2 or A.3. It blanks out, just as if it were testing a sphere.

Null testers are not inherently flawed, in spite of the poor reputation they received during the Hubble Space Telescope fiasco (Capers *et al.* 1991). However, many problems make the Dall null test difficult, including the following:

1. The testing optics must be of impeccable quality. They must have good edges and surface accuracy of approximately the precision that the tester hopes to find. A poor Dall lens cannot be used to test a good mirror.

2. The wavelength is no longer free; refractive null testing should not be done in white light. Calculations of the proper pinhole-lens separation depend critically on the index of refraction of the lens and will be valid for one color only. To use this test, one must have full information on the refractive properties of the lens. A color filter should be placed between the lamp and the pinhole to restrict the bandwidth of the light employed.

3. The Dall lens-pinhole distance must be very precisely set or the test does not null for perfect optics but for overcorrected or uncorrected wavefronts. (Testing on the Hubble telescope failed for a similar reason.)

4. The axis of the pinhole-lens assembly must be very carefully aligned on the center of the mirror.

5. The Dall null test has zonal spherical aberration that results in improper curvature when fast mirrors are made to conform to the testing apparatus. It is not an accurate null (Buchroeder 1994). See Figure A.14.

One further note that may be of interest to those tempted to try a null test is that the amateur is no longer constrained to use of the Dall test or any other variant that has appeared in the literature. If one has access to a short focal-length lens where the curvatures and material are known, it may be inserted in the commonly available ray-trace software and the theoretical spacing may be adjusted until nulling is achieved. Careful choice of curvature and geometry may lead to lower residual aberrations than in Figure A.14.

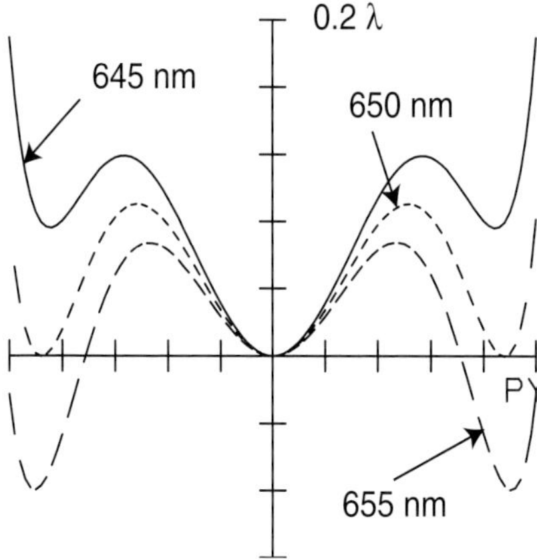

Fig. A.14 The residual aberration of a Dall null tester when used to invert the overcorrection of an 8-inch f/4 paraboloid tested near the center of curvature. The central red LED wavelength of 650 nm has 1/12th wavelength of high order spherical aberration, and a 10 nm bandpass is used to constrain the longitudinal chromatic aberration to about 0.07 mm.

A.7 The Roddier Test

The Roddier test is a quantitative star test in that it uses the defocused stellar image. It is oriented toward evaluating completed telescopes, rather than individual elements of unfinished instruments. It acts like the Hartmann test in that images of the defocused point source are gathered on either side of the focus. Unlike the Hartmann test, it puts no mask over the aperture and it is sensitive to the curvature rather than the slope. It relies on the use of an electronic imager.

To do this test, an image is taken at equal distances on either side of focus and the relative irradiances (power per unit area) are inferred on locations at the pupil. These irradiances are related to the local curvature of the wavefront. The complete wavefront is the result of an integration of all these curvature samples taken together. The mathematical technique is best learned from the original source documents (Roddier *et al.* 1988-1994). Suffice it to say that it is based on a solution of a differential equation in the interior of the pupil with starting conditions provided by the slope of the edge. Currently, amateur efforts are centered on a program, called WinRoddier2.exe, released by a French group (Lequèvre *et al.* 2007).

The proper operation of the common amateur procedure demands careful control of the amount of defocus. Users must defocus far enough that diffraction effects are no longer easily visible but not so far that sensitivity is lost. Because the test is sensitive to light intensity rather than image location, the image must be within the dynamic range and the focal plane must be calibrated for a linear response. This is most easily done with a scientific-type imager of long dynamic range. The amounts of defocus must be balanced.

While I am reluctant to recommend against using this procedure, since it is the quantitative version of the qualitative test advocated in these pages, I must warn readers that it will require a deft touch and good equipment. It appears to have been best used by those who understand all the pitfalls, namely professional astronomers. In short, the following list of procedures is recommended:

1. Use a scientific-grade imager capable of at least 10 bits of dynamic range, and preferably 12.

2. Align the telescope. Before taking a fresh set of data, align it again.

3. Carefully dark- and flat-field the imager. If the gamma power can be set, make sure that it is set to 1 (linear response).

4. If you have one, use a digitized focusing mechanism, where the amount of defocus can be read off a display.

5. Average many frames together and use a steady drive (or use an artificial point source).

6. Filter the image to narrow the bandwidth used.

7. Use multiple values of defocus and slightly different field positions to identify Zernike terms that are stable and those that are not.

8. If they are available, use two imagers and check that the results are similar. (Use the Zernike term for low- and high-order spherical aberration for the comparisons).

Appendix B
Calculation Methods

To generate the photograph-like patterns, modulation transfer functions, and other miscellaneous graphs in this book, it was necessary to use the Huygens-Fresnel principle in the context of Fourier optics. What follows is intended only as a sketch of the concepts used here. For more insight, consult the references mentioned below. The most readable is Joseph W. Goodman's *Introduction to Fourier Optics* (Goodman 1968) or Eugene Hecht's text *Optics* (Hecht 1987).

B.1 Diffraction Concepts

Chapter 4 presented a simplistic introduction to some of the ideas behind diffraction and the Huygens-Fresnel principle. For the purpose of discussion, we assumed every point on a wavefront re-radiates the wave. That convenient fiction has some difficulties, however. If each point is allowed to radiate in every direction equally, the aperture would arbitrarily reflect back into space some of the energy that reached it.

Reflection at the aperture stop is not observed, nor is it seen at any point on the wavefront. Waves don't suddenly reverse direction unless they encounter a change in the medium. The wave sum also doesn't deliver the correct answer when integrated over a situation with no aperture, just a source of light and a receiver. The sum would be over a complete sphere, and the radiators at right angles between the source and receiver have too strong a weight.

Fresnel realized that these were impediments, so he made certain approximations. His model was not derived from first principles, which probably helps explain its intense scrutiny by such worthies as Poisson. In the early 19th century, even the physical process that produces light was not yet understood.

Later research of Gustav Kirchhoff produced a version of the Huygens-Fresnel theory derived from first principles. His result, called the Fresnel-Kirchhoff diffraction formula, no longer requires that the rearward direction be ignored, though it demands a few conditions:

1. Light is modeled as a scalar wave. Polarization is not included. The mod-

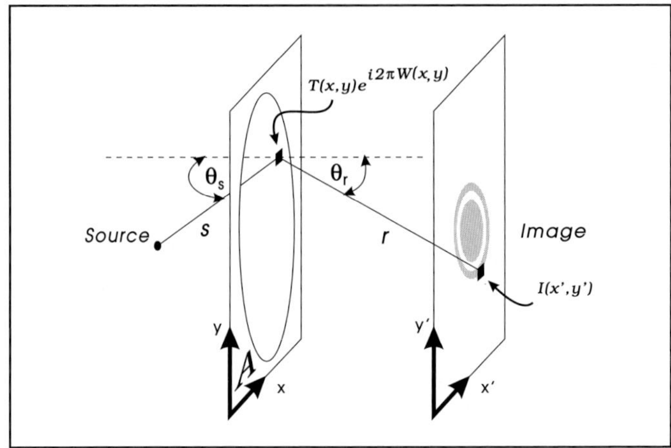

Fig. B.1 Variables appearing in the Fresnel-Kirchhoff formula. The aperture is in surface A.

el is oblivious to the vector nature of light.

2. The field values near the aperture are the same as they would be in the absence of the aperture (weighted by a simple trigonometric function).

3. The distance from the aperture is sufficient to ensure that no bound or evanescent fields are present.

One consequence of the second condition is the simultaneous specification of both the field values on the aperture in addition to their derivatives. The time-harmonic form of the wave equation[1] is a second-order differential equation for which the normal method of solution is to set either the value of the field on the boundaries or the derivative of that field. It is unusual that Kirchhoff gave both.

With its overspecified boundaries, we must regard Kirchhoff's solution as an approximation. This approximation will not affect results if the aperture does not have a lot of structure. We may rightly suspect that the Fresnel-Kirchhoff equation will be less accurate for devices like high-resolution diffraction gratings. For gently sloped optics, with most of the area many wavelengths from all edges, the approximation does not cause much trouble. More detail concerning these criticisms is available in Baker and Copson (1950).

A modified form of the Fresnel-Kirchhoff formula (using variables defined in Figure B.1) is

---

[1] This formulation is called the *Helmholtz equation*.

$$U(x', y') = \frac{1}{N} \int \int_A T(x, y) e^{i2\pi W(x, y)} \frac{e^{i2\pi(r + s)/\lambda}}{rs} \left(\frac{\cos\theta_s + \cos\theta_r}{2} \right) dx\, dy. \qquad \textbf{B.1}$$

Here, N is an arbitrary normalization and the time dependence has been suppressed. $U(x', y')$ is the value of the field at the image location (x', y'). (Uniform transmission versions of the formula are derived in Hecht 1987, pp. 461–462; Goodman 1968, pp. 37–41; Born and Wolf 1980, pp. 378–380.) The integral may be performed over the entire surface A, but the integrand is non-zero only inside the aperture pupil.

The term $T(x, y) e^{i2\pi W(x, y)}$ is called the *pupil function* and is merely a complex number in circular notation. The transmission coefficient $T(x,y)$ is its modulus[2] and the function $W(x,y)$ (in wavelengths) implies the phase. $W(x,y)$ contains aberrations like turbulence, pinched optics, defocusing, astigmatism, etc. The cosine term in brackets is the *inclination* or *obliquity factor*. This function will eliminate nonphysical backward propagation. If we place the source far away, then θ_s is 0 and $\cos\theta_s$ is 1. Set θ_r to 180°, as it would be for backward propagation, and $\cos\theta_r$ becomes −1 and the inclination factor vanishes.

In writing Equation B.1, another approximation has been implicitly made. The aberration function $W(x, y)$ affects the angles in the inclination factor slightly, and the equation doesn't contain this effect anywhere. For all practical purposes, this change in angle is extremely small. Typically, the worst wavefront tilts in this book are 30 wavelengths over 100 mm, or about 0.01°. Equation B.1 also contains an assumption of linearity.

Key elements in the Fresnel-Kirchhoff formula are the two factors e^{ikr}/r and e^{iks}/s (where k is the wave number $2\pi/\lambda$). Here, Huygens' principle of re-radiating elemental points is written in mathematical form. Each of these expressions is the time-independent part of a spherical wave function. The denominator will cause the intensity to obey the inverse square law of light (twice as far, one-fourth as bright). The numerator will ensure that the Fresnel zones will be properly painted on the aperture. The "s" spherical wave represents the propagation of the wave from the source to a point on the aperture. It is then reborn as the "r" spherical wave, which propagates to the receiving point.

The intensity is related to the energy, so it cannot be complex-valued. It is calculated from the field above as follows:

$$I(x', y') = U(x', y')U^*(x', y') = |U(x', y')|^2. \qquad \textbf{B.2}$$

Once the integral has been performed over the whole open aperture, Equa-

---

[2] *Modulus* is the absolute value of a complex number.

tion B.2 says that the intensity is known for only *one point* in the image space. To find it at any other location, we must change the values of x' and y' and evaluate the field integral again. To completely map out an entire image this way requires (to say the least) a great deal of time.

Another name for the expression $I(x, y)$ is the *point-spread function,* or PSF. The PSF determines how diffraction, obstructions, and aberrations degrade a perfectly sharp point source of light into a fuzzy disk. In the case of perfect optics filtered only by a finite circular aperture, the PSF follows the familiar Airy disk pattern.

B.2 The Fraunhofer and Fresnel Approximations

We can simplify Equation B.1 by noticing some symmetries and by casting the problem in another coordinate system. First, we place the source at a distant on-axis location so that s is a very large constant. Then we can pull the e^{iks}/s wave function constant out of the integral and bury it in the normalization constant. Because we are using a focusing lens or mirror, we pretend the receiver point is also very distant by "unbending" the wavefront (Born and Wolf 1980, pp. 382–386). In the Fraunhofer approximation (which is what this unbending is called), the important quantities are no longer the distances but the off-axis angles from the center of the aperture to the distant sensor.

This approximation is written as (Hecht 1987, p. 494)

$$U(\phi_{x'}, \phi_{y'}) = \frac{1}{N'} \int\!\!\int_A T(x, y) e^{i2\pi W(x, y)} e^{i(k\sin\phi_x x + k\sin\phi_y y)} \, dx dy \,. \qquad \textbf{B.3}$$

While not superficially looking much simpler than Equation B.1, the Fraunhofer approximation has done away with those very messy distances r and s (or at least they are now neatly tucked away). The angle ϕ_x is the angle to the image point in the x-direction, while ϕ_y is the angle toward the image point in the y-direction. The second exponential in the integrand is just the phase difference induced by the tilt angle to the image location being calculated. This term does the same thing as picking up the corner of a table and asking how much higher the table is all along its surface. Over the leg still touching the floor, the extra height is zero. At the lifted corner, it has its full value. Everywhere over the aperture, this "tilt" aberration may be easily calculated. For all cases of interest here, $\sin\phi$ is very close to ϕ, which makes it even simpler.

In the Fraunhofer approximation, the inclination factor is ignored because of the very tiny deflections of the wavefront from its spherically converging path. N' is the new normalization constant.

The Fraunhofer formula is not supposed to apply to non-zero values of defocusing aberration. However, the next term of the expansion of Equation B.1 used in deriving the Fraunhofer approximation is just this defocusing. Including this term changes the integral to the Fresnel approximation. However, an external defocusing cannot be imposed on the Fresnel approximation—that step would doubly count the defocusing. It is tidier to encapsulate all such terms into the pupil function, which is then applied to the skeletal Fraunhofer formula.

The light touch of small amounts of defocusing can perhaps be gauged by the focus shift in Table 5.1 divided by the focal length of the instrument (which is the same as the fractional change in the sagitta of the wavefront). For example, at 12 wavelengths defocusing aberration on an 8-inch (200-mm) $f/6$ telescope, $\Delta f/f = 0.0016$, a very tiny fraction.

The star test of a focused aperture does not precisely reproduce at equal distances inside and outside of focus, but the difference is small. For example, a 200-mm $f/6$ telescope defocused by 1.9 mm is listed in Table 5.1 as having a defocusing aberration of 12 wavelengths. A careful calculation shows that 1.9-mm inside focus has a more precise aberration of 12.02 wavelengths. On the outside, an equal eyepiece motion becomes 11.98 wavelengths defocusing aberration. If we were to adjust the focuser until the aberration were precisely 12 wavelengths, then the image inside focus would be ever-so-slightly smaller and brighter, and the image on the outside of focus would be incrementally bigger and dimmer. Thus, the effect is manifested as a magnification difference. Perhaps if we were to defocus very precisely with a measurement screw, we could barely detect such changes, but most star testers will never notice the difference. We must simply ensure that the fraction $\Delta f/f$ is small (Bachynski and Bekefi 1957; Li 1982; Erkkila and Rogers 1981).

B.3 Image Calculations for Symmetric Apertures

Further simplification of the Fraunhofer approximation results if the pupil function is circularly symmetric. If the integral in Equation B.3 is rewritten in circular coordinates, the angle integral can be performed to yield

$$U(r') = \frac{1}{N''} \int_{\rho = 0}^{1} T(\rho) e^{i2\pi W(\rho)} \rho J_0(\rho \pi D r'/f\lambda) d\rho \,, \qquad \text{B.4}$$

where ρ is the normalized radial coordinate of the aperture, r' is the radial coordinate on the focal plane, J_0 is the zeroth-order Bessel function, and

N'' is another normalization constant (Luneburg 1964, p. 345; Schroeder 1987, pp. 181–182).

In this book, all monochromatic circularly-symmetric images were calculated using this simplification. My program APERTURE used divided the radius into N_ρ equally-spaced points (typically 300–500) and calculated the intensity sum

$$I(\varphi) = \frac{4}{N_\rho^{\,2}} \left| \sum_{j=0}^{N_\rho} T_j e^{i2\pi(W_j + W_d)} \left(\frac{j}{N_\rho}\right) J_0\left(\frac{\varphi \pi j}{N_\rho}\right) \right|^2 . \qquad \textbf{B.5}$$

Here the value φ is the reduced image angle $(D\theta)/\lambda$, where θ is the true angle in radians, D is the diameter of the aperture, and λ is the wavelength. Reduced angle φ conveniently reaches the edge of the Airy disk for a uniform circular aperture when it has the value 1.22. W_d is the defocusing aberration $A_2(j/N_\rho)^2$ (defined in Chapter 10) in wavelengths, and W_j contains the rest of the aberrations. T_j is just the transmission coefficient at a sampled radial point j. J_0 is calculated using a subroutine adapted from *Numerical Recipes* (Press *et al.* 1986).

It was straightforward also to keep track of the encircled energy of the image as φ was increased. Because Equation B.5 is a radial sum, the intensity of the outer portions of the image had to be weighted for the increased perimeter. The normalized encircled energy increment is approximately

$$\Delta EE(\varphi) \approx \frac{\Delta\varphi \pi^2 \varphi I(\varphi)}{2\varepsilon_{TR}}, \qquad \textbf{B.6}$$

where $\Delta\varphi$ is the increment in image angle and ε_{TR} is the total energy fraction that enters the system. If the obstruction is 50%, ε_{TR} is 0.75, and the encircled energy increases enough so that it reaches unity far beyond the Airy disk.

Actually, if the encircled energy is accumulated in the simple-minded manner of Equation B.6, results are poor. The energy in focused images sharply increases just as the circle radius starts to open. Errors made here can be large. The program used Simpson's rule integration with a sophisticated starter.

The modulation transfer function (MTF) was not calculated for circularly symmetric apertures in its more straightforward form as an autocorrelation (see below). Such a calculation is too slow. Instead, once $I(\varphi)$ had been carried out far beyond the Airy disk, it could be used to infer the transfer function. The image is merely the convolution of a perfect sinusoidal target with the point spread function. For spatial frequency ν, the integral is (as adapted from Schroeder 1987, p. 204)

$$\text{MTF}(\nu) = \frac{1}{M_0} \int_0^\infty I(\varphi) J_0(2\pi\varphi\nu)\varphi \, d\varphi. \qquad \textbf{B.7}$$

The number M_0 is a normalization at zero spatial frequency.

A problem with this calculation is not apparent from casual inspection. The upper limit of the integral assumes that the intensity is known out to an arbitrarily high angle. We can calculate the intensity out to any angle, of course, but there must be energy beyond that angle that we know nothing about. High angles are represented by very low spatial frequencies. Thus, at first glance it would seem that this energy would be automatically taken into account by forcing the normalization M_0 to be the integral evaluated at $\nu = 0$.

This approximation throws a wrench into the works. A modulation transfer function calculated this way does have the proper value (i.e., 1) at a spatial frequency of 0, but at higher spatial frequencies, the calculated transfer function is too high. The inescapable conclusion is that the truncation of the angle causes a numerically diminished MTF at low spatial frequencies. In other words, if we have not gathered that distant energy in our intensity calculation, we had better not take tally of it in Equation B.7. Therefore, the computational algorithm is slightly modified:

$$\text{MTF}(\nu) = \frac{EE(\varphi_{\max})}{\text{MTF}(0)} \sum_{m=0}^{N_\varphi} I(m)\left(\frac{m\varphi_{\max}}{N_\varphi}\right) J_0\left(\frac{2\pi m \varphi_{\max}\nu}{N_\varphi}\right). \qquad \textbf{B.8}$$

Here MTF(0) is just the sum done for $\nu = 0$, and m is the summation index. $EE(\varphi_{\max})$ is the fractional encircled energy as far as the point-spread function integral is done. N_φ is the number of angles in the intensity summation. The effect on the MTF is to induce a small downward curl at low spatial frequencies. In most of the figures appearing in this book, the intensity function was carried out to angles where the lower part of the MTF could be approximated by a straight line to the known intercept of unity at the origin. Usually, the encircled energy at the most distant angle of a focused pattern exceeded 99%.

B.4 Image Calculations in ASYMM for Nonsymmetric Apertures

Although the field calculations of Equation B.4 are long, they are a great deal shorter than the double integral required for a nonsymmetric pupil function. Others have noted that generating these patterns is tedious (Allred and Mills 1989); before computer time became inexpensive, they were not often attempted at all. Each aperture is modeled by a square grid of 129 × 129 points, with points farther than halfway from the center set to zero. Within the aperture are about 13000 points. The image plane is a square grid of points that cannot be truncated, as at the pupil. The integral in Equation B.3 must be done for each of these 16000 image locations. For each nonsymmetric image frame in the first edition, the integrand of Equation B.3 needed to be evaluated about 210 million times, and for the polychromatic 256x256 image and pupil of the second edition, repetitions had to stretch into the billions.

The image plane may be mapped out in one grand sweep using another technique. If a two-dimensional discrete Fourier transform were taken of the complex pupil function, the result would be (after rearrangement and further processing) a complete image. The fast Fourier transform (FFT) is useful to reduce the computational load (Brigham 1988).

Once such an FFT is finished calculating, the array has to be reorganized and the image intensity extracted. The additional processing to sample and compress the image would take more computer time. For the self-written program, it was decided that the extra effort required to process the image by simple integration of Equation B.3 was not a severe burden, and was more than compensated by the complete transparency of the code.

One other interesting note is that the larger, polychromatic, image used in the second edition compensated for the increase in computer speeds since the first edition was written in 1992. Each frame of the polychromatic images required over a half hour to compute on a 2.7 GHz computer.

The algorithm that was used at the core of the program ASYMM was a discrete version of Equations B.2 and B.3. It had the form

$$I_{m,n} = \frac{1}{N^2}\left|\sum_{t,u=-64}^{64} T_{t,u}e^{i2\pi(W_{t,u}+W_d)}e^{i\pi\varphi_{\max}(mt+nu)/64^2}\right|^2 \qquad \text{B.9}$$

where W_d is the defocusing aberration, $W_d = A_2(t^2 + u^2)/64^2$ and values $W_{t,u}$ contain the rest of the aberrations. N here is the number of pupil points, i.e., 12,853. This sum is done for each image point in a grid that runs indices m and n from -64 to 64.

Asymmetric pupils have complex transfer functions. The full description of the performance of asymmetric apertures is called the *optical transfer function* (OTF) and has the following form:

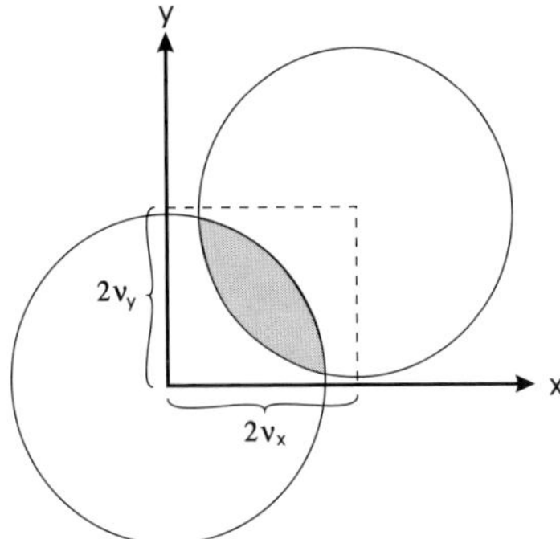

Fig. B.2 The MTF as the overlap with a shifted pupil.

$$\text{OTF}(v) = \text{MTF}(v)e^{i\Psi(v)}.$$

B.10

The imaginary exponent $\Psi(v)$ is the *phase transfer function*. For small aberrations the real part of the OTF is much larger than the imaginary portion. Thus, phase is commonly neglected. The modulus $\text{MTF}(v)$ is the quantity usually equated with the optical quality of the system.

The effect of a small imaginary part of the OTF on the bar pattern image is to shift it sideways a tiny amount but not enough to reverse it completely. Reversal is adequately handled by a negative MTF.[3]

Clearly, we cannot use the circularly symmetric formulation of Equation B.7 to calculate the OTF of an asymmetric pupil. Instead, we may use a tidy formulation described well in a number of places (Born and Wolf 1980, p. 485; Parrent and Thompson 1969, p. 22; Luneburg 1964, p. 356). The OTF is calculated as the autocorrelation of the pupil function:

$$\text{OTF}(v_x, v_y) = \frac{\iint P(x, y)P^*(x - 2v_x, y - 2v_y)\,dx\,dy}{\iint |P(x, y)|^2\,dx\,dy},$$

B.11

where the integral is taken over the aperture plane. Here P is a shorthand notation for the pupil function defined in the text following Equation B.1. The aperture coordinates are defined so that the perimeter is at $x^2 + y^2 = 1$, and the spatial frequency fractions are similarly normalized. One can envision the integral above as the overlap between the pupil function and the pupil function moved sideways by the spatial frequency fractions as in Figure B.2. For non-aberrated pupils, the OTF is the dark area divided by

---

[3] The MTF as defined in Equation B.10 is always positive, but in this work the sign component of the phase transfer function is sometimes appended to the MTF when the OTF is mostly real. The resulting "MTF" is smoother and tells more about the optical performance.

the uncovered area of the whole aperture.

The actual algorithm took into account the sampling of the aperture on a rectangular grid. Since the offsets have to land on sampled positions, the OTF was calculated only for integer offsets along the x and y directions and a 45° offset with equal shifts. The algorithm for a shift in only the x direction was

$$\text{OTF}(\nu_x) = \frac{\sum_{t=s}^{64} \sum_{u=-64}^{64} T_{tu} e^{i2\pi W_{tu}} T_{t-s,u} e^{-i2\pi W_{t-s,u}}}{\sum_{t,u=-64}^{64} T_{tu}^2} \qquad \text{B.12}$$

where indices s (for "shift"), u, and t run from -64 to 64 and $\nu_x = (s+64)/128$. The maximum spatial frequency corresponds to $\nu_x = 1$. Here the pupil array is measured from its center.

Encircled energy can be defined for an asymmetric aperture, but no such calculations appear in this book.

B.5 Use of Programs

While preparing the first edition, I was forced to write the programs AP-ERTURE and ASYMM for the symmetric and asymmetric pupil functions because, quite literally, there were no commonly available software products that would calculate the star test patterns. I also wrote a driver to translate the numerical values into Postscript^{TM} pictures that could be rendered into grayscale images by a high-end printer. This procedure was tedious, especially in 1992. Since that time, the commercial optical design program ZEMAX has appeared, which allows the calculation of either a FFT point-spread function, or the more accurate Huygens one. It also will perform the calculation over many colors. The particular advantage of using ZEMAX is that it allows a real modeled system on the front end.

One disadvantage of using any code written out-of-house is that you are not entirely aware of its core operations, approximations, and as-yet undetected bugs. For example, the present version (2007) of ZEMAX has a peculiar way of rendering true colors. Anything resembling the photopic response of the eye comes out bright green. It seems that the normalization or the CIE color-curve contributions of the individual wavelengths are not yet predictable. An attempt to artificially generate a white image seems to be weighted as a sort of flat curve, giving red and blue colors excessively strong contributions. Still there is no guarantee that the curve will work at other values of defocus and aberration. Until a routine reproduces the fairly color-free appearance of the perfect defocused reflector at a number of

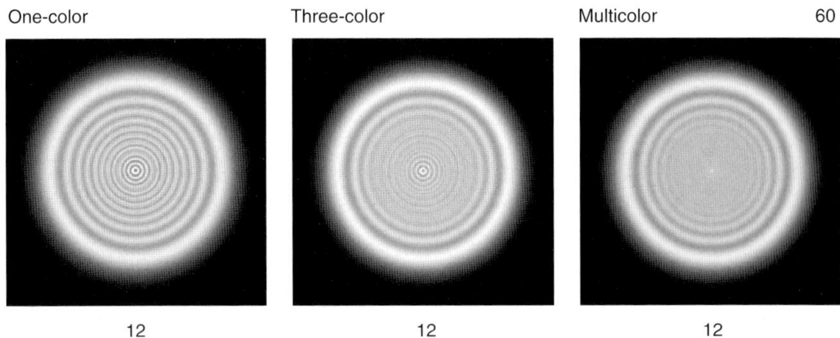

Fig. B.3 Fidelity increases as wavelength sampling increases from monochromatic on left, through multicolor on right. The three-color image in the middle uses the visually balanced 510, 560, and 610 nm wavelengths of Bruce Walker (Walker 2000).

defocus values, such true color computations of the star test should not be done. This is not so much a failing of ZEMAX, but a difficulty in creating a realistic color image from nothing. Most color schemes are designed for reproduction of an existing color image using a three-color device like a computer monitor. A seventeen-wavelength creation is necessary to reproduce the effects of diffraction but overdoes the colors. As a consequence of this and other reasons related to the difficulty of shepherding the figure through all the intermediate steps of the printing process, I have chosen not to calculate true color images. Instead, I retreated to the grayscale polychromatic image, which passed the tests mentioned in the next section.

It may be tempting to calculate the polychromatic contributions of just the red, blue, and green components of the color image. In some situations, though, such a calculation does not sample the spectrum sufficiently. Figure B.3 shows the progression from monochromatic to three-color to multicolor of an unobstructed, perfect aperture. Most new figures in this book were prepared with a multicolor photopic distribution. Figure B.4 shows the sampling and weights that were used.

B.6 Verification of Numerical Procedure

As you may expect, the actual implementation of the algorithms is very messy compared to the sparse presentation in Section B.1 through B.3. Four methods were used to check that the coding was properly done:

1. Checks were made between APERTURE, ASYMM, and ZEMAX for circularly-symmetric apertures,

2. A comparison of APERTURE with an exact procedure was made,

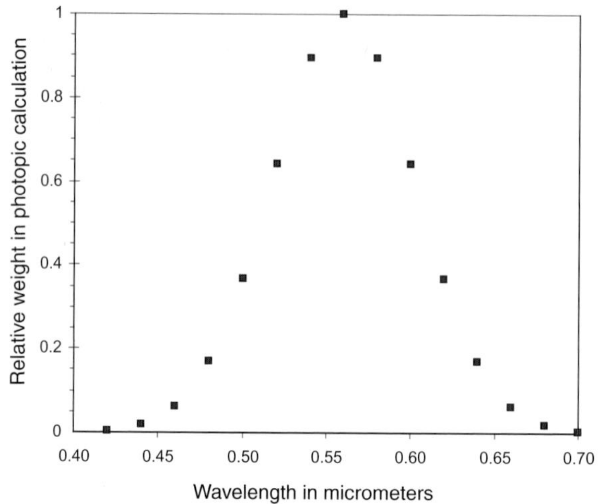

Fig. B.4 Wavelengths and weights used in polychromatic calculations.

3. Reproductions were generated of complicated patterns appearing in the literature, and

4. Calibration was done of contributions of multiple colors in ZEMAX.

B.6.1 Comparison of the Three Programs

Even though all three programs have a pedigree that traces back to Equation B.3, very little resemblance between them can be easily seen. They use different routines and are written in different languages. We can calculate the same situations using these separate procedures and verify that the answers are indeed about the same.

Taking a circularly-symmetric pupil and cranking it through the slow, direct-integration programs ASYMM and the slow Huygens PSF option of ZEMAX, one should obtain the same answer as APERTURE does. Tests were done for a number of pupils, and good comparisons resulted. Three such images appear in Figure B.5; the point spread functions appear in Figures B.6 and B.7.

Because APERTURE is circularly symmetric and greatly oversampled compared with ASYMM and ZEMAX, I have used it as the solid comparison curve in each. The curves are virtually identical except for a slight structuring of the ASYMM output caused by coarse sampling. A possible reason for image differences is discussed below in Section B.7.

Had these programs generated markedly different patterns, we would not have known whether any was correct. But because three quite dissimilar routines generate more or less the same answer, they are either coded correctly or an unlikely accident has occurred—an error that exhibits itself

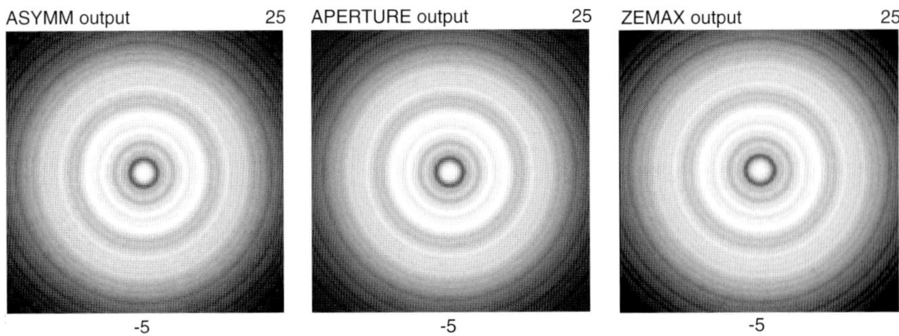

Fig. B.5 Comparison images of APERTURE, ASYMM, and ZEMAX on $+\frac{1}{4}$-wavelength spherical aberration of a 25% obstructed pupil.

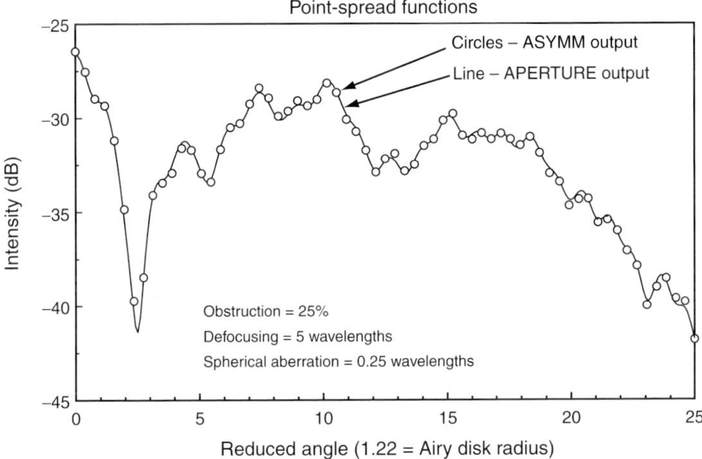

Fig. B.6 Point-spread functions that were used to generate the previous figure for the ASYMM and APERTURE contributions along one radius.

identically in each procedure. Lacking evidence to the contrary, we will assume that they are correct for now and proceed with other checks. Note that this verification does not confirm the original theory contained in Equation B.3. It merely says that the programs all seem to be calculating Equation B.3 correctly. The physics of the star test have been verified experimentally with the modern advent of CCDs and phase-retrieval methods, inferring an estimate of the wavefront shape at the pupil from changes in the defocused image (e.g., Fienup et al., 1993, Roddier *et al.*, 1988– 1994). These algorithms have been compared with results of other testing methods and give good agreement. They are exceedingly difficult and prone to systematic error, however, and should not be attempted by the unwary. For more information, see the entry on Roddier's method in

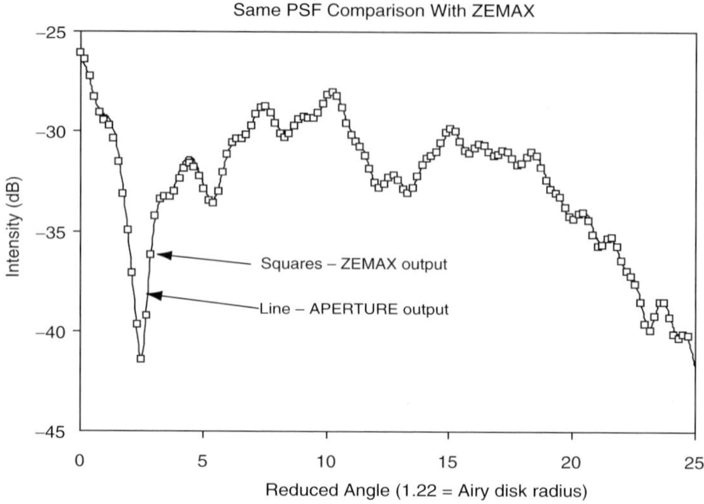

Fig. B.7 APERTURE results compared with ZEMAX.

Appendix A. Their brightnesses are difficult to measure quantitatively (Taylor and Thompson 1958, Burch 1985).

B.6.2 A Numerical Comparison with an Analytic Solution

Any numerical procedure should reproduce the simple systems for which the answer is known analytically. Unfortunately, few diffraction problems have been solved in closed form. From our point of view, an analytic theory of a circular aperture containing at least one aberration would be acceptable.

A solution to an otherwise perfect circular aperture with defocusing aberration appears in Chapter 8 of Born and Wolf. It involves what are called the *Lommel functions* to perform a version of the integral in Equation B.4 (Born and Wolf 1980, pp. 438–439). Two solutions are each written as an infinite series having its own region of applicability, one inside geometric shadow and one outside of it. These series were programmed into a MathCad™ document and each was graphed in the region where it was supposed to work (an 8-wavelength defocused example is shown in Figure B.8). Results were indistinguishable from the results of APERTURE. A variety of tests went up to 12 wavelengths defocusing aberration and were limited to Bessel functions up to order 80.

Again this verification is reassuring, but incomplete. It tests the focus component of the aberration function (which appears in a separate term of both algorithms) but leaves the other aberrations unverified. However, the central loops of ASYMM and APERTURE do not distinguish between defo-

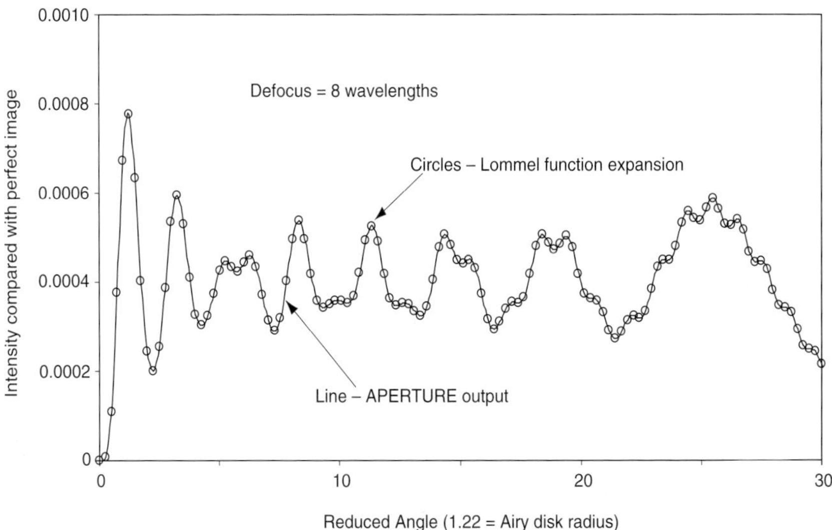

Fig. B.8 Lommel function expansion of defocus aberration compared to `APERTURE`.

cusing and other aberrations. If a mistake were being made, it would have to be in the preparatory statements of both programs. Furthermore, any supposed error would have to leave the defocusing aberration untarnished in both procedures.

B.6.3 Comparison with Published Patterns

Perhaps the most severe test of the approximation here was the comparison of patterns generated by `ASYMM` and `ZEMAX` with published contour plots for coma and astigmatism. The coma contour plots were figured analytically by R. Kingslake (1948), but they are most commonly available as reprinted in Born and Wolf (6th ed., 1980). The astigmatism contour plot was published by Nienhuis and Nijboer (1949).

The expressions for aberration differ somewhat from the primary aberrations used to derive these contour plots, but if we trace through the details, we find that 3.2 and 6.4 wavelengths of primary coma translate to 2.13 and 4.27 wavelengths of peak-to-valley coma aberration as defined at Zernike diffraction focus. In other words, coma as defined here is a third the coefficient appearing in Born and Wolf. Likewise, the amount of astigmatism used there was 0.64 wavelength, which translates to 1.28 wavelengths peak-to-valley astigmatism. The patterns as drawn by `ASYMM` and `ZEMAX` appear in Figure B.9. They match the overall shapes of the contour plots very well (axes were relabeled in Born and Wolf pp. 478, 480).

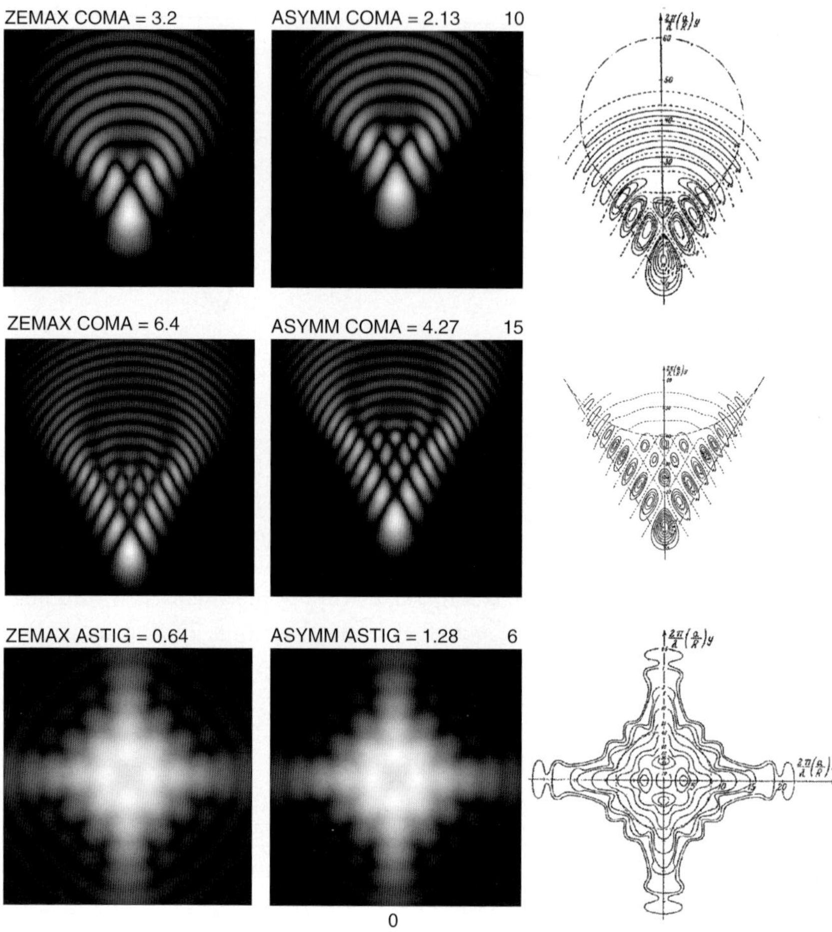

Fig. B.9 Reproduction of coma and astigmatism image patterns. Left images are not at the same scale as the contour plots to the right. (Coma contours appear with permission of R. Kingslake and the astigmatism contour appears courtesy of Elsevier Science Publishing.)

B.6.4 Calibration of ZEMAX Grayscale Polychromatic Images

One of the things we would like to know (rather than assume) is that the grayscales are added properly in polychromatic images. To properly test ZEMAX's addition of colors, we will have to split up the colors of a single image, and see if they appear in the correct amounts.

This operation is done in Figure B.10. Here the images are split with a mild dispersive prism (with index of refraction at the d Fraunhofer line

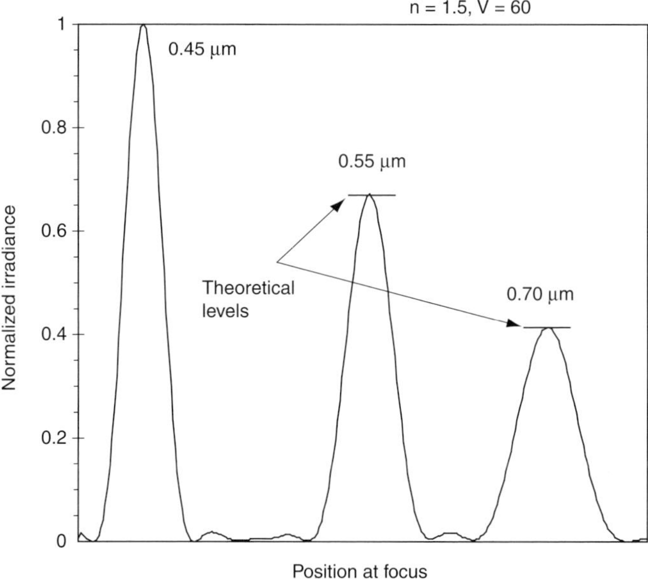

Fig. B.10 The expected values of the amplitudes for the 550 nm and 700 nm Airy disks are 0.67 and 0.41 respectively. Generated by ZEMAX.

of 1.5 and Abbe number at the same line of 60) and weighted equally in the image. The Airy disk would be tightest for the 450 nm line, so the image is normalized there. The expected maximum intensity scales as the square of the ratio of wavelengths, so $(0.45/0.55)^2 = 0.67$ and $(0.45/0.70)^2 = 0.41$.

We can see that the ZEMAX model does a good job of predicting these levels.

B.7 Numerical Limitations on Programs

Although APERTURE is also a numerical model, it is ASYMM which has the greater opportunity to fail. Experience showed that the 129 points on the aperture's diameter had difficulty simulating defocusing aberration exceeding 10 wavelengths. Along each diameter under such conditions, a minimum of about 5 samples occur within a Fresnel zone. The errors have a chance to average to zero only because many such diameters are added together statistically.

ASYMM simulates an aperture with an approximate circular pattern 64 points in radius. These points are distributed in a rectangular pattern. The edge of such a sampled aperture is necessarily ragged. We can perhaps estimate how much error is induced by inspecting Figure B.11, which shows the differences between APERTURE and ASYMM on a defocused

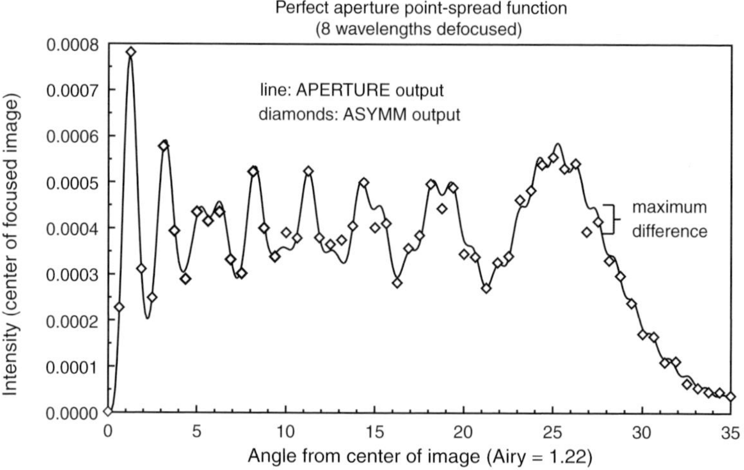

Fig. B.11 Limited sample-size errors in the 129x129 pupil of ASYMM.

unaberrated aperture expanded until the errors are obvious. Because APERTURE can be set to sum over a finer one-dimensional grid (here it is 500 points), it suffers less corruption. The model has smooth edges in APERTURE because the angular integral is done analytically.

The maximum difference is about 0.000075, or 1 part in 13,300. Thus, we are seeing less than 1 erroneous sample of the 12,853 points on the aperture contributing to the intensity.

Think about the way those ragged edges are working for a moment. If the point is outside the aperture, the model ignores it. Each point represents a square area around its feet. Thus, if this little tile is more than halfway outside the radius, its contribution is neglected. If it extends less than halfway outside the boundary, its full contribution is counted, even for the area that stretches outside. A $\frac{1}{13,000}$ error is a very small mistake and is perhaps better than the program deserves. One suspects that errors as large as 4 parts in 13,000 appear occasionally. Such an error would be 0.0003. Since this intensity is about that of an image defocused 9 wavelengths, we should anticipate that the accuracy beyond defocus values of 8 to 10 wavelengths is lessened in APERTURE.

In practical use, ASYMM did not fail from rough pupil edges. Figure B.12 shows a comparison between two perfect images, each defocused by 8 wavelengths. One was produced with ASYMM and the other with APERTURE. They were both printed at extremely low contrast to show the feathery appearance induced by the coarse edges of the ASYMM pupil. This spurious detail was much dimmer than 0.0003. In all real cases, contrast

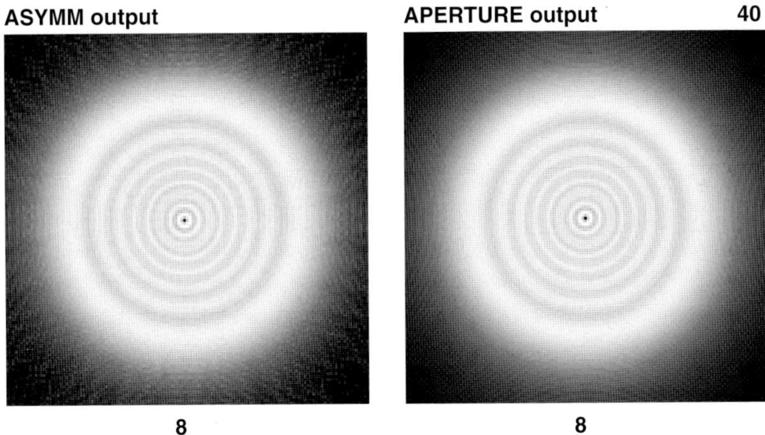

Fig. B.12 Extremely low-contrast image patterns showing tiny errors in ASYMM calculation.

was high enough so that such wrinkling was nearly invisible. Note also the comparison of the performance the Lommel function expansion and APERTURE of Figure B.8, which plots identical defocus.

The final problem with the sparsely sampled pupil of ASYMM is grating interference. If the phase shift between adjacent contributing elements is one wavelength, they once again act as if they are in phase. For our 128-point diameter pupil, this condition occurs at $D\theta/\lambda = 128$.

This effect is an artifact of pretending that the wonderfully smooth surface could be represented by a 129×129 square grid of points. In FFT terminology, we can call such a phenomenon "aliasing." It is an effect of changing the continuous Fourier transform to a discretely sampled one.

If we haven't defocused far enough to divert significant light to distant portions of the image space, no trouble results. In other words, the image and the twin image (over at a reduced angle of 128) aren't interfering because they don't throw much light that far from their centers. Nevertheless, a limit exists as to how far the image can be defocused until interference becomes severe. We must know that limit so that it can be intelligently avoided.

Figure B.13 shows the pattern out to $D\theta/\lambda = 128$ for three cases: defocusing aberrations of 8, 12, and 16 wavelengths. Clearly, 16 wavelengths is too far, and 12 is questionable because of dark gray tendrils between the images. Only the frame depicting 8 wavelengths seems to show the diffraction disks as isolated.

We should only expect the images to be free of interference if we can

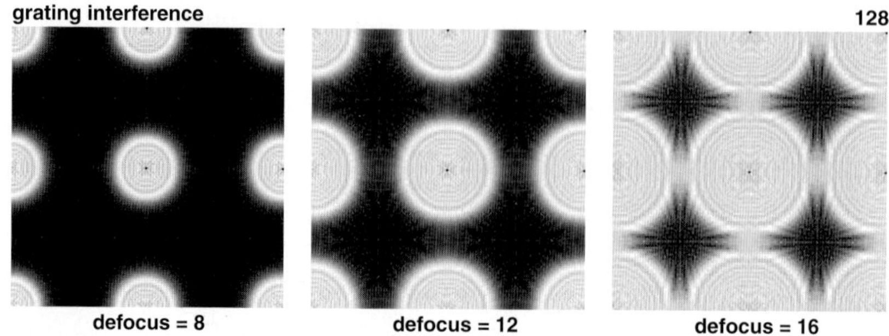

Fig. B.13 Interference between side-order images in ASYMM. Note that the reduced angle is very large at 128.

fit a whole defocused disk into the darkness between the images. The geometric radius of an image in reduced angle is 4 times the number of wavelengths of defocus. This result may be derived by considering

$$f - f' = 8F^2 \Delta n \lambda, \qquad \textbf{B.13}$$

where F is the focal ratio and Δn is the number of wavelengths of focusing difference. We can calculate the geometric radius of the image:

$$\text{radius} = \frac{(D/2)(f-f')}{f} = \frac{f-f'}{2F} = \frac{8F^2 \Delta n \lambda}{2F} = (4\Delta n)F\lambda. \qquad \textbf{B.14}$$

Since the radius of the Airy disk is $1.22F\lambda$, we can identify $F\lambda$ as the conversion factor from reduced angle to radius. Thus $4\Delta n$ is the value of the reduced angle at the edge of geometric shadow. In the present case, 8 wavelengths defocusing aberration means the geometric shadow starts at a reduced angle of $4 \times 8 = 32$, or 26.2 times the radius of the Airy disk.

Eight wavelengths should be barely enough to fit a whole disk between the disk and the interference order centered at a reduced angle of 128. A geometric radius would appear at the center and edge with a diameter in the middle: $32 + 64 + 32 = 128$. Of course, the image appears to be somewhat smaller than the geometric radius. A whole disk would drop between the two with a tiny bit of rattle.

Occasionally, ASYMM was asked to calculate an image 10 wavelengths out-of-focus. Some interference from the side order was noticeable in a few cases, but the damage was not severe.

All patterns with a defocusing aberration beyond 10 wavelengths were calculated with the program APERTURE (which is far less sensitive to sampling error) or with the optical design model ZEMAX (which allows more than 128x128 sampling on the pupil). The ZEMAX models in this book which were calculated at more than 5 wavelengths of defocusing aberration were done so mostly with 256-point diameter pupils.

B.8 A Note on Apodization

Figure 9.14 showed the response of a stepwise transmission function roughly simulating the action of a Gaussian transmission function. Although the effects of this apodizer may be readily calculated with the APERTURE routine, it may also be analytically calculated for stepwise dependence by extending an expression usually employed in the modification to the Airy pattern caused by central obstruction (Schroeder 1987). If the pupil is broken up to a set of annuli of constant transmission coefficient, each zone of the pupil contributes the Equation B.1 amount to the field

$$U_\eta(\rho) \; = \; 2T_\eta\left(r_{\eta o}\frac{J_1(r_{\eta o}\pi\rho)}{\pi\rho} - r_{\eta i}\frac{J_1(r_{\eta i}\pi\rho)}{\pi\rho}\right), \qquad \text{B.1}$$

where ρ is the radius in reduced units at the focal plane, $r_{\eta o}$ is the outer radius in the normalized pupil plane (at the edge of the aperture, $r = 1$) and $r_{\eta i}$ is the inner radius in that plane. J_1 is the first-order Bessel function. The total field is the sum of all these annular contributions,

$$U(\rho) \; = \; \sum_\eta U_\rho(\rho) \;, \qquad \text{B.2}$$

and for the final intensity value, the field is squared. Equation B.1 and B.2 are just an extension of Babinet's principle to partial transparency. The advantage of using this method was is that it may be calculated quickly in a spreadsheet. This formula has been confirmed by duplicating the predicted apodization in APERTURE (Suiter 2001).

B.9 Difficulties in Printing

The dynamic range of book reproduction is not nearly the same as the dynamic range of the human eye. When the eye receives the wrong light level, it adjusts. Paper can't do that. The halftone process divides grayscales into dot patterns. An ideal dynamic range for such a process allows a grayscale with 256 levels of intensity.

Clearly, the most desirable way of proceeding would be to use an unchanging scale for the printing process. This procedure is impossible with only 256 levels. We saw above that much interesting detail appears with an intensity of 0.0001, particularly with large values of defocusing aberration. Thus, no uniform scale can cover both in-focus images (intensity = 1) and defocused images (intensity $\cong 0.0001$). We would need a gray scale with 10,000 intensity levels. Also, an image that is self-luminous and one that is reflective have subtle differences. Real stars are viewed in a dark field, with the eye's variable sensitivity tracking the lowered intensity of a defocused image. Images on paper must be lit by a lamp. When paper images get darker, the eye does not follow. The bright corners of the paper are still in the field of view, and the extra illumination subverts the tracking.

Coupled with this problem are the non-linearities associated with printing. Often the dark end of the scale is darker than anticipated due to the phenomenon of ink spot spreading. The brightest parts of the image are too bright because very small isolated dots fail to pick up any ink at all. The net result is an increase in contrast and a decrease in dynamic range.

There is no choice but to arbitrarily follow the decreased range of a roughly linear printing scale. In the images throughout this book, the brightness of the frame varies considerably. Focused images are compact; they seem to be drawn with just a spot of white coloring. On the other hand, defocused images are rendered much brighter than they actually appear when viewed through the telescope. Both contrast and brightness are under subjective control.

When you look at any image in this book, concentrate on the shape rather than the absolute brightness. The illumination of such an image is the least significant (and most deceptive) of its properties.

B.10 Viewing ZEMAX Rendering

For the first edition, all of the star-test figures were generated with a hand-written Adobe Postscript^{TM} driver that more or less followed a magnitude scale, matching the peculiarities of human vision. In ZEMAX graphics, the selection is between a purely linear or a purely logarithmic scaling in the Huygens point-spread function figures. A long logarithmic scale is useful on the focused image if used with a darkening gamma in Adobe Photoshop, but a logarithmic scale is far too washed out for use in defocused images (most of the interesting detail is in the topmost few levels).[4] A linear scale is better for defocused images but gamma adjustment is less helpful in recovering a realistic eye response.

The result is that the defocused images as generated for real telescopes are a trifle more contrasty than they are in real life.

---

[4] Recently (2008) ZEMAX was modified to allow modification of the depth of the logarithmic rendering. This would have made the renderings appearing in the book more realistic but, alas, the modification came too late to affect this edition.

Appendix C
Diagonal Calculations

Ralph Dakin published expressions for minor axis and offset in *Sky & Telescope* years ago, but these expressions are arranged for convenience of calculation rather than exposition. Hence, the equations are a little mysterious and subject to error when published in secondary sources. This situation was not helped by an unfortunate typographical error in the original article. The mistake was corrected a few months later, but the repair was often missed and incorrect expressions progressed further in amateur literature (Dakin 1962, *erratum* 1963).

These quantities are not difficult to derive, and something can be learned from seeing the analytic geometry logic, so they are derived anew here. The formulas that follow are superficially different than Dakin's expressions, but they produce the same answers. The differences probably are caused by the way the coordinate systems are set up. Such modifications often lead to ostensibly different formulas.

Telescope makers are not expected to actually use these formulas, as they are not exact[1] and diagonals are available in such a limited set of sizes and shapes. Use the approximations to minor axis unless very fast systems are contemplated or if the answers are very close to changing which diagonal is available.

C.1 Derivation of Minor Axis and Offset

Figure C.1 displays an "unfolded" Newtonian and defines various quantities useful in this derivation. These variables are f, the focal length; D, the diameter of the mirror; s, the sagitta (or depth) of the mirror; T, the distance from the focal plane to the center of the tube; and L, the diameter of the fully-illuminated region at the focal plane. In this derivation, I will determine the values of the minor (short) axis of the diagonal and the sideways

---

[1] They are not exact for several reasons, but far and away the greatest is that the ratio of the major axis to the minor axis is somewhat greater than 1.414 for all aperture ratios less than infinity (Thiebaux 2006). Perspective error causes stretching of the major axis. Dakin's calculation is very close for the more constrictive major axis, however, and all real diagonals are made with ratios of major to minor axis of 1.414. Thus, Dakin's calculation along the long axis leads to larger non-vignetted answers.

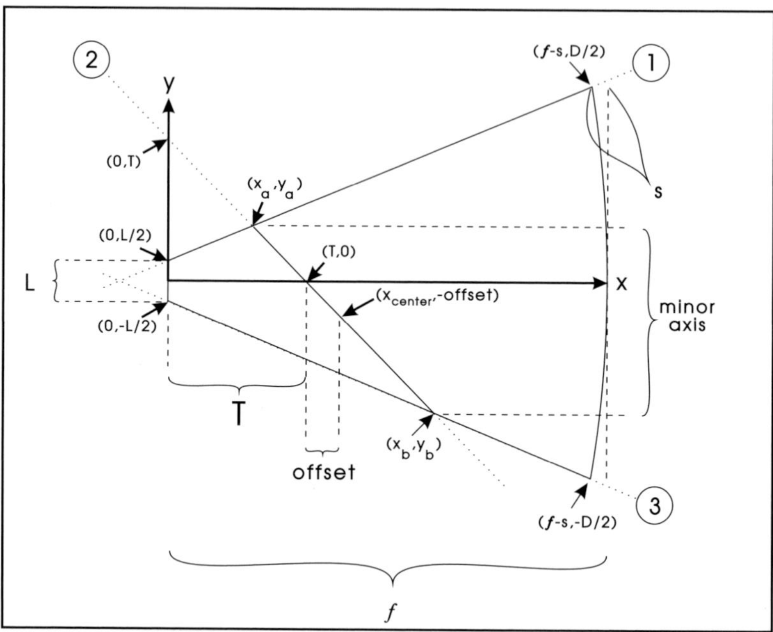

Fig. C.1 The coordinate system and quantities of interest in sizing and properly placing the Newtonian diagonal. The minor axis is slightly overestimated by the approximation that minor axis is major axis divided by 1.414 for all realistic aperture ratios.

or down offset of its center. The coordinate system is defined with its origin at the center of the focal plane. The *x*-axis extends through the center of the mirror, and the *y*-axis extends laterally away from the optical axis.

Three lines of Figure C.1 are significant. Lines #1 and #3 follow the edges of the fully-illuminated field to the edges of the mirror. Line #2 is directed along the surface of the diagonal. We know it is at 45° and therefore strikes the two axes at the indicated points having coordinates $(0,T)$ and $(T,0)$. Similarly, we know the coordinates of two points along each of the lines #1 and #3, and we can use this knowledge to obtain the defining equation of those lines as well.

Once the equations of the lines are known, the points of intersection at the far edges of the diagonal can be calculated. These intersections have coordinates x_a, y_a and x_b, y_b. These points are all that are required to derive both the minor axis and the offset. The minor axis is

$$MA = y_a - y_b,$$

C.1

and the offset is

$$\text{Offset} = x_{\text{center}} - T = \frac{x_a + x_b}{2} - T.$$ C.2

Lines have the general form $y = mx + b$, where b is the y-intercept and m is the slope. Line #2 is the easiest to determine, and by inspection one can assign it the equation

$$y_2 = -x_2 + T.$$ C.3

Line #1 has a y-intercept of $L/2$, and the slope can be obtained by inserting the known coordinates of the point at the upper corner of the mirror:

$$\frac{D}{2} = m_1(f - s) + \frac{L}{2}.$$ C.4

Thus,

$$y_1 = \frac{D - L}{2(f - s)}x_1 + \frac{L}{2}.$$ C.5

Similarly, the equation of line #3 becomes

$$y_3 = -\frac{D - L}{2(f - s)}x_3 - \frac{L}{2}.$$ C.6

We can rename the quantity $(D{-}L)/2(f{-}s) \equiv n$. All three equations can be summarized as

$$y_1 = nx_1 + \frac{L}{2}, y_2 = -x_2 + T, y_3 = -nx_3 - \frac{L}{2}.$$ C.7

By setting $y_1 = y_2$, we can determine the coordinate

$$x_a = \frac{T - (L/2)}{1 + n},$$ C.8

and we insert that expression back into the line #2 equation to derive

$$y_a = -\frac{T - (L/2)}{1 + n} + T.$$ C.9

Similarly, the coordinates of the other intersection point are

$$x_b = \frac{T + (L/2)}{1 - n} \text{ and } y_b = -\frac{T + (L/2)}{1 - n} + T.$$ C.10

The expression for the minor axis (Equation C.1) is just the difference between the two y-values:

$$\text{MA} = \frac{T+(L/2)}{1-n} - \frac{T-(L/2)}{1+n} \text{ or MA} = \frac{L+2nT}{1-n^2}. \qquad \textbf{C.11}$$

The offset (Equation C.2) is only a little more complicated:

$$\text{Offset} = \frac{1}{2}\left(\frac{T-(L/2)}{1+n} + \frac{T+(L/2)}{1-n}\right) - T \text{ or Offset} = \frac{T+(nL/2)}{1-n^2}(-T). \quad \textbf{C.12}$$

The final expressions of Equations C.11 and C.12 only require an expression for the sagitta. A parabola of focus f is determined by the equation $x = f-(y^2/4f)$, and the exact sagitta is obtained by evaluating the shift in x on axis and at the y-value of the edge, $D/2$:

$$s = \frac{D^2}{16f}. \qquad \textbf{C.13}$$

Please note that Equation C.12 is not the same as the wavefront sagitta that will be calculated in Appendix E, but the surface sagitta, which is half as large.

C.3.1 Test Case

Before proceeding further, it would be prudent to check these expressions against Dakin's results. One such calculation appears in Appendix G of Texereau (1984). With

$$\begin{aligned}
s &= 0.0833333 \\
n &= 0.0789913 \\
T &= 6.3 \text{ inches, and} \\
L &= 0.43 \text{ inch,}
\end{aligned}$$

one readily calculates (from Equations C.11) that MA = 1.43423950 inches. This number compares well with the result of 1.43423955 using Dakin's formulas as reprinted in Texereau. Offset = 0.056646217 inch there, and 0.056646220 here. The slight differences are caused by numerical truncation error. (The calculations in Texereau were carried out to more decimal places for comparison.)

C.2 Minor Axis and Offset Approximations

The nearly exact expressions are not that difficult to evaluate with the common availability of calculators and computers. Spreadsheet programs are especially handy for calculations but, as these expressions are not exact, some people may prefer the more compact form of an approximation.

Notice that in Equations C.11 and C.12, the factor $1/(1-n^2)$ can be

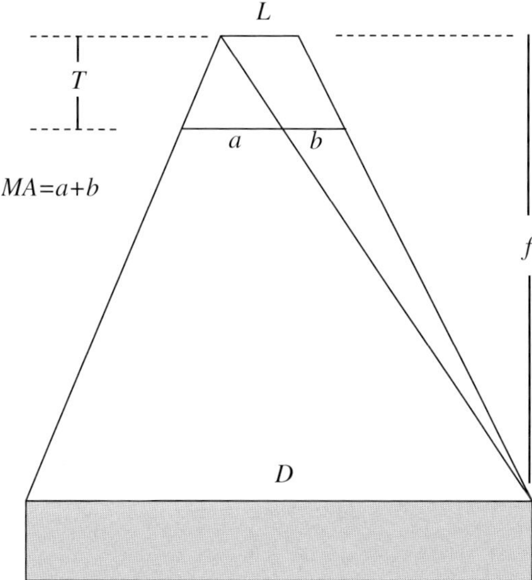

Fig. C.2 Simplification from which another derivation of minor axis may be inferred.

approximated by $1 + n^2$. With most mirrors, the quantity $(D - L)/2(f - s) \equiv n$ is very close to $(D - L)/2f$. Capital F is the aperture ratio, so

$$\text{Offset} = \frac{T + \frac{nL}{2}}{1 - n^2} - T = \frac{T + \frac{nL}{2}}{1 - n^2 \dots} - T \cong \frac{T}{4F^2} + \frac{L}{4F}\left(1 - \frac{2T}{f}\right) \qquad \textbf{C.14}$$

where the smallest terms have been ignored.

The same expansion could be done with the equation for MA, again ignoring exceptionally tiny terms, to obtain the answer

$$\text{MA} \cong L + \left(\frac{D - L}{f}\right)T. \qquad \textbf{C.15}$$

The test case above was recalculated with Equations and C.15 and the answers are MA = 1.424 inches and Offset = 0.057 inch, quite close to the Dakin numbers. Of course, with a very small, fully illuminated region of the focal plane, these values approach MA = T/F and Offset = $T/4F^2$.

An easy yet satisfying alternate derivation of minor axis is possible by inspection of Figure C.2. Here we have cut the system the other way and have the minor axis as the sum of a and b, neglecting niceties like sagitta and offset. Noting two triangle relationships, $a/T = D/f$ and $L/f = b/(f - T)$, we may write the expression $MA = D(T - L)/f + L$. Putting in the test case parameters, we derive a minor axis of 1.41 inches. We might make a further

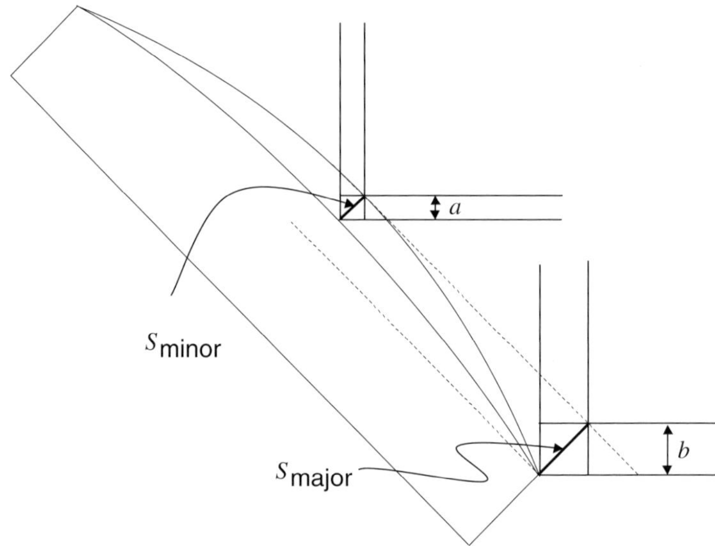

Fig. C.3 S_{minor} and S_{major} are the sagittae of the minor and major axes respectively; a and b are the projections of these sagittae along the incoming and outgoing beam directions. Please note that the minor and major axes referred to are only the part of the diagonal that is struck by the beam emanating from a single point of the object, not the whole physical extent of the diagonal.

approximation $MA = a + L$ or $MA = T/F + L$. Our example (at $f/6$) works out to $6.3/6 + 0.43 = 1.48$ inches. One is a slight underestimate, the other an overestimate.

C.3 Tolerance for Spherical Deformation in the Diagonal

Figure C.3 shows a diagonal with an exaggerated spherical curvature from the side. This diagonal is meant only to represent that portion of the real diagonal under a single beam bundle, or that portion of the diagonal that may affect the wavefront coming from a single star or other point source. It is the part of the diagonal that affects the optical path difference. Another nearby point on the image may partake of a large portion of the same surface, but it is slid sideways somewhat, depending on its angular distance away. If the diagonal is near enough to the eyepiece and the two points at opposite ends of the field-of-view, these two portions of the diagonal surface need not overlap. Two such optical paths need not be in phase at all.

A wavefront reflected from the middle of the diagonal and the edge of the minor axis is delayed by $2a$. A wavefront reflected from the diagonal center and the edge of the major axis is delayed a net $2b$. If this is a spherical deformation and the diagonal is the usual type of a cylinder cut at 45 degrees, $2b = 4a$. The wavefront reflected from such a spherical curvature is a compound of spherical curvature and an astigmatism saddle-point, because it bends away from the observer by $2a$ in one direction and $4a$ in the other. There are slight additional aberrations cause by perspective effects, which result in the distortion of the spot diagram in Figure C.4.

If the telescope is refocused to the wavefront centered on the pure saddle point (by subtracting $3a$), the greatest delay is $\pm a$, or a total amplitude of $2a$. An expression relating this delay to the sagitta across the minor axis (S_{minor}) is:

$$2a = \frac{2S_{minor}}{\sqrt{2}} = \sqrt{2}\, S_{minor} \; . \qquad\qquad \textbf{C.16}$$

Remembering that for astigmatism the amplitude of the aberration is about 3/8 wavelength (more precisely to reduce the Strehl ratio to 0.8, this delay may be rewritten 0.37)

$$\sqrt{2}\, S_{minor} = 0.37\lambda \qquad \text{for Strehl} = 0.8 \qquad \textbf{C.17}$$

so, the mirror only need be flat across the ray bundle portion (that part of the mirror with the incident beam from one point on the image) to 0.262 wavelength. For spherical curvature, ¼ wavelength is good enough, ⅛ wavelength is excellent and ¹⁄₁₀ wavelength nearly perfect. Keep in mind that this number applies only to the part of the diagonal within the beam, so that diagonals for which these numbers apply to the whole optical piece are even better.

For other aberrations, perhaps we may not do such convenient refocusing. We could take the most stringent criterion as a halving of this tolerance, but a more reasonable limit is to say that aberrations tend to add in a root-mean-square sense, and to divide the criterion above by another factor of $\sqrt{2}$:

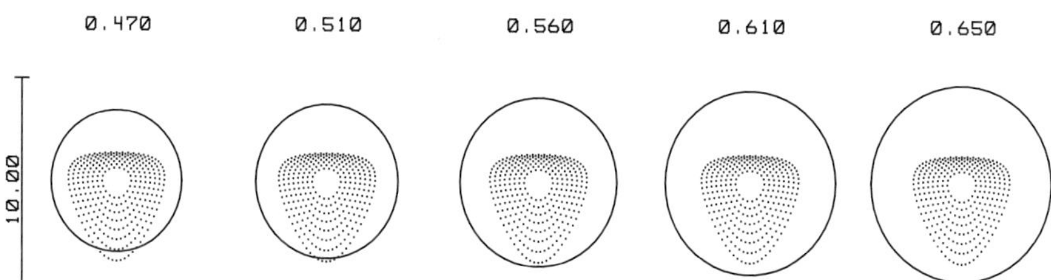

Fig. C.4 Spot pattern of a spherical diagonal. This is mostly astigmatism, but enough coma and trefoil is induced by perspective to change the pattern from a strictly symmetrical one.

$$S_{\text{minor}} = \frac{\lambda}{5.4}.$$

C.18

Or, the diagonal must be flat over a beam bundle region to within $\frac{1}{5.4}$ wavelength to be well within the Strehl = 0.8 tolerance; and $\frac{1}{11}$ wave is excellent.

We may test the derivation of the spherical deformation by inserting the parameters into the ZEMAX model. First, we need a more convenient expression. Recalling the expression for the Strehl = 0.8 generated astigmatism:

$$\sqrt{2} S_{\text{minor}} = \frac{r_{ill}^2 \sqrt{2}}{2R} = \frac{\lambda}{2.7},$$

C.19

where r_{ill} is the semi-minor axis of the illuminated region ($MA/2$) and R is the radius of the spherical diagonal. If we note that the approximation above for a minimal-sized diagonal is $MA=T/F$, and we take the wavelength as 560 nm, then we can rewrite this radius as a convenient numerical expression

$$R = 852.3 \left(\frac{T}{F}\right)^2 [\text{mm}]. .$$

C.20

Table C.1 **Zernike expansion of slightly spherical diagonal**

```
Listing of Zernike Standard Coefficient Data

Title: 250-mm f/6 Newtonian w/radius 949300 on diag

Wavelength                        : 0.5600 µm
Peak to Valley (to centroid)      : 0.40052807 waves

From integration of the rays:
RMS (to centroid)                 : 0.07959102 waves
Variance                          : 0.00633473 waves squared
Strehl Ratio (Est)                : 0.77873447

Z  1  -0.00023087 : 1                                    piston
Z  2   0.00000000 : 4^(1/2) (p) * COS (A)                tilt x
Z  3   0.01806775 : 4^(1/2) (p) * SIN (A)                tilt y
Z  4   0.00017851 : 3^(1/2) (2p^2 - 1)                   defocus
Z  5   0.00000000 : 6^(1/2) (p^2) * SIN (2A)             astig 45 deg
Z  6  -0.07621307 : 6^(1/2) (p^2) * COS (2A)             astig 0   deg
Z  7   0.00640878 : 8^(1/2) (3p^3 - 2p) * SIN (A)        coma y
Z  8   0.00000000 : 8^(1/2) (3p^3 - 2p) * COS (A)        coma x
Z  9  -0.00276182 : 8^(1/2) (p^3) * SIN (3A)             trefoil 30 deg
Z 10   0.00000000 : 8^(1/2) (p^3) * COS (3A)             trefoil 0
Z 11   0.00024198 : 5^(1/2) (6p^4 - 6p^2 + 1)            3rd order SA
```

Note that Strehl ratio is approximately 0.8.

Appendix D
Labeling of Diffraction Patterns

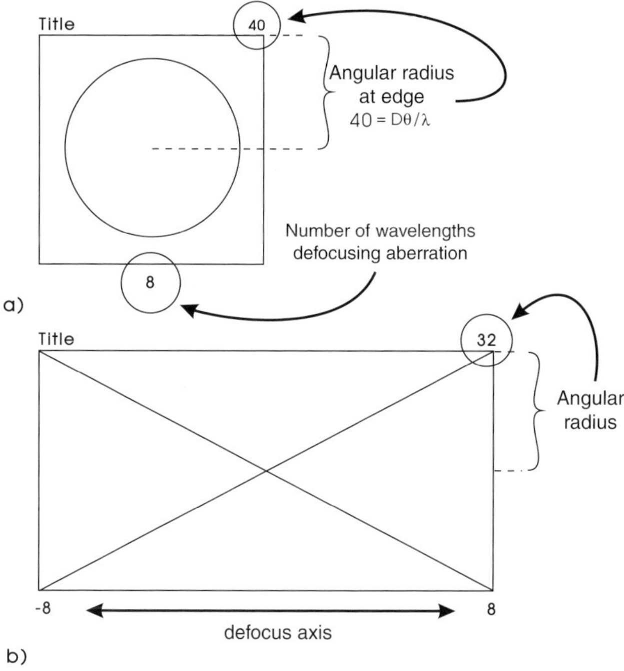

Fig. D.1 Image notation: a) labels used on diffraction patterns, b) labels used on longitudinal slice patterns.

Figure D.1a shows how the typical image pattern appearing in this book is labeled, and Figure D.1b indicates the slight modifications to annotate slice patterns. In the upper left corner of any given pattern is a title, if one is present. In the upper right corner of a frame is a number that gives the reduced angular radius $(D\theta)/\lambda$ from the edge to the center of the square box. In the case of the square diffraction pattern frame, this angle also applies to the horizontal axis. If the pattern had the number 1.22 here, the Airy disk of a focused perfect aperture would just be contained within the box (see Figure D.2). Most of the focused patterns in the text have limits

Fig. D.2 A perfect Airy disk is barely contained within a box with a reduced angular dimension of 1.22 from center to edge. The bright corners are the first diffraction ring.

of 5 or 10, allowing us to inspect the ring structure as well as the central spot. Out-of-focus patterns are viewed at lower magnification, which accounts for the large numbers appearing in their upper right corners.

The centered number at the bottom of one of the square patterns is the number of wavelengths of defocusing aberration. The numbers at the lower corners of a slice pattern denote the value of defocus at each end of the box, with the defocus changing steadily between them. Longitudinal slice frames are defocused an equal distance inside and outside focus, with the best focus near the center. In all slice frames, the objective lens or mirror is to the left.

In some cases, clutter is reduced by labeling only one frame in a row or column. If a defocus value is missing, the value appears at the bottom of the column. If the angular coordinate is missing on a particular frame, it can be found on a row entry on the far right.

For composite figures with a focused image frame appearing to straddle two lower columns, the *inside* focus behavior is the column to the left (negative defocus values), while the *outside* focus behavior (positive) is in the middle column. Such figures often feature the behavior of perfect optics in a column to the far right. These unaberrated columns give only one side of focus; the other side is identical. See Figure D.3.

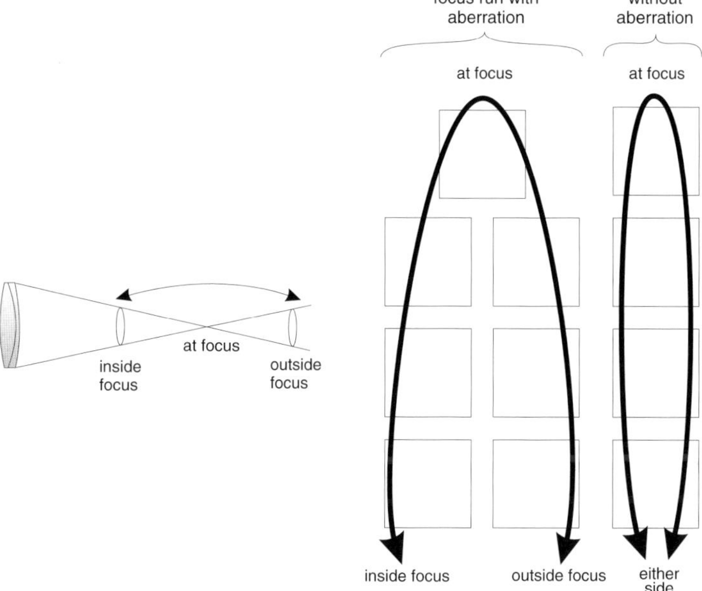

Fig. D.3 A focus-run pattern shows how the frames "fold over" the region of focus. The perfect pattern to the right need not show both inside- and outside-focus images because they are identical.

Appendix E
Eyepiece Travel and
Defocusing Aberration

This book uses the generic unit of defocusing aberration when referring to distance out of focus. However, the most convenient way of thinking about defocus is in terms of eyepiece travel. Here, we wish to determine the amount of difference between two spheres at the aperture, one centered on focus f and the other centered on a defocus position f'. We will then relate that very tiny distance at the aperture (the defocusing aberration) to the relatively large amount we have moved the eyepiece between f and f'. This situation is depicted in Figure 4.9.

The derivation proceeds from the difference between the two wavefront sagittae, or how much the wavefront is "cupped" at the aperture. Another common sagitta involves surface shape (or the shallow depth of the mirror itself). It is half the wavefront sagitta. Don't confuse surface sagitta with wavefront sagitta.

If the distance to a focus position is f and if D is the diameter of the aperture (See Figure E.1), then by the Pythagorean theorem, the sagitta

$$s = f - \sqrt{f^2 - \left(\frac{D}{2}\right)^2}.$$

E.1

If the focal length is much greater than the aperture diameter, this equation may be approximated by performing a Taylor expansion. Such an approximation results in

$$s \cong \frac{D^2}{8f}.$$

E.2

Taking the difference between two different wavefront sagittae in Equation E.2,

$$s - s' = \frac{D^2}{8f} - \frac{D^2}{8f'} = \frac{D^2}{8}\left(\frac{1}{f} - \frac{1}{f'}\right) = \frac{D^2}{8}\left(\frac{f'-f}{ff'}\right).$$

E.3

Next we demand that the quantity $s - s'$ be thought of as a certain

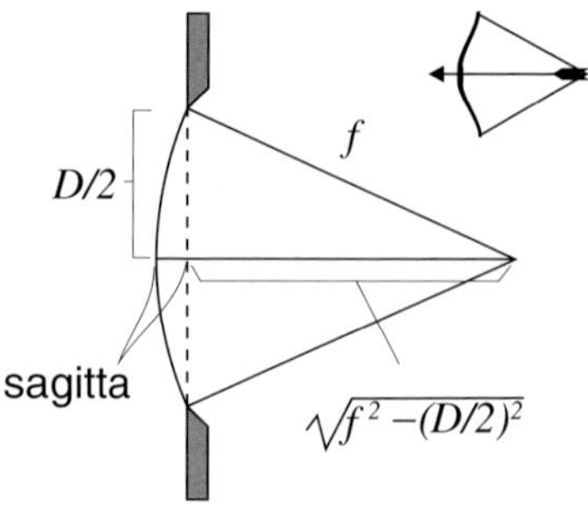

Fig. E.1 Geometry of the sagittal relation. Focal length *f* is the radius of curvature of the wavefront.

number of wavelengths "$n\lambda$" out of focus, where λ is the common symbol for wavelength. Noticing that ff' is to a very high degree of precision just equal to the average focal length squared, then

$$s - s' \equiv \Delta n\lambda = \frac{(f-f')}{8F^2},\qquad\text{E.4}$$

where F is the focal ratio. The final change in focus as measured by advancing or pulling back the eyepiece is

$$f - f' = \Delta f = 8F^2\Delta n\lambda.\qquad\text{E.5}$$

As an aside, we can derive an expression for the depth of the shallow bowl in a mirror by noting that the focal distance f in Equation E.2 can be replaced by the radius of curvature $R = 2f$, or

$$s_d \cong \frac{D^2}{16f}.\qquad\text{E.6}$$

Appendix F
Glitter in a Shiny Sphere

Figure F.1a shows a reflection in a small sphere. In its most general form, it is roughly shaped like a kidney bean, longer in the azimuthal direction than in the radial direction. Since the long axis of the bean is the limiting factor, an expression for this angle is derived first.

We wish to calculate u, the long axis of the glitter. If the glitter is far enough from the center and not near the edge (i.e., θ_r is somewhere near 90 degrees), its approximate length is given by

$$\frac{u/2}{x} = \tan\left(\frac{\phi}{2}\right), \qquad\qquad \text{F.1}$$

with x being the radius of the circle containing the glitter seen in perspective and ϕ being the angle of the source of light. If we view the situation from the side (Figure F.1c), the radius x is obtained from the great circle radius of the sphere R and the light diversion angle θ_r as in

$$x = R\sin\left(\frac{\theta_r}{2}\right), \qquad\qquad \text{F.2}$$

with the complete expression being

$$u = 2R\sin\left(\frac{\theta_r}{2}\right)\tan\left(\frac{\phi}{2}\right). \qquad\qquad \text{F.3}$$

Thus, for a 1-inch sphere ($R = 0.5$ inch) and a reflection of the Sun ($\phi = 0.5° = 0.0087$ radians), a reflection angle (θ_r) of 90° yields a glitter point 0.003 inches long. The result for a typical θ_r, the gleam of a solar reflection will be smaller than $\frac{1}{300}$ the diameter of the sphere.

Similarly, we can obtain an expression for the short axis (v) of the bean. This dimension is obtained by perturbing Figure F.1c around the average reflection angle,

$$v = R\left[\sin\left(\frac{\theta_r + \phi/2}{2}\right) - \sin\left(\frac{\theta_r - \phi/2}{2}\right)\right]. \qquad\qquad \text{F.4}$$

For the example situation, v is 0.0015 inch. The glitter point is about twice as wide as it is long.

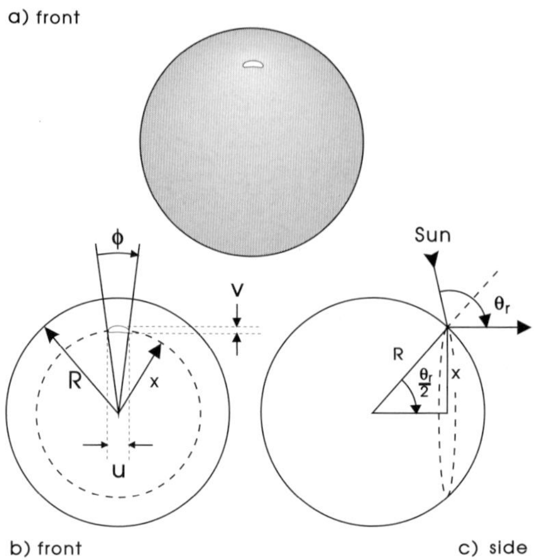

Fig. F.1 The variables used to determine the size of the glitter reflected in a sphere.

For a nearly centered bounce with the gleam appearing near the middle of the sphere ($\theta_r = 0$) and for a very small source angle ϕ, the expression for v is approximately

$$v \cong \frac{R\phi}{2}.$$

F.5

If we want to figure the size of the round gleam where the 1-inch reflective sphere is placed opposite to the Sun, we could use this approximation (the angle must be in pure number format, i.e., radians):

$$v = \frac{0.5 \text{ in } (.0087 \text{ radians})}{2} = 0.0022 \text{ in.}$$

F.6

The glitter size shrinks to about ¹⁄₄₅₀ the diameter of the sphere at its smallest aspect. The source of light is nearly behind the viewer's head and the reflection is approximately centered in the shiny sphere for this favorable condition to occur. For a more complete treatment plus an interesting article on the general topic of reflection in spheres, see Berry 1972.

Appendix G
Specific Designs Used in the Main Text

In this Appendix, I will give surface parameters and discuss the optical performance of amateur-sized telescopes more complicated than the Newtonian. Star-test patterns for most of these designs have already appeared in the main text, where they were used as examples. These are theoretical instruments modeled in ZEMAX, and although these designs are straightforward, none of these telescopes has been built. As a consequence, I wish to emphasize the following point. *All who attempt fabrication of any of these designs do so at their own risk.* I have attempted to design these telescopes as well as I can, and I believe them to be good starting points for actual working instruments and for discussion of star-test results. However, they use "book" values for indices of refraction and little thought has been directed at ease of manufacture. Refractive optical elements should not be attempted without specific glass-melt data. Some sort of optical design program must be used to slightly adjust the radii for the difference between book indices and the actual glass. Also, these designs tend not to be oriented towards the fussy amateur telescope maker or high-end manufacturer, but are often left spherical in "good enough" form the way a consumer builder might do it. Last of all, they are not oversize to account for lens cells and the like.

If any of these designs are close to those of existing commercial instruments, they do so coincidentally. No reverse engineering was done. In some cases, I looked at published specifications and inferred an approximate design from general behaviors. The most sophisticated measurement was done using a classroom ruler.

This appendix features the following designs:

1. 152-mm f/15 Fraunhofer achromat.
2. 152-mm f/8.4 ED triplet apochromat.
3. 152-mm f/12 aluminized-spot Maksutov-Cassegrain.
4. 152-mm f/12 separated-secondary Maksutov-Cassegrain.
5. 200-mm f/10 spherical-mirror Schmidt-Cassegrain.

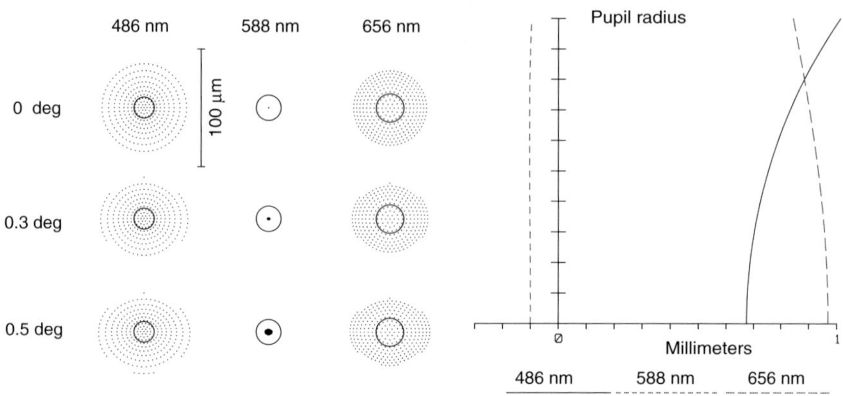

Fig. G.1 Spot diagram with Airy disk (left) and longitudinal aberration (right) of 152-mm achromat.

G.1 152-mm *f*/15 Fraunhofer-type Achromat

The surface summary appears in Table G.1. The two glasses selected for this objective are Schott BAK1 and F2, respectively the crown and flint elements (see Chapter 12 on chromatic aberration).[1] The Fraunhofer is an achromat that usually features slightly different curvatures on either side of a central airspace. The benefit of such a complication is the simultaneous correction of coma with spherical aberration. The spot diagram in Figure G.1 displays a residual amount of astigmatism off-axis.[2]

Table G.1

Surface Summary of 152-mm *f*/15 Achromat [mm]

Surf #	Radius	Thickness	Glass	Semi-Diameter
STOP 1	1449.65	17.780	BAK1	76.20
2	−722.50	0.254		76.20
3	−724.78	12.700	F2	76.20
4	−5856.61	2269.149		76.20
IMAGE	−850.90			19.98

This is a conventional achromat with about 1-mm defocus out of 2286 mm at the C and F lines (656 and 486 nm respectively). Its nearest focus is about 553 nm. As with all achromats, the reason it works so well is because the response of the human eye is low at the edges of the visual spectrum.

---

[1] These glasses are the old versions of the modern low-lead formulations. Changing the glasses to the new materials would require only a minor adjustment of the curves.

[2] To those following the ray traces with their own optical design programs, the figures above and later on in this Appendix may be found at trifling differences of the back focal lengths quoted in the prescriptions. This was done to elucidate points presented in the figures.

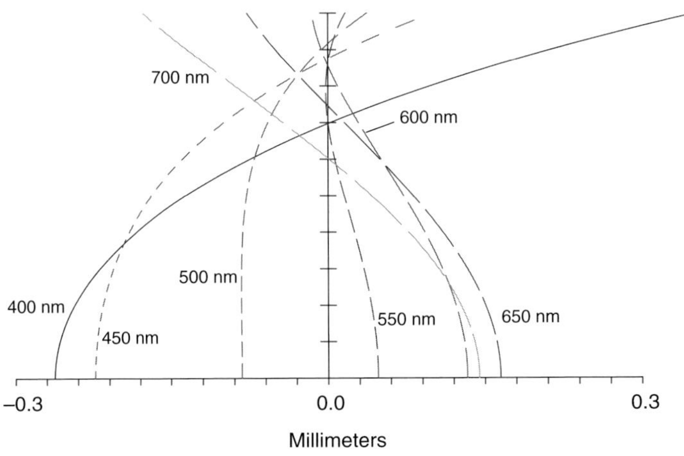

Fig. G.2 Longitudinal aberration of 152-mm f/8.4 ED apochromat from 400 to 700 nm.

Indeed, Walker (2000, p. 5), states that the visual spectrum may be approximated by three equally-weighted lines at 510, 560, and 610 nm. The wavefront everywhere between 510 and 610 nm is within about ¼ wavelength. This objective is assumed to be coated, because the interior of the airspace allows a ghost reflection only 22.8 mm out-of-focus. This may be adjusted to about twice this value by changing the separation. Such pitfalls reinforce the warning in the first paragraphs to make these designs at your own risk.

G.2 152-mm *f*/8.4 ED Triplet Apochromat

This apochromat takes advantage of superior color correction to shorten the focal length to an easily mounted 50 inches. It has one airspace, and the other contacting surface must have an optical couplant such as oil or gel. It uses a simple crown, Schott BK7, to sandwich both sides of the ED glass, Ohara S-FPL53. The surface summary is in Table G.2; the longitudinal aberration is shown in Figure G.2. The chromatic focal shift in the 71 percent zone was displayed in Figure 12.8. It is less than 120 μm.

Table G.2
Surface Summary of 152-mm f/8.4 Apochromat [mm]

Surf #	Radius	Thickness	Glass	Semi-Diameter
STOP 1	613.11	7.620	BK7	76.20
2	321.11	0.051		76.20
3	315.55	20.320	S-FPL53	76.20
4	−439.83	7.620	BK7	76.20
5	−1552.22	1266.218		
6	−466.73			

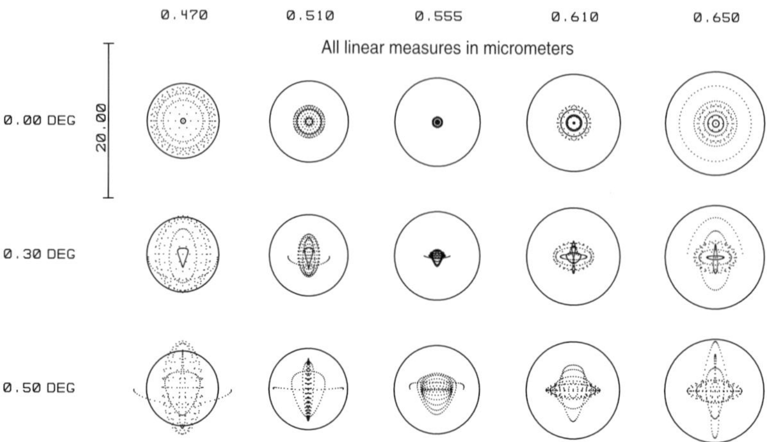

Fig. G.3 Spot diagram of the 152-mm *f*/8.4 ED apochromatic triplet.

From the longitudinal aberration plot we can see that the primary contributor to optical error is no longer chromatic defocus but spherochromatism. In fact, the chromatic focus shift in Chapter 12 is overly optimistic in that it shows no more than about a 0.12 mm focal shift, out of which the total focal length of 1250 mm is only one part in 10000. (Recall from the longitudinal aberration diagram that the achromat had a focal shift of 1 mm out of a 2286 mm focal length.) If we were to actually build such an ED objective, we might smugly state that we had improved our chromatic correction nearly five times, but the residual aberration of spherochromatism prevents improvement of the root-mean-square focus over the whole pupil beyond about one part in 5000. Nevertheless, this lens is a significant improvement over a simple achromat, as the spot diagram in Figure G.3 shows.

G.3 152-mm *f*/12 Aluminized-spot Maksutov-Cassegrain

The focal length of this telescope is chosen to be about 1830 mm to match several commercially available instruments. Maksutovs in general and Maksutov-Cassegrains with aluminized spot secondaries in particular are often made with a measurable component of higher-order spherical aberration (see Section 11.1), and it helps if they are planned to be of high aperture ratio. For example, this *f*/12 all-spherical telescope would work a lot better if it were forced to beyond *f*/15. The surface summary is in Table G.3 and the layout is shown in Figure G.4.

The design uses a high-index glass for the corrector to generate less

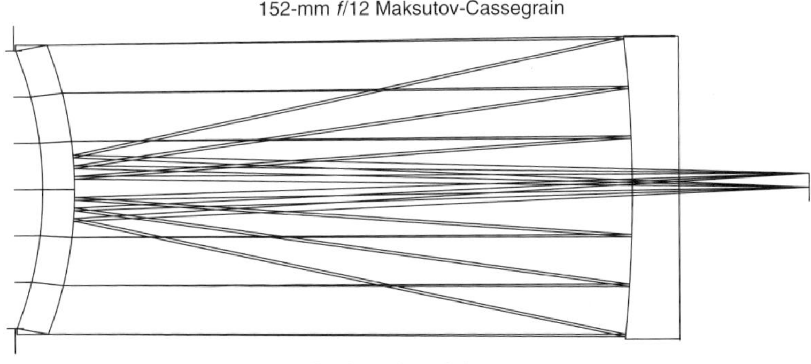

Fig. G.4 Layout diagram of 152-mm ƒ/12 aluminized-spot Maksutov-Cassegrain.

geometric aberration. Since chromatic errors are so much smaller than geometric errors in the Maksutov design, using a flint is not as foolish as it looks. It allows a thinner meniscus, and this one is thick enough at 18 mm, even with the high-index glass. The downside is a much-diminished wedge tolerance.

Spherical-surface Maksutovs do not scale well; what is possible and common at 90 mm becomes difficult to do at 152 mm. The aluminized spot secondary cannot be coupled with a fast (for Cassegrain telescopes) aperture ratio without doing something peculiar to the design. This problem was somewhat relieved in this theoretical case with a high-index corrector. Any maker who was interested in actually fabricating such a telescope would more likely retouch the surface or redesign the instrument.

Longitudinal aberration and the sagittal and meridional wavefront cross-sections are shown in Figure G.5. The wavefront can be refocused to have a minimum on-axis peak-to-valley error of ⅛ wavelength at 560 nm or a maximum aberration-only polychromatic Strehl ratio of about 0.95. The dominant on-axis residual is higher-order spherical aberration. Even though this aberration is always present in Maksutovs that have not been retouched, it may in some instances be ignored.

Table G.3
Surface Summary of 152-mm ƒ/12 Maksutov-Cassegrain [mm]

Surface		Radius	Thickness	Glass	Semi-Diameter
Stop	1		16		76.2
	2	−201.065	18	SF11	79.5
	3	−213.741	314.434		79.5
	4	−773.655	−314.434	MIRROR	82.24
	5	−213.741	414.364	MIRROR	20
	6	−190.8			7.97

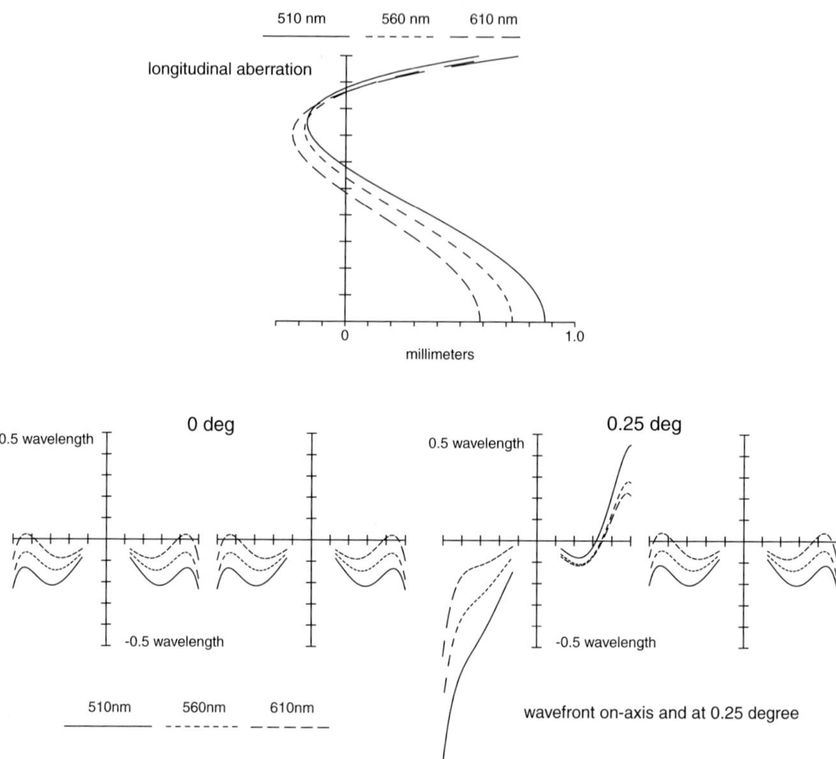

Fig. G.5 The longitudinal aberration and optical path difference (wavefront) in the 152-mm all-spherical aluminized spot Maksutov-Cassegrain.

Because the curvature of the secondary mirror is locked to the rear curvature of the corrector, the optical designer does not have the freedom to cure coma. Off-axis coma and astigmatism diminish the Strehl ratio to 0.8 beyond a field radius of 0.24 degree. Still, this telescope encompasses nearly the whole Moon inside a diffraction-limited region. The spot diagram of Figure G.6 shows the image deteriorating with distance from the center.

It also shows something else. There is a thin scatter of rays well outside the Airy disk. Ray theory fails in diffraction-limited systems, but when some rays are diverted this far outside the Airy disk, the ray approximation becomes increasingly valid.

The reason for this diversion of rays is the rolled region of the wavefront from about the 85-percent zone out to the edge. It has an amplitude of only about ⅛ wavelength and doesn't seriously affect the Strehl or the P-V wavefront measures, but diverts a moderate amount of light in a halolike pattern much resembling that of a turned edge.

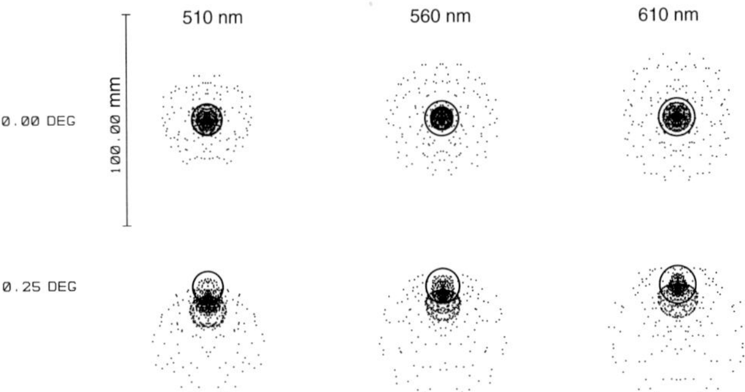

Fig. G.6 Spot diagram of 152-mm Maksutov-Cassegrain with indicated Airy disk.

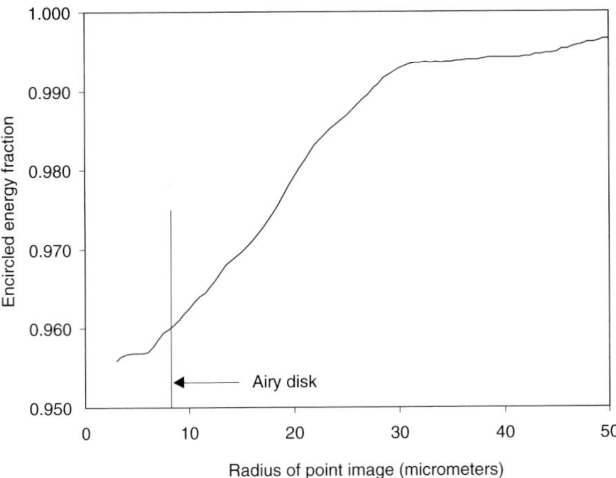

Fig. G.7 Encircled energy fraction of 152-mm Maksutov-Cassegrain. Note that this is a full wave theory calculation.

Like the turned edge, the aberration is at the perimeter of the disk where it is most heavily weighted. Figure G.7 shows the fraction of diffraction encircled energy compared with obstructed, but otherwise perfect, optics.

About 4% of the energy is excavated from the diffraction spot (radius about 8.2 micrometers). Of this, 3.5% is deposited in a halo of radius less than 30 μm (about four Airy radii). The remaining half-percent consists of a huge, dim glow out to about 70 μm. Depicting this tiny optical difficulty in such a manner puts the trouble in an overly harsh spotlight. People may get the impression that this is an unacceptable feature of Maksutovs, when it is an error that can be more or less ignored, much in the same way that slight turned edge can be ignored in most situations. This aberration, however, *will* show itself in a star test and other tests.

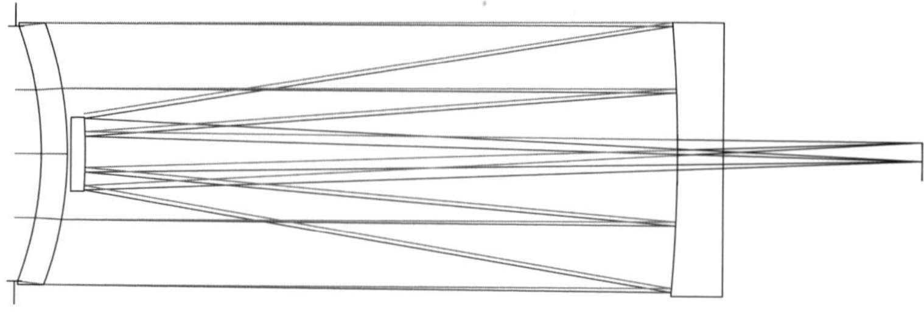

Total Length: 554.7 MM

Fig. G.8 . Layout of the 152-mm separate secondary Maksutov-Cassegrain

G.4 152-mm *f*/12 Separated-secondary Maksutov-Cassegrain

As mentioned before, the residual errors in a purely-spherical aluminized-spot Maksutov become too severe for all but miniature apertures. The 152-mm described in the previous section, with its up-and-down loops of zonal aberration, is close to overstepping the bounds of respectability. This minimum aberration also has little tolerance for other manufacturing errors. The 1-mm longitudinal aberration compares unfavorably with the 1-mm longitudinal aberration seen in the 6-inch *f*/15 achromat, since the achromat's aberration is much less than this toward the center of the spectrum.

Table G.4
152-mm *f*/12 Maksutov-Cassegrain with Separated Secondary [mm]

#		Radius	Thickness	Glass	Semi-Diameter
Stop	1		16.000		76.2
	2	–214.485	16.000	BK7	78.40
	3	–223.740	372.674		78.40
	4	–970.00	–361.494	MIRROR	81.24
	5	–359.208	511.484	MIRROR	23.00
	6	–355.451			7.98

We can repair the difficulty somewhat by going to long aperture ratios or by careful hand-figuring, but here we choose to separate the secondary from the rear surface of the corrector. An example result is the design presented in Table G.4, with layout in Figure G.8. Note that this telescope is inherently longer, which helps the design chores considerably.

The aluminized-spot corrector in the previous section had to maintain enough curvature to deliver the required magnification, but the natural tendency of the optimized secondary is to flatten out. The result was a tension between two incompatible goals. Separating these two elements allows

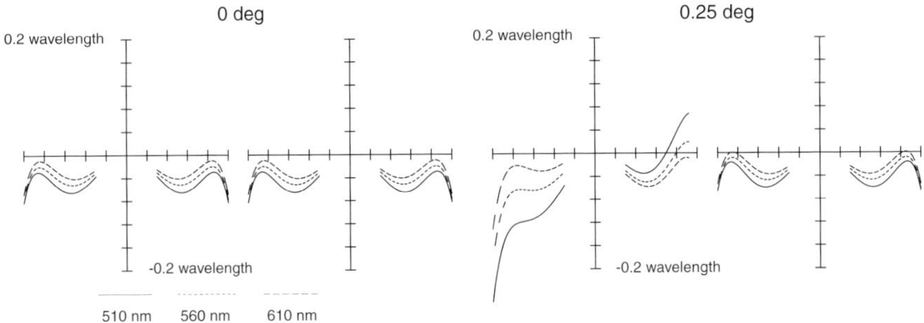

Fig. G.9 Wavefront of the 152-mm *f*/12 separate-secondary Maksutov-Cassegrain.

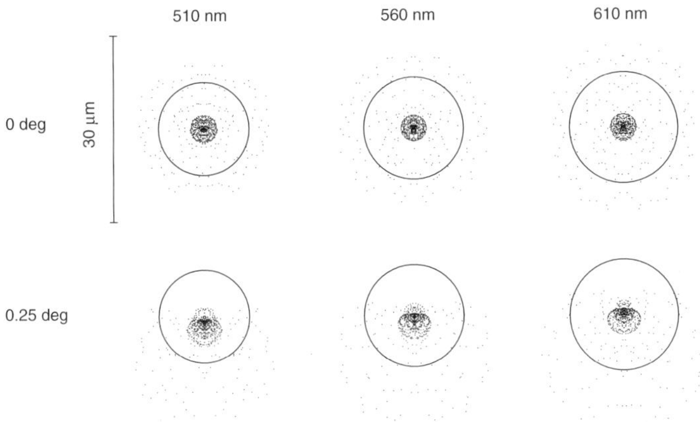

Fig. G.10 Spot diagram of 152-mm f/12 separate-secondary Maksutov-Cassegrain.

each to achieve the optimum. In the design of Table G.4, the corrector is made of a thinner piece of the lighter and less expensive crown glass BK7.

Figures G.9 and G.10 show the definite advantages of separating the secondary. Refocused on-axis error has diminished to ⅟₂₀ wavelength P-V and the refocused polychromatic Strehl ratio is above 0.993. This is very similar to the plot in Figure G.5, but the scale is smaller. Coma has been mostly corrected and there is only a slight amount of astigmatism. The longitudinal aberration (not plotted) is about the same shape as for the aluminized-spot Maksutov-Cassegrain, except it is has a much smaller amplitude of about 0.5 mm.

G.5 200-mm *f*/10 Spherical-mirror Schmidt-Cassegrain

The reason this telescope is designed with spherical mirrors is so that it will resemble the vast majority of Schmidt-Cassegrains out in the real world.

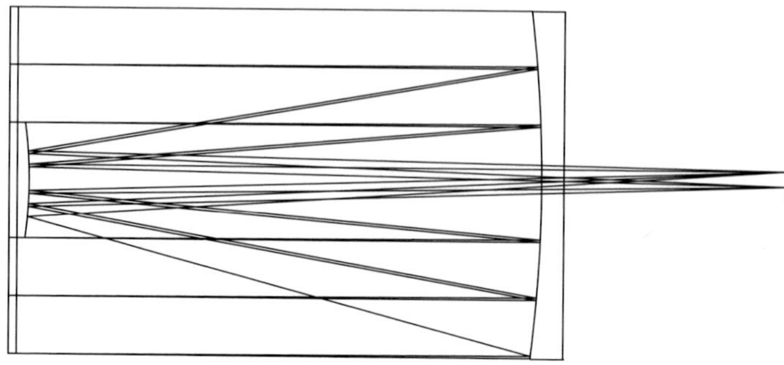

Total Length: 475 mm

Fig. G.11 Layout of a commercial-type 200-mm spherical-mirror Schmidt-Cassegrain.

People have envisioned improved designs, such as the one presented in Rutten and van Venrooij (1984) and the design in Section G.6.3, but few are in production. One manufacturer recently (2005) has been making a variation that presumably has a spherical main mirror and a hyperboloidal secondary to achieve better off-axis performance, but any randomly selected Schmidt-Cassegrain is much more likely to be a spherical mirror design with only an aspherical corrector plate.

Table G.5

Surface Summary of Commercial-style 200 mm Schmidt-Cassegrain [mm]

	Radius	Thickness	Glass	Semi-Diameter	C(2)	C(4)
Stop 1		5.000	BK7	101.60	7.81×10^{-6}	-6.70×10^{-10}
2		320.000		101.62		
3	−812.800	−312.622	MIRROR	101.60		
4	−231.067	462.622	MIRROR	33.00		
5	−208.011			10.75		

The design appears in Table G.5 and the layout in Figure G.11. The new columns that appear in this prescription are the non-spherical terms C(2) and C(4), which comprise the coefficients of r^2 and r^4 additions to the flat front surface of the BK7 corrector plate (units: μm). The coefficient C(4) contains all of the optical fireworks; C(2) merely adds a mild focusing power to the corrector to reduce chromatic aberration. The sag diagram appears in Figure G.12. A maximum thickness of around 41 wavelengths of 560 nm light must be removed from a 76 mm radius.

This telescope shares with other Cassegrain designs an intense field curvature that is exacerbated by the secondary magnification factor of 5. The primary has a natural aperture ratio of *f/2*, which is extended to *f/10* by this strong magnification.

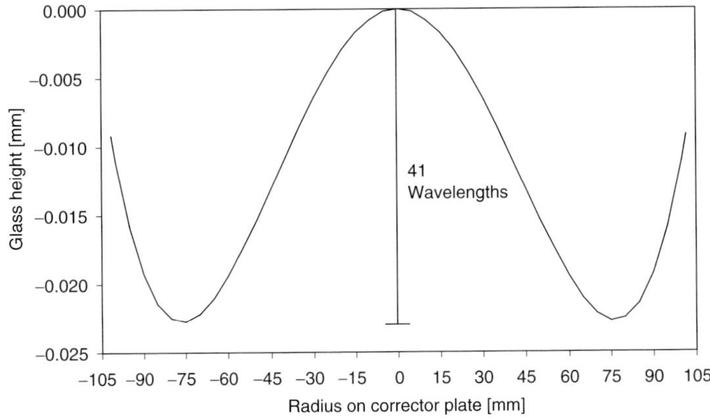

Fig. G.12 . Contour across diameter of front of S-C corrector plate.

The spot diagram (Figure G.13) reveals that although on-axis imaging is good, coma is uncorrected. The image at the edge of the Moon is four times as bad as diffraction will allow. The diffraction-limited field is only 4.2 arcminutes in radius, a figure that is common in fast Newtonians but less often seen in Cassegrains. This is not a peculiarity of the particular design. To do better off-axis imaging, another surface must be aspherical.

Figure G.14 displays the longitudinal aberration of the S-C. Note that the chief aberrations at the central wavelength are a balance of 3rd and 5th order spherical (4th and 6th order on the wavefront), but spherochromatism varies more rapidly across the spectrum than in the Maksutov-Cassegrains. Here, however, both aberrations start smaller.

On axis, it is clear that if a good corrector plate can be made—by no means an easy process (see the sag chart in Figure G.12)—then you can

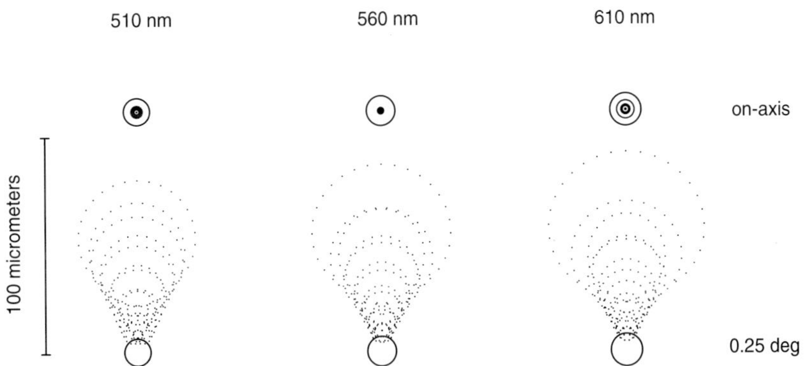

Fig. G.13 Spot diagram of 200-mm Schmidt-Cassegrain.

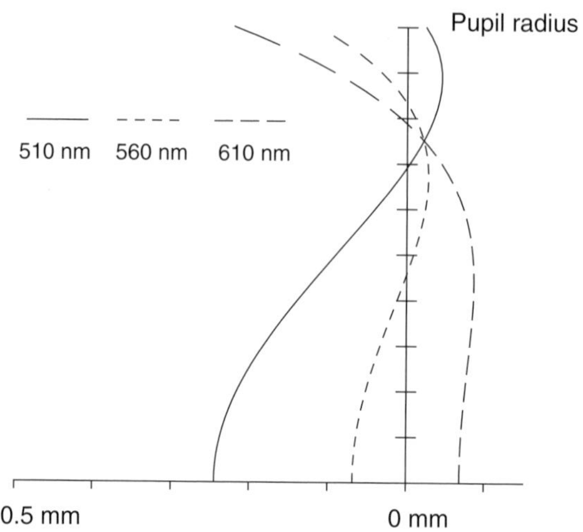

Fig. G.14 Longitudinal aberration of commercial-type Schmidt-Cassegrain

achieve an impeccable image in the 200-mm aperture.

G.6 Other Designs

Throughout the book and this Appendix I refer to other designs than the ones in the previous sections. I do not want to give the impression that these designs have not been modeled. However, these designs may be considered to be only slight variations, requiring less exposition. Here are the prescriptions of the following telescopes:

1. 152-mm *f*/12 all-spherical Maksutov-Cassegrain with aluminized spot having different mixes of aberrations.

2. 152-mm *f*/12 conic-primary Maksutov-Cassegrain with aluminized spot.

3. 200-mm *f*/8 improved Schmidt-Cassegrain.

4. 150-mm *f*/10 digital-imaging apochromat.

G.6.1 Another 152-mm *f*/12 All-spherical Maksutov-Cassegrain

The Maksutov telescope of Section G.3 has a strong effective turned edge which may be slightly reduced by taking advantage of the obstruction. Re-balancing fourth- and sixth-order wavefronts allows the designer to reduce the sharp curvature in the edge zone at the cost of more activity in the center, but the center is largely covered by the obstruction. It still has an aluminized spot on the rear of the corrector in lieu of the preferred separate secondary. It also still uses a high-index SF11 corrector. I still recommend that makers go either to a smaller aperture, a separated secondary, or aspherical optics if they want an *f*/12 Maksutov-Cassegrain.

Table G.6

152-mm *f*/12 All Spherical Maksutov-Cassegrain with Different Aberration Mix [mm].

		Radius	Thickness	Glass	Semi-Diameter
Stop	1	−203.459	18.000	SF11	76.205
	2	−216.178	315.620		79.366
	3	−777.771	−315.620	MIRROR	83.114
	4	−216.178	414.580	MIRROR	23.000
Image	5	−220.00			7.955

G.6.2 150-mm *f*/12 Conic-primary Maksutov-Cassegrain with Aluminized Spot

This design is a compromise between the Dall-Kirkham Cassegrain (attained by allowing the meniscus thickness to go to zero) and the all-spherical Maksutov. It has a much reduced on-axis error, yet does not allow off-axis coma to grow as explosively as it does in the pure Dall-Kirkham. The disadvantage is that it requires an aspheric curve on the primary mirror, but this is only a weak prolate spheroid, such as would be found on the way toward parabolizing a Newtonian mirror. Peak-to-valley wavefront error refocuses to $\frac{1}{16}$ wavelength, and at diffraction focus the aberration-only polychromatic Strehl ratio is over 0.98. At $\frac{1}{8}$ degree off-axis, Strehl ratio drops to about 0.92, and at $\frac{1}{4}$ degree, the Strehl ratio is about 0.7. The parameter labeled "conic" is the common usage where a sphere is zero and a paraboloid is −1. It still has the thick 18 mm corrector, but makes a small concession to manufacturability by changing the corrector material to BK7, or ordinary crown glass.

Table G.7

Surface Summary of Aspheric Maksutov-Cassegrain [mm]

		Radius	Thickness	Glass	Semi-Diameter	Conic	Comment
Stop	1		16		75		
	2	−207.498	18	BK7	77.4		
	3	−217.952	300.281		77.4		obs ~ 30%
	4	−759.771	−300.281	MIRROR	80.0	−0.30	
	5	−217.952	420.069	MIRROR	22.0		
Image	6	−228.253			7.9		0.25 deg

G.6.3 200-mm *f*/8 Improved Schmidt-Cassegrain

This design is shortened somewhat to accommodate its use in imaging. It has a Strehl ratio above 0.8 to beyond 0.4 degree on the curved field. Its performance is very similar to a Ritchey-Chrétien (RC) Cassegrain, even though the details of the design are different from the RC telescope. It has a great deal of field curvature, as does the RC, so care should be taken when using flat photographic film. It will cover a 10-mm square digital imager without a field-flattener, however.

Table G.8
Surface Summary of Improved Schmidt-Cassegrain [mm]

		Radius	Thickness	Glass	Semi-diameter	Conic	C(2)	C(4)
Stop	1				100			
	2		5	BK7	100			
	3		302.10		100		7.8×10^{-6}	-7.8×10^{-10}
	4	−800	−293.04	MIRROR	102.85			
	5	−281.71	293.04	MIRROR	33	−1.16		
	6		130.02					
Image	7	−164						

G.6.4 152-mm *f*/10 Fluorite Digital-imaging Apochromat

In Chapter 12, I used this design to exemplify a digital-imaging 6-inch *f*/10 apochromat. It is a calcium fluoride triplet with the same glass (N-BAK4) on either side of the fluorite crystal.

Incidentally, the *f*/10 film-emulsion photographic triplet also mentioned in Chapter 12 is a lengthened version of the *f*/8.4 apochromat appearing earlier in this Appendix. Its design doesn't appear here because it is just a trivial scaling and stopping down.

Table G.9
512-mm *f*/10 Digital-Imaging Apochromat [mm]

Surface	Radius	Thickness	Glass	Diameter
1	591.23	9	N-BAK4	152.4
2	358.65	18	CaF2	152.4
3	−1091.09	9	N-BAK4	152.4
4	−2914.30	1496.32		152.4
5	−560.00			13.29

Appendix H
List of Common Symbols

Some symbols don't appear here at all because they are used in different ways in different locations (a good example is N). Some that do appear here occasionally are used in alternate ways. Most symbols are used only in the way defined in this Appendix, but if readers are unsure in any case, the immediate context takes precedence.

A_n	nth order coefficient of aberration.
c	speed of light propagation.
CCD	acronym for charge-coupled device.
CD	acronym for Compact Disc.
D	diameter of the objective lens or mirror.
DAC	acronym for digital-to-analog converter.
Δf	focus shift or change in focus.
e	2.71828….
E	energy.
EER(θ)	encircled energy ratio at angular radius θ.
f	effective focal length of the objective.
F	the focal ratio as used in formulas.
f, f', etc.	various focus positions.
f/#	the "f-number" or focal ratio.
h	Planck's constant.
i	$\sqrt{-1}$.
i_s	the Strehl ratio.
$I(x', y')$	intensity at image point (x', y').
J_n	the nth-order Bessel function.
k	$k = 2\pi/\lambda$, called the "wave number."
λ	the wavelength of light.
MA	the minor, or short axis of a Newtonian diagonal.
MTF	acronym for modulation transfer function.
Mult	the multiplier of focal lengths in the distance of the source N.
ν	frequency (time or spatial).
OTF	the optical transfer function.
φ	reduced image angle $D\theta/\lambda$.

π 3.14159… .

$P(x,y)$ the pupil function at position (x,y).

R radius of curvature.

r_{Airy} radius of Airy disk.

RFT   acronym for richest-field telescope.

ρ radial coordinate of aperture (range 0 to 1).

RMS   acronym for root-mean-square deviation of wavefront.

S or s   the sagitta in all but Appendix B.

σ_{RMS} RMS deviation of wavefront in wavelengths.

SNR   acronym for signal-to-noise ratio (S/N).

T distance from focal plane to center of tube. Ch. 6 and App. C.

$T(\rho)$ or $T(x, y)$ transmission coefficient as a function of aperture radius.

$U(\rho)$ or $U(x', y')$ complex scalar field value at point (x', y').

$W(\rho)$ or $W(x, y)$ aberration over the pupil in wavelengths.

x or y usually the aperture coordinates.

x' or y' usually the image coordinates.

Glossary

Aberration Any deviation from a spherically converging wavefront after the optical system has finished processing light is called an aberration. Aberration is commonly defined in one of two ways. The most convenient ray-optics usage expresses it as a deviation from the point of geometric focus. The wave optics convention defines it as deviation from a perfect converging sphere. These two seemingly different definitions are closely related. The slope of the wavefront determines the direction of ray propagation in a homogeneous medium.

Accommodation A biological adjustment to changing light or focusing conditions.

Achromat Literally meaning "without color," the term is used in astronomical optics to indicate a two-element refractor lens corrected for the dispersion of glass (chromatic aberration) over the color range of interest. The term achromat is a misnomer. For most doublets, only two colors are deliberately brought to a common focus, but the resulting folded spectrum offers much less dispersion than exists in simple lenses. (See *refractor, apochromat, secondary spectrum, dispersion*, and *chromatic aberration*.)

Airy disk (Named after 19th Century scientist Sir George Airy.) The appearance of the central peak in the focused diffraction pattern superficially resembles a disk. The illusion is made stronger since the peak is surrounded by a zero-intensity region. The disk has a soft edge, so the effective radius of the central diffraction spot varies with the brightness of the image. Brighter spots look bigger. (See *resolution*.)

Analytic geometry Originated by Pierre de Fermat and René Decartes in the early 17th century and further developed by later mathematicians, analytic geometry is a way of reducing geometrical relationships to algebra.

Angle A division of a circle. Four measures are common in this book: radians ("natural" or unitless), degrees, arcminutes, and arcseconds. The measure denoted in the upper left hand corner of the star-test plots is a reduced angle valid for all instruments — the angle in radians times the ratio D/λ. (See *resolution*.)

Annealed If materials are melted and allowed to cool below their solidification temperature too quickly, they will exhibit permanent strains. Improperly annealed glass deforms under the influence of grinding, and over time, to make the surface shape uncontrollable. Almost never appearing in glasses made specifically for optical use, it often occurs in glass disks adapted for precision optics, but cast with another purpose in mind.

Aperture The opening through which parallel light enters a telescope. Aperture is typically measured as the diameter of the most restrictive opening of the telescope before the light is focused. Some confusion may exist with the abbreviated photographic terminology that uses the term "aperture" interchangeably with "focal ratio." Aperture is not used that way here.

Aperture stop A physical obstruction (usually a circular hole) placed somewhere in front of the telescope. The function of the aperture stop is to limit the aperture in a specific plane. (See *field stop.*)

Apochromat If three colors are simultaneously focused, the lens is said to be apochromatic. The rigorous definition also includes corrections for spherical aberration and coma. Because three colors within the spectral range of interest are held to a common focus, secondary spectrum is markedly reduced. (See also *achromat, refractor, secondary spectrum,* and *chromatic aberration.*)

Apodization Shading the aperture to diminish diffraction rings. This term has come to refer to modifying the transmission and phase of the aperture to achieve any sort of diffraction pattern change.

Arcminute $\frac{1}{60}$ of a degree or $\frac{1}{21,600}$ of a full circle. (See also *arcsecond, degree, radian,* and *angle.*)

Arcsecond $\frac{1}{60}$ of an arcminute, $\frac{1}{3600}$ of a degree, or $\frac{1}{1,296,000}$ of a full circle. 1 arcsecond $= 4.848 \times 10^{-6}$ radians. (See *arcminute, degree, radian,* and *angle.*)

Aspheric surface In optics, a concave or convex surface that superficially resembles a sphere but has another shape. Most commonly applied to conic sections.

Astigmatism An aberration that results from two radii of curvature oriented at right angles to one another on a wavefront. (See *aberration.*)

Attenuation A diminishing of wave intensity that includes diffraction, scattering, spreading of the beam, and absorption.

Bandwidth/bandpass The frequency range passed by a given filter.

Barlow lens A diverging lens placed near focus to increase the effective focal length of an instrument without appreciably increasing the telescope's physical size.

Bench test An indoor, laboratory-style test generally used during the fabrication of optics. Because of the necessarily compact nature of the testing geometry, most bench tests of astronomical instruments use secondary references or complicated data reduction procedures.

Catadioptric A mixed lens-mirror system that makes up the objective or main optical group.

Caustic A region where rays of light cross each other and pile up. Focus is a caustic, but this word also refers to the one-dimensional focusing that occurs with astigmatism or even the quasi-focusing that occurs in systems suffering from aberrations. In geometric optics, caustics are regions where the ray-tracing formalism breaks down.

Chromatic aberration Color errors caused by the dispersion of light in glass. Bringing the colors to different focus points has two forms of damaging effect on the image: all colors except those in focus are imaged as expanded disks, and magnifications vary with color. (See *refractor, achromat, dispersion,* and *apochromat.*)

Collimation Here, collimation is used to indicate achieving accurate alignment. More properly, it refers to the generation of a flat wavefront, but good alignment and a good wavefront are usually inseparable.

Coma An aberration that most often occurs when some optical systems are tilted. Rare in quality refractors, it is very common in reflectors and catadioptric systems. Coma results in a fan-like shape of the image. (See *aberration.*)

Corrector (See *Schmidt corrector* and *meniscus.*)

Correlation An expression of the similarity of two functions as they are offset from one another. If the functions are precisely the same, the maximum correlation is one.

Criterion (resolution) In astronomical optics, this term refers to a certain separation in resolved objects (as in "the stars were separated just at the Rayleigh criterion"). Colloquially, it has been also used for stating optical quality, as in "the ¼-wavelength Rayleigh criterion," although this terminology is properly replaced by "the ¼-wavelength limit." (See *Rayleigh tolerance* and *Rayleigh criterion.*)

Curtate cycloid A cycloid is the path followed by a point on the rim of a wheel. A curtate cycloid is the path of a point nearer the axle.

Dawes criterion A separation angle of about $1.02\lambda/D$. It occurs between the loose Rayleigh criterion and the tight Sparrow criterion.

Decibel A change in intensity of sound (or any signal) by a factor of 10 was originally termed "a bel." Since this change was too coarse, bels were never used. People much preferred the finer "decibel" measure. The dB level of I_1 referenced to I_2 is $10 \log_{10}(I_1/I_2)$.

Defocus The amount of defocus as used here is more precisely the defocusing aberration measured in wavelengths of light. This number is not to be confused with defocus distance, or how far one must move the eyepiece to obtain defocusing aberration. (See Table 5.1.)

Degree $1/360$ of a full circle The reason early mathematicians probably used such a peculiar number is because it can be divided into so many whole number portions: 180, 120, 90, 72, 60, 45, 36, 30, 20, 18, 16, 15, 12, 10, 8, 6, 5, 4, 3, and 2. This measure was very helpful before the decimal number system was invented (with its compact algorithm for long division). Also equal to $\pi/180$ or about $1/57.3$ radian. (See *arcminute, arcsecond, radian,* and *angle.*)

Dielectric A non-conducting material that becomes polarized in an electric field (i.e., that develops "electric poles" similar to "magnetic poles"). Many such materials are transparent to visible light. Glass, polyester, air, and quartz are all examples of dielectrics.

Diffraction-limited Used conventionally as an equivalent for the $\frac{1}{14}$-wavelength RMS wavefront deviation limit of Maréchal. Colloquially, it is used to mean the same thing as the $\frac{1}{4}$-wavelength Rayleigh limit, but that usage is only true for broadly-varying wavefront deformations such as correction error.

Diopter A measure of lens focusing strength. A lens having a strength of 2 diopters has a focal length of $\frac{1}{2}$ meter, 3 diopters has a focal length of $\frac{1}{3}$ meter, etc. Thus, the human eye, which is focused at a distance of about an inch, has a native strength of roughly 45 diopters. Focusing corrections typical in eyeglasses—1 to 4 diopters—are minor adjustments.

Dispersion Frequency or color dependence of optical effects. Dispersion in refractive materials leads to desirable effects in spectroscopes and undesirable effects when white-light images are the goal. The phenomenon of chromatic aberration in lenses is a difference of focal lengths for light at various portions of the spectrum. It results in the breaking of white light into rainbow colors.

Doublet Two lenses placed in close proximity that act as if they were a single unit. Used to correct various single-lens defects, especially chromatic aberration. (See *achromat* and *chromatic aberration*.)

Dynamic range The intensity range over which a sensor or emitter is approximately linear. Dynamic range can be written in decibels.

ED glass Glass made to emulate the low dispersion characteristics of calcium fluoride, called "fluorite." When used with normal glasses, even nominal "crown" glasses become the flintlike elements.

Effective *Effective* is applied to another quantity such as focal length or focal ratio. Complicated multi-element systems like Cassegrain telescopes or the use of a Barlow lens on a telescope demand some sort of leveling terminology to make their description comparable. "Effective focal length," for example, is the same as the focal length if the complicated optics were replaced by a single thin lens. "Effective focal ratio" is determined by examining how precipitously the light cone converges to focus. Cassegrain telescopes are much shorter than their effective focal lengths and aperture ratios indicate.

Encircled energy ratio In this book, the encircled energy ratio (EER) is the normalized energy fraction enclosed at a certain radius compared to the normalized energy fraction of a perfect unobstructed aperture at the same radius. By normalized, it is meant that both the numerator and denominator of this ratio go to unity far from the center of the image. If both the calculated aperture and comparison aperture are perfect, differing only by obstruction, then EER accounts only for changes to the image shape and not absolute brightness of the image.

Evanescent waves Waves that can only occur at the interface between media of different optical characteristics. They cannot propagate away from the interface and are bound to that surface.

Field stop A mask defining the field near focus. It is usually visible in the eyepiece as a sharply defined circle. This circle is not the edge of the main lens or mirror, but is contained in the eyepiece itself. It can be seen in most eyepieces by removing the eyepiece and turning it upside down. It is the sharp-edge opening through which the lenses can be seen. (See *aperture stop.*)

Figure (Noun) The shape of an optical surface. (Verb) To perform polishing operations to achieve the proper shape.

Focal length If an infinitely distant target is imaged by a lens or mirror, the focal length of the optical element is the distance from the lens or mirror such that the sharpest image of the target is found. Here, if no other indication is given, the focal length refers to the effective focal length of the objective system, rather than the eyepiece.

Focal ratio (Also called aperture ratio or f-number) The ratio between the effective focal length and the aperture is the focal ratio. It is written as $F = f/D$, where F is the focal ratio, D the diameter of the aperture, and f the focal length. By convention, a 6-inch telescope with a focal length of 48 inches is referred to as an "$f/8$" system because $^{48}\!/_6$ equals 8. F is used in formulas as a simple number (such as "8").

Fourier transform A mathematical procedure used to derive the frequency content of a function.

Foucault test A test using an obscuring edge placed near a point or slit focus. It is no longer commonly used by professionals, who prefer tests that rely on interference.

Fraunhofer lines Prominent dark lines in the solar spectrum labeled by the alphabet.

Fresnel zone A conceptual device to keep track of the phase sign. At locations where the wave is above its average value, it is in a positive Fresnel zone, and where the wave is below its average value, it is in a negative Fresnel zone. Thus, a given Fresnel zone applies only at an instant of time and is only an approximation of what is really happening.

Gaussian function A function of form Ae^{-x^2/w^2}, with amplitude A and exponential decay width w. Random deviations often follow a Gaussian distribution, as does the well-known "bell curve."

Geometric shadow In the light-as-particles ray-tracing approximation, the geometric shadow consists of the regions beyond the cone extending from the aperture and passing through focus.

Grit Loose particles of abrasive used in grinding optical surfaces. Grit sizes range from coarse sandlike grains to powdered finishing abrasives.

Hyperboloid A three-dimensional surface having a blunted cone shape.

Incoherent When two waves are added, they are said to be incoherent if they have no phase relationship with one another. At one instant, they may add constructively, and at the next, subtract destructively. Coherent light is usually derived from the same atomic transition or the same cavity resonance (in lasers), incoherent light from unrelated transitions. Severely restricting the geometry (as in focusing light on one or more slits) will often achieve approximate spatial coherence of even temporally incoherent light. (See *interference.*)

Index of refraction (See *refractive index.*)

Intensity As used here, intensity is proportional to the wave value squared. Actually, this number is not intensity as defined radiometrically, but the misuse has become customary.

Interference When intensity is calculated from the sum of two coherent waves, it is figured something like this equation: $I = (W_1 + W_2)^2 = W_1^2 + W_2^2 + 2W_1W_2$. The term on the end of the equation is called the interference term. If $W_2 = -W_1$, this term completely cancels the first two. If $W_2 = W_1$, the intensity is doubled. For incoherent light, this term can be anything at a given instant, but it averages over longer times to zero.

Iris An aperture stop that is adjustable in diameter. The eye has an iris and most cameras have a leaf-type iris used to adjust the aperture ratio. An iris is typically placed in or near a pupil in an afocal beam of light. (See *aperture stop.*)

Knife-edge test Colloquial expression for the Foucault test.

Lap A contraction of "lapping tool," it is a disk coated with pitch and powdered polishing agent in a slurry of water. Laps are used to polish optics. Typically, they are crossed by trenches ("channels") that have been cut into the lap to ease conformance to the optical element being worked.

Longitudinal As used here, an orientation along the axis of the instrument; perpendicular to transverse. The conventional image as viewed in an eyepiece is a transverse slice. A longitudinal slice is usually in a plane passing through the centers of the objective and eyepiece.

Magnification Literally, the object size divided by the image size. The usefulness of this term breaks down when we talk about very distant, very large objects. In astronomical telescopes, it is much more common to speak of *angular magnification*. Angular magnification refers to the angle subtended by the object in the eyepiece divided by the angle subtended without optical aid. Thus, if we look at the half-degree Moon in binoculars with a magnification of 7, we should see an image that extends an apparent 3.5 degrees.

Magnitude The magnitude difference between two stars is $-2.5 \log_{10}(I_1/I_2)$. Thus, if one star is 10 times brighter than another, it is only 2.5 magnitudes brighter. Magnitude is similar to the decibel scale used in electronics or sound and the Richter scale of earthquakes. For historical reasons, it increases with dimmer stars.

Meniscus A lens with the same sign optical curvature on each side. The meniscus used in Maksutov telescopes is a lens with very little optical power that is used to correct spherical aberration.

Micrometer A unit of measure equal to 10^{-6} meters. Twenty years ago, this unit was in common use as the "micron."

Microripple A surface roughness originating from a correlated area on the order of 1 mm across. Microripple is usually of very small amplitude.

Model, modeling Fitting a phenomenon to a mathematical system that may or may not be physically derived. The fit can be empirical, with no scientific basis, but the best and most extensible models are usually derived from fundamental theory.

Modulation transfer function MTF predicts the ability of an optical system to preserve light-dark contrast in periodic targets with finer and finer bar spacing.

Newtonian telescope The parabolic reflector was first announced by Sir Isaac Newton in 1672. It was not useful for astronomical purposes until John Hadley made the first approximately paraboloidal surface in 1721.

Normalized An integral that is normalized has been reduced to a value of one in an ideal case. Thus, an integral of, say, energy over the aperture is multipled by a constant to yield a value of 1.00. Such a procedure is useful in comparisons of imperfect apertures to perfect ones.

Objective The main image-forming element or group of elements in a telescope. Often, it is useful to equate "objective" with "non-removable" optics in the system.

Oblate spheroid A conic surface of revolution that is flatter in the middle. An oblate spheroidal mirror focused at infinity has more spherical aberration than a sphere.

Optical transfer function Full complex form of the spatial-frequency transfer function. Its absolute value is the modulation transfer function. Gives the contrast and shift in position of a sinusoidal bar pattern.

Overcorrection A form of low-order spherical aberration in which marginal rays cross beyond the focus of central rays. In Newtonian telescopes, an overcorrected mirror is hyperboloidal. (See *undercorrected*.)

Paraboloid A theoretical curve between prolate spheroids and hyperboloids. Represents the ideal in making Newtonian telescope mirrors.

Paraxial focus The focus of rays incident at or near the center of the objective and parallel to the axis of the instrument. Seldom the same as "best" focus.

Photodetector A sensor capable of detecting one or a few individual photons.

Pit An unpolished crater left over from the grinding process remaining in a polished surface.

Polychromatic Having many colors. White light is polychromatic. In this book a polychromatic image does not mean it is printed in color, but that the calculations of many colors are superimposed.

Primary aberrations Pure low-order forms of aberrations. Examples include coma, spherical aberration, and astigmatism, etc. (Also called *Seidel aberrations*.)

Primary ripple Coarse, quasi-periodic roughness having a spacing about the same as the channels in the lapping tool. (See *lap*.)

Prolate spheroid A conic surface of revolution intermediate between a sphere and a paraboloid.

Pupil function A mathematical construct that defines transmission and phase on the exit pupil. It allows theoretical calculations of the effects of obstruction and apodization by manipulation of the transmission component and the effects of aberration by manipulation of the phase component. In this work it is usually graphed as a three dimensional circular surface with phase in the vertical direction.

Quantum mechanics A name applied to the wave theory of matter. When wave-like particles are caught in a potential well (such as an electron in the Coulomb field of an atom), the requirement that the waves precisely fit in these wells demands that only certain energy states be occupied. Thus, energy can only be added or subtracted to such systems in discrete steps (called quanta).

Radian The measure of angle equal to moving one unit along the perimeter of a circle with unit radius. A radian is the "natural" measure of angle in that it is unitless. The word "radians" is actually a placeholder. A frequency of 2π radians/second is the same thing as 2π /second. An angle of 2π means that one full cycle of a circle has been traversed. (See *angle, arcminute, arcsecond,* and *degree.*)

Radiator An elemental source of secondary waves used in the Huygens-Fresnel theory of diffraction.

Rayleigh criterion When a double star is separated by an amount equal to the radius of either star's Airy disk, the separation is said to be just at the Rayleigh criterion. Most observers are able to resolve stars separated by less than this amount, but they do so more by the shape of the pair than the darkening between them.

Rayleigh tolerance or limit Visual optics that satisfy Rayleigh's limit produce wavefronts that can be enclosed by concentric shells with radii differing by ¼ wavelength of yellow-green light. (See *RMS.*)

Refractive index The ratio of the speed of light in empty space to the speed of light in a material. Examples: water has a refractive index of 1.3 and most glasses are somewhere over 1.5. Refractive index varying with wavelength is a conventional way of describing dispersion.

Resolution A measure of optical quality that depends solely on whether two equally bright incoherent points or bands of light are distinguishable. (See *modulation transfer function* and *optical transfer function.*)

Retouch To test and apply individual figuring strokes repeatedly until the wavefront is flat or spherical. Although this must be done on any large optical surface, certain small consumer instruments are unretouched to lower costs.

RMS Stands for "root mean square," or the square root of the averaged deviations squared. In telescopes, RMS is used as a measure of surface quality of optical elements, with lower values for more perfect optics.

Sampling When closely-spaced discrete values are used to represent continuous functions, individual values are called *samples*. For example, a weather gauge that takes barometric pressures every five minutes is said to be sampling at five minute intervals. A CCD chip measures a sampled image plane at a spatial frequency equal to the pixel spacing.

Scattering Diffraction from randomly distributed optical defects or obstructions.

Schiefspiegler A German word meaning "oblique reflector." A schiefspiegler is a telescope that avoids the additional diffraction of central obstruction by using tilted, round mirrors.

Schmidt corrector A plate placed in the incoming parallel beam to introduce an equal and opposite amount of spherical aberration to that produced by the rest of the optical system. The typical curve defining a Schmidt corrector is a 4th-order radial polynomial.

Secondary spectrum The residual chromatic aberration that exists in the bright portions of the spectrum (among the deliberately corrected wavelengths) even after a good attempt has been made at fixing color error in achromats. (See *chromatic aberration, achromatic, apochromatic,* and *dispersion.*)

Seeing The minimum point-spreading of the image due to turbulence, typically given in arcseconds. A site with "1-arcsecond" seeing is much worse than a site with "⅓-arcsecond" seeing.

Seidel aberrations The earliest formal description of spherical aberration, coma, astigmatism, field curvature, and distortion using polynomials and trigonometric functions. Also called the "primary" third order aberrations, or the first non-trivial components of a generalized aberration expansion.

Shading (See *apodization.*)

Signal-to-noise ratio (SNR) A comparison between the amount of interesting information to the amount of non-interesting information, as it is transferred through a system. Because the noise power is usually so much less than the signal power, the SNR is usually given in decibels. Higher SNRs are desirable. (See *decibels.*)

Sparrow criterion If the separation of two points of light is set to where the diffraction structure creates a flat isthmus bridging them, then they are said to be separated by the Sparrow criterion.

Speckles Little points of light surrounding the image, produced from rough surfaces or turbulent media. Speckles are caused by interference.

Spherical aberration When the converging wavefront is subtracted from a sphere centered on best focus, any remainder described by radial polynomials of low order is called "spherical aberration."

Stigmatism Perfect imaging of one surface (object surface) on another (image surface). Not quite the same as "not having astigmatism," though that is true also. In homogeneous space, the only way stigmatism can be achieved near stellar focus is by a spherically converging wave.

Strehl ratio The ratio of the maximum central brightness of an aberrated aperture's image to what it would be if the aberration were removed. A Strehl ratio of 0.8 is associated with the ¼-wavelength Rayleigh tolerance. In this book the Strehl ratio is formally defined as the limit of the normalized encircled energy as image radius approaches zero, divided by the same for the perfect unobstructed aperture, thus allowing the effect of obstruction to be calculated. This value is the same as the integral of the 2-dimensional modulation transfer function where the maximum spatial frequency is normalized to unity.

Substrate A telescope mirror is a layer of metal about 100 nanometers thick. The glass holding it is properly termed the "substrate."

Superposition If the combined effects of two waves are no more complicated than the simple sum of the waves, the system is called linear, and the net effect is called a linear superposition of the two waves. In other words, no effect arises from one wave's influence on the other. Intense light beams do not obey superposition in nonlinear media. All optical phenomena in this book are assumed to be linear. (See *interference*).

Telephoto A rear lens element that lengthens the effective focal length. Differs from a Barlow lens primarily in that it cannot be removed and was specifically designed to work with a given optical system.

Undercorrection As used here, a type of low-order spherical aberration in which marginal rays cross nearer to the objective than the central rays.

Wave function The three-dimensional function describing a wave, generally including its amplitude, phase, and direction of propagation.

Wavelet Here, this term refers to the tiny subsequent waves emitted by the Huygens-Fresnel radiators. It also refers to a mathematical method used in signal processing, but that usage does not apply in this book.

Wedge A prismatic aberration of refractor lenses. Although wedge can result from a lens that is actually thicker on one side, it can also result from a decentered symmetric lens.

Zonal Spherical Aberration Another name for higher-order spherical aberration.

Zones A contraction of "zonal defects," they are a special case of spherical aberration. "Zones" can be thought of as localized circular corrugations on the surface.

References

Allred and Mills 1989 Daniel B. Allred and James P. Mills, "Effect of Aberrations and Apodization on the Performance of Coherent Optical Systems. 3: The near field," *Applied Optics*, vol. 28, no. 4, pp. 673–681, 15 Feb. 1989.

Bachynski and Bekefi 1957 M.P. Bachynski and G. Bekefi, "Study of Optical Diffraction Images at Microwave Frequencies," *Journal of the Optical Society of America*, vol. 47, no. 5, pp. 428–438, May 1957.

Baker 1992 Lionel Baker, *Selected Papers on Optical Transfer Function: Measurement*, SPIE Milestone Series Volume MS 60, SPIE Optical Engineering Press, 1992.

Baker and Copson 1950 Bevan B. Baker and E.T. Copson, *The Mathematical Theory of Huygens' Principle,* Oxford Clarendon Press, Oxford UK, 1950.

Barakat 1962 Richard Barakat, "Solution of the Luneberg Apodization Problems," *Journal of the Optical Society of America,* vol. 52, no. 3, pp. 264–275, March 1962.

Bell 1922 Louis Bell, *The Telescope,* McGraw-Hill, 1922, Dover reprint, 1981.

Berkovitz 1968 M.A. Berkovitz, "Edge Gradient Analysis OTF Accuracy Study," *Proceedings of SPIE* vol. 13, pp. 115–123, 1968. (Reprinted in Baker 1992 above).

Berry 1972 M.V. Berry, "Reflections on a Christmas-tree Bauble," *Physics Education*, vol . 7, no. 1, January 1972.

Berry 1979 M.V. Berry, "Diffractals," *Journal of Physics A: Mathematical and General,* vol. 12, no. 6, pp. 781–797, 1979.

Berry and Blackwell 1981 M.V. Berry and T.M. Blackwell, "Diffractal Echoes," *Journal of Physics A: Mathematical and General* vol. 14, pp. 3101–3110, 1981.

Berry 1992 Richard Berry, "How to Build a Portable Artificial Star," *Sky & Telescope* (Gleanings for ATMs), p. 572, November 1992.

Beyer and Clune 1988 Louis M. Beyer and Laverne C. Clune, "Intensity and Encircled Energy for Circular Pupils Obscured by Strut Supported Central Obscurations," *Applied Optics,* vol. 27, no. 24, 15 Dec. 1988.

Born and Wolf 1980 Max Born and Emil Wolf, *Principles of Optics,* 6th ed., Pergamon Press, 1980. (also the seventh edition, Cambridge 1999).

Brigham 1988 E. Oran Brigham, *The Fast Fourier Transform and its Applications,* Prentice-Hall, Englewood Cliffs, NJ, 1988.

Buchdahl 1970 H.A. Buchdahl, *An Introduction to Hamiltonian Optics,* pp. 232–234, Cambridge University Press 1970 (Dover reprint, 1993).

Buchroeder 1994 R.A. Buchroeder, private correspondence, 1994.

Burch 1985 D.S. Burch, "Fresnel Diffraction by a Circular Aperture," *American Journal of Physics,* vol. 53, no. 3, pp. 255–260, March 1985.

Cagnet et al. 1962 M. Cagnet, M. Francon, J.C. Thrierr, *Atlas of Optical Phenomena,* Springer-Verlag, Göttingen and Heidelberg 1962.

Capers et al. 1991 Robert S. Capers, Eric Lipton, and staff writers "The Looking Glass—How a Flaw Reflects Cracks in Space Science" March 31 to April 3, 1991, *The Hartford Courant,* Hartford Connecticut.

Ceragioli 2003 Roger Ceragioli "A Survey of Refracting Systems for Astronomical Telescopes," on the web at http://alice.as.arizona.edu/~rogerc/, 2003–2005.

Ceravolo et al. 1992 Peter Ceravolo, Terence Dickinson, and Douglas George "Optical Quality in Telescopes," *Sky & Telescope,* vol. 83, no. 3, pp. 253–257, March 1992.

Ceravolo 1994, "Build Your Own Interferometer," *Sky & Telescope*, Jan. 1994.

Ceravolo 1994b *Interferometry and Telescopes*, published on web, 1994–2003, Ceravolo Optical Systems.

Conrady 1957 A.E. Conrady, *Applied Optics and Optical Design*, Dover Publications, Part One–1957, Part Two–1960.

Cornejo and Malacara 1970 A. Cornejo and D. Malacara, "Ronchi Test of Aspherical Surfaces, Analysis, and Accuracy," *Applied Optics,* vol. 9, no. 8, Aug. 1970, pp. 1897–1901.

CRC 1973 *Handbook of Chemistry and Physics*, 54th Ed., ed. by Robert C. Weast, The Chemical Rubber Company Press, 1973.

Danjon and Couder 1935 André Danjon and André Couder, *Lunettes et Télescopes,* Éditions de la Revue D'Optique Théorique et Instrumentale, Paris, 1935. [in French]

Delvo 1985 Pierino Delvo, "Point-Diffraction Interferometry Made Easy" *Sky & Telescope* (Gleanings for ATM's), February 1985.

Dakin 1962 R.K. Dakin, "Placing and Aligning the Newtonian Diagonal," *Sky & Telescope* (Gleanings for ATM's), pp. 368–369, Dec. 1962. *erratum* p. 114, Feb. 1963.

di Francia 1952 G. Toraldo di Francia, "Super-Gain Antenna and Optical Resolving Power," *Supplemento al Volume IX,* Serie IX del Nuovo Cimento, no. 3, 1952.

Dolph 1946 C.L. Dolph, "A Current Distribution of Broadside Arrays which Optimizes the Relationship between Beam Width and Side-Lobe Level," *Proceedings of the Instute of Radio Engineers,* vol. 34, p. 335, June 1946.

Edberg 1984 Stephen J. Edberg, "Apodizing Screens: A Critical Evaluation," *Telescope Making #24,* p. 12, Fall 1984.

Erkkila and Rogers 1981 John H. Erkkila and Mark E. Rogers, "Diffracted Fields in the Focal Volume of a Converging Wave," *Journal of the Optical Society of America,* vol. 71, no. 7, pp. 904–905, July 1981.

Fienup et al. 1993 J.R. Fienup, J.C. Marron, T.J. Schulz and J.H. Seldin, "Hubble Space Telescope Characterized by Using Phase Retrieval Algorithms," *Applied Optics 32*, 1747–1768, 1 April 1993.

Goodman 1968 J.W. Goodman, *Introduction to Fourier Optics,* McGraw-Hill, 1968.

Gordon 1984 Rodger W. Gordon, "Apodizing and Metzger," Letters, *Telescope Making #24*, p. 47, Fall 1984.

Greer 2000 Bryan Greer, "Understanding Thermal Behavior in Newtonian Telescopes" *Sky and Telescope*, Sept. 2000.

Greer 2004 Bryan Greer, "Improving the Thermal Properties of Newtonian Reflectors," May 2004 (Part 1) and June 2004 (Part 2).

Gregory 1957, John Gregory, "A Cassegrainian-Maksutov Telescope Design for the Amateur", *Sky and Telescope*, March 1957.

Harrington 1987 Steven Harrington, *Computer Graphics, A Programming Approach,* 2nd ed., McGraw-Hill, New York, 1987.

Harvey and Ftaclas 1995 J.E. Harvey and C. Ftaclas, "Diffraction Effects of Telescope Secondary Mirror Spiders on Various Image-Quality Criteria," *Applied Optics*, vol. 34, no. 28, pp. 6337–6349, Oct. 1995.

Hecht 1987 Eugene Hecht, *Optics,* 2nd ed., Addison-Wesley, Reading, MA, 1987.

Hufnagel 1993 Robert E. Hufnagel, "Propagation through Atmospheric Turbulence," Chap. 6 in *The Infrared Handbook*, ed. by William L. Wolfe and George J. Zissis, Infrared Information Analysis Center, Environmental Research Institute of Michigan, 4th printing, 1993.

Ingalls 1976 *Amateur Telescope Making Book One,* 4th ed., edited by Albert G. Ingalls, Scientific American, Inc., 1976 (originally published 1935). Book Two, 1978 (orig. 1937) and Book Three, 1953 by the same publisher.

Ingalls 1996 *Amateur Telescope Making, 1, 2, and 3,* republished and rearranged with slight editing by Willmann-Bell, Inc. 1996. [These are generally distinguished by use of arabic numerals on the newer set.]

Jacquinot 1958 Pierre Jacquinot "Apodization" (Appendix E) *Concepts of Classical Optics,* by John Strong, Freeman and Co., 1958.

Jacquinot and Roizen-Dossier 1964 P. Jacquinot and B. Roizen-Dossier, "Apodisation," [alternate spelling] *Progress in Optics Vol. III,* ed. by E. Wolf, p. 29, North-Holland, 1964.

Kestner 1981 Bob Kestner, "Grinding, Polishing, and Figuring Thin Telescope Mirrors," Part 1–Grinding, *Telescope Making #12,* pp. 30–35, Summer 1981 (Part 2: *TM#13*; Part 3: *TM#16*).

King 1955 Henry C. King, *The History of the Telescope,* Charles Griffin and Co., 1955, Dover reprint, 1979.

Kingslake 1948 R. Kingslake, "The Diffraction Structure of the Elementary Coma Image," *Proc. Phys. Soc.*, vol. 61, p. 147, 1948.

Kingslake 1978 R. Kingslake, *Lens Design Fundamentals*, Academic Press, Inc., 1978.

Kinsler et al. 1982 Lawrence E. Kinsler, Austin R. Frey, Alan B. Coppens, and James V. Sanders, *Fundamentals of Acoustics*, 3rd Ed., John Wiley and Sons, New York, 1982.

Leonard 1954 Arthur S. Leonard, in "The Amateur Astronomer" column of *Scientific American,* ed. by Albert G. Ingalls, p. 104, June 1954.

Lequèvre et al. 2007 www.astrosurf.com/tests/roddier/projet.html, the Roddier-test site with a download of WinRoddier2.exe (in French).

Li 1982 Yajun Li, "Dependence of the Focal Shift on Fresnel Number and f Number," *Journal of the Optical Society of America,* vol. 72, no. 6, pp. 770–774, June 1982.

Luneburg 1964 R.K. Luneburg, *Mathematical Theory of Optics,* University of California Press, Berkeley and Los Angeles, 1964. (Luneburg's name was misspelled "Luneberg" in the 1944 first edition; unfortunately, this spelling was propagated in the literature.)

Mackintosh 1977 *Advanced Telescope Making Techniques Vol. 1,* 1977. Willmann-Bell, Inc. reprint.

Mahajan 1981 Virendra N. Mahajan, "Zernike Annular Polynomials for Imaging Systems with Annular Pupils," *Journal of the Optical Society of America,* vol. 71, no. 1, Jan. 1981.

Mahajan 1982 Virendra N. Mahajan, "Strehl ratio for Primary Aberrations: Some Analytical Results for Circular and Annular Pupils," *Journal of the Optical Society of America,* vol. 72, no. 9, Sept. 1992.

Maksutov 1944 D. D. Maksutov, "New catadioptric meniscus systems ," *Journal of the Optical Society of America* vol. 34, 270–284 (1944).

Malacara & Welford 2007 D. Malacara and W.T. Welford "Star Tests" *Optical Shop Testing* 3rd ed., Wiley, 2007 (Also see Welford 1978).

Mallick 1978 S. Mallick, "Common-Path Interferometers" in *Optical Shop Testing,* ed. by Daniel Malacara, Wiley 1978.

Mandelbrot 1983 B.B. Mandelbrot, *The Fractal Geometry of Nature,* W.H. Freeman, San Francisco, 1983.

Maréchal 1947 André Maréchal, "Études des Effets Combinés de la Diffraction et des Aberrations Géométriques sur l'Image d'un point Lumineux," *Revue d'Optique,* vol. 26, no. 9, pp. 257–277, 1947.

Maurer 1991 Andreas Maurer, "Measuring Resolution Indoors," *Sky & Telescope* (Gleanings for ATM's), Sep. 1991.

Millies-Lacroix 1976 Adrien Millies-Lacroix, "A Graphical Approach to the Foucault Test," *Sky & Telescope,* Feb. 1976, p. 127 *et seq.*

Mobsby 1974 E.G.H. Mobsby, "A Ronchi Null Test for Paraboloids," *Sky & Telescope* (Gleanings for ATM's), Nov. 1974.

Muirden 1974 James Muirden, *The Amateur Astronomer's Handbook,* Crowell, New York, 1974.

Murata 1965 Kazumi Murata, "Instruments for the Measuring of Optical Transfer Functions," *Progress in Optics V,* ed. by E. Wolf, North Holland, 1965.

Newton 1730, Sir Isacc Newton, *Optics*, Dover Publications Reprint, 1979.

Nienhuis and Nijboer 1949 K. Nienhuis and B.R.A. Nijboer, "The Diffraction Theory of Aberration. Part III: Ggeneral Formulae for Small Aberrations: Experimental Verification of the Theoretical Results," *Physica,* vol. 14, no. 9, pp. 590–608, Jan. 1949.

Osterberg and Wilkins 1949 Harold Osterberg and J. Ernest Wilkins, Jr., "The Resolving Power of a Coated Objective," *Journal of the Optical Society of America,* vol. 39, no. 7, p. 553, July 1949.

Park 1974 David Park, *Introduction to Quantum Mechanics* 2nd ed., McGraw-Hill, Inc., 1974.

Parrent and Thompson 1969 George B. Parrent, Jr. and Brian J. Thompson, *Physical Optics Notebook*, Society of Photo-Optical Instrumentation Engineers (SPIE), Redondo Beach, CA, 1969.

Peitgen and Saupe 1988 *The Science of Fractal Images,* ed. by H. Peitgen and Deitmar Saupe, Springer-Verlag, New York, Berlin, 1988.

Peltier 1965 Leslie C. Peltier, *Starlight Nights,* pp. 215–216, Sky Publishing, Cambridge, Massachusetts, 1965.

Peters and Pike 1977 W.T. Peters and R. Pike, "The Size of the Newtonian Diagonal." *Sky & Telescope* (Gleanings for ATMs), p. 220, March 1977.

Press et al. 1986 William H. Press, Brian P. Flannery, Saul A. Teukolsky, and William T. Vetterling, *Numerical Recipes: The Art of Scientific Computing (Fortran),* Cambridge University Press, Cambridge, 1986.

Prugna 1991 F.D. Prugna, "A Monte Carlo Approach to the Ronchi Test," *Sky & Telescope* (Astronomical Computing), April 1991.

Pugh & Winslow 1966 Emerson M. Pugh, George H. Winslow, *The Analysis of Physical Measurements*, Addison-Wesley 1966.

Roddier 1981 F. Roddier, "The Effects of Atmospheric Turbulence in Optical Astronomy," *Progress in Optics*, ed. by E. Wolf, vol. XIX, pp. 281–376, North-Holland 1981.

Roddier et al. 1988 F. Roddier, C. Roddier, and N. Roddier "Curvature Sensing: a New Wavefront Sensing Method," *Proc. SPIE,* Vol. 976, 203–209, 1988.

Roddier et al. 1990 C. Roddier, F. Roddier, A. Stockton, A. Pickles, and N. Roddier, "Testing of telescope optics: a new approach," *Proc. SPIE,* Vol. 1236, pp. 756–766, 1990.

Roddier et al. 1994 C. Roddier, J.E. Graves, M.J. Northcott, and F. Roddier, "Testing Optical Telescopes from Defocused Stellar Images," *Proc. SPIE* Vol. 2199, pp. 1172–1177, 1994.

Ronchi 1964 Vasco Ronchi, "Forty Years of History of a Grating Inteferometer," *Applied Optics,* vol. 3, no. 4, pp. 437–450, April 1964.

Rowe 2003 David Rowe, *FringeXP ver. 2*, appearing on the web Feb. 2008 at cer-avolo.com/fringe/FringeXP/FringeXP.htm, 2003.

Royce 2001 Robert Royce "Testing Telescope Mirrors and Statement of Standards," on the web at www.rfroyce.com/testmethod.htm, *circa* 2001.

Russell 1995 Mark D. Russell, "Telescopic Performance on the Planets," *Sky & Telescope*, Mar. 1995, p. 90 *et seq.*

Rutten and van Venrooij 1988 H.G.J. Rutten and M.A.M. van Venrooij, *Telescope Optics: Evaluation and Design,* Willmann-Bell, Inc. 1988. Reset in 1999, with different pagination.

S&T 1990 "Hubble's Flaw Pinpointed," *Sky & Telescope,* November 1990, p. 470.

Schroeder 1987 Daniel J. Schroeder, *Astronomical Optics,* Academic Press, Inc., 1987.

Schultz 1980 S. Schultz "The Macalester Four-Goal System of Mirror Making and the Ronchi Test," *Telescope Making #9,* Fall 1980.

Sidgwick 1980 J.B. Sidgwick, *Amateur Astronomer's Handbook,* 4th ed., Enslow Publishers, Hillside New Jersey, 1980. (First published 1955.)

Sinnott 1990 "HST's Magnificent Optics... What Went Wrong?" *Sky & Telescope,* Oct 1990, p. 356.

Sinnott 1991 R.W. Sinnott, "Focus and Collimation: How Critical?" *Sky & Telescope* (Astronomical Computing), May 1991.

Smartt and Strong 1972 R.N. Smartt and J. Strong "Point Diffraction Interferometer" (Abstract) *Journal of the Optical Society of America,* vol. 62, p. 737, 1972.

Smartt and Steel 1975 R.N Smartt and W.H. Steel, "Theory and Application of Point-Diffraction Interferometers," *Proc. ICO Conf. Opt. Methods in Sci. and Ind. Meas.,* Tokyo, Japan 1974. *J. Appl. Phys.* vol. 14, (1975), Suppl. 14–1.

Stevick 2006 http://bhs.broo.k12.wv.us/homepage/alumni/dstevick/weird.htm. This long-running website entitled "Weird Telescopes" by David Stevick existed much earlier than 2006.

Strong and Plitnick 1992 William J. Strong and George R. Plitnick, *Music Speech Audio,* Soundprint, Provo UT, 1992.

Suiter 1983 D. Suiter, "Star Testing Your Telescope," *Astronomy,* April 1983.

Suiter 1986a D. Suiter, "Modifications to the Diffraction Image Caused by the Surface Roughness in Mirrors," *Telescope Making #28,* p. 24, Fall 1986.

Suiter 1986b D. Suiter, "Changing the Aperture Pupil," *Telescope Making #29,* p. 34, Winter 1986/87.

Suiter 1987 D. Suiter, "Letter to a Beginner," *Deep Sky* vol. 5, no. 1, Spring 1987.

Suiter 1988 D. Suiter, "Testing Paraboloidal Mirrors," *Telescope Making #32,* Spring 1988, p. 4.

Suiter 1990 D. Suiter, "Test Drive Your Telescope," *Astronomy,* pp. 56–61, May 1990.

Suiter 2001 H.R. Suiter "Apodization for Obstructed Apertures" published on the web at http://home.digitalexp.com/~suiterhr/TM/TM.htm.

Suiter and Zmek 2003 "A Dialogue on Spider Diffraction", *The Best of Amateur Telescope Making Journal Vol. 1* (originally ATMJ #11, ca. 1999), Willmann-Bell, Inc. 2003.

Taylor 1983 H. Dennis Taylor, *The Adjustment and Testing of Telescope Objectives,* 5th ed., first published 1891, Adam Hilger Press, 1983.

Taylor and Thompson 1958 C.A. Taylor and B.J. Thompson, "Attempt to Investigate Experimentally the Intensity Distribution near the Focus in the Error-Free Diffraction Patterns of Circular and Annular Pupils," *Journal of the Optical Society of America,* vol. 48, no. 11, pp. 844–850, November 1958.

Terebizh 1990 V.Yu. Terebizh, "A New Ronchi Null Test for Mirrors," *Sky & Telescope* (Gleanings), Sept. 1990.

Texereau 1984 Jean Texereau, *How to Make a Telescope,* 2nd ed., (first published 1951), Willmann-Bell, Inc., 1984.

Thiebaux 2006 Martial L. Thiebaux, private communication 2006.

Twyman 1988 F. Twyman, *Prism and Lens Making,* 2nd ed., (originally published 1952), portion on roughness abstracted from work by Texereau, Adam Hilger, Bristol, 1988.

van de Hulst 1981 H.C. van de Hulst, *Light Scattering by Small Particles,* p. 107, Dover Publications, 1981 (originally appeared in 1957).

Van Nuland 1983 J. Van Nuland, "More on Apodizing," Letters, *Telescope Making #21,* p. 4, Winter 1983.

Walker 2000 Bruce H. Walker, *Optical Design for Visual Systems*, vol. TT45 SPIE Press, 2000.

Walker 1977 Jearl Walker, *The Flying Circus of Physics With Answers*, probs. 4–13, 4–15, Wiley and Sons, New York, 1977.

Welford 1960 W.T. Welford, "On the Limiting Sensitivity of the Star Test for Optical Instruments," *Journal of the Optical Society of America,* vol. 50, no. 1, pp. 21–23, Jan. 1960.

Welford 1978 W.T. Welford, "Star Tests," ch. 11 of *Optical Shop Testing,* ed. by D. Malacara, pp. 351–379, John Wiley and Sons, New York 1978.

Yoder et al. 1953 P.R. Yoder, Jr., F.B. Patrick, and A.E. Gee, "Permitted Tolerance on Percent Correction of Paraboloidal Mirrors," *Journal of the Optical Society of America*, vol. 43 no. 8, pp. 702–703, Aug. 1953.

Zmek 1993 W.P. Zmek, "Rules of Thumb for Planetary Scopes–I," *Sky & Telescope* (Telescope Making), July 1993; "Rules of Thumb for Planetary Scopes–II," *Sky & Telescope*, Sept. 1993.

Index